Praise for A Quiet Evening

He is the travel writer's travel writer, conjuring prose with more humanity than Bruce Chatwin, more insight than Jan Morris and more humour than Patrick Leigh Fermor. And as this new collection brilliantly reveals, he is a better writer than all three. This book is perfect.
Sara Wheeler, *Financial Times*

For those who have yet to discover the strange greatness of Norman Lewis, this is an excellent place to start. The Sage of Enfield, with his sniffer-dog's nose for raffish broken-down places, may serve as a tonic for 2025.
Ian Thomson, *Spectator*

What observation! What majesty of style! What laconic humour! For me, Lewis has been the discovery of the year.
Matthew Parris, '*What* Spectator *writers read in 2024*', *Spectator*

Masterfully selected and introduced by John Hatt, *A Quiet Evening* covers the breadth of Lewis's extraordinary oeuvre in thirty-six articles of descriptive rigour and often devastating insight.
Colin Thubron, *Books of the Year, Times Literary Supplement*

Lewis is the master whose work all travel writers (and journalists) should aspire to. As good as Orwell
Henry Porter, *X*

… he found, like an anthropologist, that 'the discipline of writing compelled me to see more, to penetrate more deeply'. What he was brilliant at recording was the way of life of a vanishing world.
James Owen, *The Times*

He is probably the best travel writer we have ever had.
Jason Goodwin, *Country Life*

Every one of the thirty-six selections in the anthology is unforgettable. Rightly, Graham Greene called Lewis 'one of the best writers, not of any particular decade, but of our century'. No twentieth-century author has blended the comic, the tragic and the lyrical so perfectly in their work.
John Gray, *New Statesman*

One of the most outstanding English travel writers of his generation…with a prose style of wit and unobtrusive scholarship
Paul Clements, *Irish Times*

I thought I might read the first fifty pages and then some selected pieces to give me a good flavour. In fact, once I'd read the first pages I was gripped and read right through … I cannot stress enough just how good this book is. Do yourself a favour.
Desmond Clifford, *Nation Cymru*

Norman Lewis, probably the greatest travel writer of the 20th century, is not afraid of tackling big issues, including war and genocide. His output underlines that the best travel writing is seldom a contrived trip, but undertaken by someone willing to immerse themselves for a considerable time in unfamiliar cultures.
Mike Phipps, *Labour Hub*

Each story in this book is like a wonderful destination, which you leave longing only to return.
Neal Ascherson, *London Review of Books*

Was there a finer, more underappreciated travel writer in the 20th century than Norman Lewis? John Hatt's collection of his writings is a sumptuous reminder of what a treasure the man was. Look out for priceless encounters with Hemingway in Cuba and a cannibal in Indonesia.
Justin Marozzi, Books of the Year, *Spectator*

A Quiet Evening

The Travels of Norman Lewis

SELECTED AND INTRODUCED BY
JOHN HATT

ELAND
London

First published by Eland Publishing Ltd in 2025
This paperback edition first published in 2025

Copyright © The Estate of Norman Lewis
Copyright introduction and prefaces © John Hatt, 2024

The right of Norman Lewis to be identified as the author
of this work has been asserted in accordance with the
Copyright, Designs and Patents Act 1988

ISBN 978 1 78060 155 7

A full CIP record for this book is available from the British Library

Cover image: *A village in the Mang Yang district of Vietnam*
taken by Norman Lewis

Text set in Great Britain by James Morris
Printed in Spain by GraphyCems, Navarra

At first I believed in pure travel, and that it was necessary never to have a purpose....
Later I found that the discipline of writing compelled me to see more, to penetrate more deeply, to increase my understanding, and to discard a little of my ignorance.

Norman Lewis
From *The Changing Sky*, 1959

Contents

Introduction

AFTER FOUNDING ELAND IN 1982 with the purpose of reviving exceptional travel literature, my earliest ambition was to bring Norman Lewis to a wider readership. Therefore Eland's first publication was *A Dragon Apparent*, his wonderful account of Vietnam, Laos, and Cambodia before those nations were engulfed in the Vietnam war.

I went on to republish what I considered to be Norman's four best books. After passing the Eland baton to Rose Baring and Barnaby Rogerson, I stayed in close touch; but, because they managed the company with such love and skill, I took a back seat. All the same, I recently beseeched them to allow me to assemble a new anthology of Norman's best articles. They generously agreed, and this is the result.

Why was I so keen to publish this collection? The main reason was my perplexity and sadness that so much of Norman's magical writing was languishing unread. Despite previous collections, published over several decades, many of Norman's keenest fans were unaware of the articles that are published here. Why were they not better known?

One reason could be the difficulty of selling anthologies. A related and wider question is why Norman, revered even by so many other travel writers, was insufficiently honoured during his lifetime. There are several reasons, quite apart from the bad luck of having an unmemorable name.

Too often he was published without sufficient enthusiasm. The publisher, Collins, for instance, didn't seem to realise that they were publishing someone whom Graham Greene has described as 'one of the greatest writers of our century'. When I purchased the world English language rights in perpetuity for *A Dragon Apparent*, Norman's stock had fallen so low that they were sold to me for £250. And when his agent sold me those rights, he never told me that they were also available for Norman's masterpiece, *Naples '44*.

Norman's modesty was another reason for his lack of fame. He didn't care to be part of any literary coterie, and he was too dignified to tolerate the degradation of the typical publicity bandwagon. But he paid a penalty. His brilliant *Voices of the Old Sea*, for instance, was published in 1984, soon after Eric Newby's *On the Shores of the Mediterranean*. Unlike Newby's celebrated *A Short Walk in the Hindu Kush*, the latest book was hardly more than a potboiler, but it was universally reviewed. By contrast, Norman's book at first failed to receive even a single review. Though Eland wasn't the publisher, and though I was a nobody in the publishing world, I was so perturbed that I wrote to every literary editor to complain. A few reviews finally surfaced.

When eventually I purchased the rights to republish *Naples '44*, I was determined that this marvellous book, so deserving of becoming a classic, should get better known. For the first time, I hired a publicist who helped promote the book, getting Norman newspaper interviews and broadcasts, including an appearance on *Desert Island Discs*. Before hiring the publicist I had asked Norman if he would go along with her plans. He replied, 'I very rarely keep promises, but this time I really will try.' He kept his promise.

For this anthology I have had the advantage of being able to select all the best articles from Norman's entire life. None of his articles are without magic, but the necessity of earning a living meant that previous anthologies always contained some articles that were workaday rather than dazzling. None of these have been included.

Two sources have been invaluable for researching the prefaces for each article. The internet, unavailable for any previous collection, has often been a source of useful information, some of it recondite. Norman would have relished some of the discoveries, including the material about Catherine the Great's pubic hair. An even more invaluable source has been Julian Evans's *Semi-Invisible Man*, his remarkable and thorough biography of Norman. I have pillaged it mercilessly.

Having selected the articles, a decision had to be made about their order. The obvious solution was to arrange them chronologically. But this concept didn't really work, as Norman quite often wrote the article a long time after the event. Instead, where it makes sense, I have created clusters of subjects. Some of the clusters are geographical, but some reflect his passion for indigenous people, or his horror at their oppression, or his admiration for their resistance to the modern world's homogeneity.

In particular, the three major articles on the dreadful plight of South America's indigenous peoples are gathered together. The subject matter makes gruelling reading, but I felt it would be wrong to intersperse them with anything less serious. Otherwise, I have endeavoured to make the book as enjoyable as possible. Long articles are mostly interspersed with shorter ones, and the most serious with those that are more light-hearted. I had in mind that, when reading Gerald Brenan's great book, *South from Granada*, I came to a dense chapter called 'A Chapter of History', but I persevered – without skipping – aware that the next chapter was called 'Almeria and its Brothels'.

When I handed Eland over to Rose Baring and Barnaby Rogerson, I pleaded with them not to place introductions or prefaces in the front of the books, especially not those written by famous people. And most especially not if the introductions gave any subjective views about the quality of the book. I believe that readers should come fresh to a book, and any additional information of interest can be placed at the end. But rules are made to be broken; and having been allowed editorial control, I am taking advantage to explain my enthusiasm for this conclusive collection of Norman's articles.

Three ingredients are necessary for an author to be great. To use a well-worn metaphor – from the era of gambling on fruit-machines – you need all three lemons in a row to hit the jackpot. With two lemons, the author may write an adequate book but, for a book to be great, all three lemons must swing into view. What are these three lemons?

The first one is obvious: content. Many travel writers manage only one good book, at the most. And, if it is a successful book, their subsequent ones are all too often the result of discussion with a publisher who is eager for a subsequent success. The writers may have no genuine communication with the people they travel among – perhaps because they are constantly on the move, or because they don't speak the language, or because their curiosity isn't sufficient. Too often, their prime reason for travelling is only to produce a book.

Norman was the very opposite: his addiction to travel preceded any idea of writing about his journeys. In his foreword to *The Changing Sky* he wrote, 'Travel came before writing. There was a time when I felt that all I wanted from life was to be allowed to remain a perpetual spectator of changing scenes. I managed my meagre supply of money so as to be able to surrender myself as much as possible to this addiction; and charged with a wonderful ignorance I went abroad by third-class train, country bus, on foot, by canoe, by tramp steamer, and by Arab dhow ...When I began to write, it was probably, at least in part, an attempt to imprison some essence of the experiences, the images which were always slipping, fading, dissolving, taking flight.'

The second lemon is the possession of a literary gene – something that is, so to speak, God-given, rather than learnt. Even a combination of cleverness and hard work can't guarantee writing that sings. Take academics: although mostly clever, they can rarely produce prose that is a pleasure to read. And if one of their number succeeds in writing a bestseller, the others get crazed with jealousy. Of course, Norman was a painstaking writer, who wrote almost every morning of his life, but he had an *innate* skill that was dazzling. One sign of a born-writer is their use of metaphor – a metaphor that manages to vividly sum up a concept, without being strained. Consider Norman's description of the Bolivian government, which 'had not been above sending planes and tanks and killing a hundred or so, but for the moment ... had fallen like a digesting crocodile into a kind of watchful inactivity.'

The third lemon, as intangible as the literary gene, is the character of the writer. So much of writing is ultimately autobiographical, whether camouflaged or overt. In Norman's case his wisdom, passion and wit shine through. A revealing test of character is the relationship between travel writers and their photographers, which can often be fraught, especially during testing journeys. It is no exaggeration to say that – with the sole exception of Lord Snowdon – Norman formed warm and strong bonds with all his accompanying photographers.

One aspect of character is what one might call 'soul'. Soul overlaps with 'empathy' but it is something more. Perhaps it also overlaps with another hard-to-define word, 'duende', so cherished by Lorca, Norman's favourite poet. I can think of travel writers who produce elegant, impeccable prose, but their cold heart or mundane brain results in a lack of soul, thus producing books that are adequate, sometimes admirable, but never great. Thirty years after publication they will mostly be forgotten. But Norman did have 'soul', and when, for instance, he writes about genocide in South America, the reader can sense that this isn't mere journalism, but that he is motivated by a fiery passion.

So the three lemons for writing to be great are content, inherent writing skill, and the author's character. These are essentials. But let me mix the metaphor – as Norman would never have done – and claim that the three lemons are the ingredients of a cake, and thus add two cherries to decorate its icing.

The first cherry is Norman's magnificent sense of humour. It is easy to miss because his humour is often subtle, deadpan, dry, and sometimes nearly invisible; but it is there on almost every page, even the darkest ones.

The second is his acute power of observation. Norman's son, Gawaine, who travelled with him along the Sumatran coast, was astounded by what his father noticed on a journey that Gawaine had considered unfruitful, even boring. I had a similar experience. Norman once told me that he was going on a holiday with his wife, Lesley, to Andalusia. Because I had recently made a similar trip, I advised them to visit an unusual nobleman's castle that was open to the public. When Norman returned, he described his visit to the castle with such vividness that I felt ashamed by how much I had failed to notice. Ian Fleming (who features in this anthology) spotted Norman's power of observation, and in a BBC conversation with Raymond Chandler said, 'Norman Lewis has got an extraordinary visual eye. Photography is one of his main hobbies and he's got this astonishingly clear eye for detail and situation. A very remarkable man.'

Another striking aspect of this anthology is Norman's unbeatable range. Can any other travel writer compete with someone who interviews Castro's executioner, investigates bandits in Sardinia, exposes genocide in Brazil, issues one of the earliest warnings of a climate calamity, lampoons tourism in Panama, provokes Ernest Hemingway in Havana, and spends whole summers among the fishermen in a still-traditional Ibiza?

In an unashamedly personal introduction, I can't resist ending on a personal note. Julian Evans interviewed me for Norman's biography, recording some of my observations. Although I have no memory of the following, I am so grateful to have this opportunity to repeat what I once said.

'I had these long, long telephone calls with Norman. He was the master of the fruitful telephone conversation, there would always be interesting or wise things to come out of it. And the thing was, I might as well just say it now, in my life he was one of the most remarkable people I ever met. I think that often one's interest in somebody or affection, or whatever, can be gauged by how often you say to yourself, "Oh I'd like to tell some particular person that; it would interest them or make them laugh or they would be able to top it with something interesting themselves." And very, very often I thought I must tell Norman something. He was completely unlike anybody I had ever known, and a tremendous addition to my life.'

John Hatt
Cumbria, 2024

A Quiet Evening in Huehuetenango

Norman Lewis first visited Guatemala in 1946. In this article, written nine years later, he arrived soon after a CIA-backed coup overthrew the elected government.

He mentions that his reason for the visit was to gather background material for a novel. Indeed soon afterwards he wrote The Volcanoes Above Us, *published in 1957. It was enthusiastically reviewed, in several cases by famous writers such as V. S. Pritchett and Cyril Connolly. It may have been Lewis's most successful novel, and Russian readers especially took it to heart, buying more than two million copies.*

Guatemala consistently inspired Lewis. The hero of The Volcanoes Above Us *contends that 'For anyone who has lived in Guatemala, other countries, by contrast, are lacking in savour'.*

First published in the *New Yorker*, July 13, 1956

IN THE BLEAK DEPTHS of an interminable English winter, I was suddenly seized with an almost physical craving to write a novel having as its background the tropical jungles and volcanoes of Central America. Having succeeded in persuading my publishers that this would be a good thing from both our points of view, I boarded a plane at London Airport one morose evening in January, and two days later I was in Guatemala City. I chose Guatemala because I had been there before and knew something about it, but also because all that one thinks of as typical of the Central-American scene – primitive Indians, Mayan ruins, the wrecks of grandiose Spanish colonial cities – is found there in the purest concentration.

For three weeks I did my best to absorb some of the atmosphere of life in seedy banana ports of the Caribbean and the Pacific, where bored men in big hats still occasionally pull guns on each other. I went hunting in jungles said to abound with jaguars and tapir without shooting anything more impressive than a species of giant rat. I talked with wily politicians of the country, survivors of half-a-dozen revolutions, and took tea with exiled fellow-countrymen on isolated coffee plantations, who had lived so long among the Indians that they sometimes stopped in mid-sentence to translate their very proper English sentiments from the Spanish in which they now thought.

My final trip was to the far north of the country, the remote and mountainous area beyond Huehuetenango, which lies just south of the Mexican state of Chiapas and is reached after three hundred miles of infamous roads and stupendous scenery. Here under the Cuchumatanes, the ultimate peaks of Guatemala, even the onslaught of the Spanish conquistadors faltered and collapsed. And here the mountain tribes were finally left in peace, to live on in the harsh but free existence of the Stone Age, touched only by the outward forms of Christianity, consoled in secret by the ancient gods, and rejecting with all their might the overtures of Western civilisation.

In the early afternoon of the fourth day, my taxi, driven by a town Indian from Guatemala City called Calmo, reached the top of the 12,000-foot pass overlooking the valley of Huehuetenango. We stopped here to let the engine cool and noticing that the trees in this windswept place were covered with orchids, I astounded Calmo by suggesting we should pick some. 'Flowers?' he said. 'Where? They don't grow at this height!' I stumbled, weak and breathless from the altitude, up the hillside towards an oak, loaded with vermilion-flowered bromeliads. 'Ah,' he said, 'you mean the *parasitos*. Well, certainly, if you like, sir. When you said flowers, I didn't realise... We call these weeds – tree-killers.' Calmo was not only an intrepid driver, but a qualified guide supplied by the State Tourist Office. He spoke a version of English which so effectively stripped the meaning from his remarks that I steered him back to Spanish whenever I could. For the rest, he was gentle, sad-looking and pious, dividing his free time between visits to churches and – although well into middle age – running after women.

We got into Huehuetenango at four in the afternoon, and it turned out to be an earthquake town, with corrugated-iron roofs on fine churches, squat houses iced over with multicoloured stuccoes, and a great number of pubs having such names as 'I Await Thee on Thy Return'. We went into one of these, each of us carrying an armful of orchids, Calmo probably hoping that no one he knew would see him bothering himself with such contemptible weeds. The woman who brought the beer had a Mayan face, flat-featured but handsome, and full of inherited tragedy. Calmo told her in his most dignified way, 'This I say with all sincerity. I want to come back to this place and marry you.' The woman said, 'Ah bueno,' shaking off the compliment as if an invisible fly had settled on her cheek. She wore a massive wedding ring, and there were several children about the floor.

After that, Calmo wanted to go into the cathedral to pray for success in that week's lottery. The cathedral had just been freshly decorated for the pre-Lenten festival with huge bouquets of imitation flowers, their stiff petals varnished, and dusted over with powdered glass. Indians were lighting candles among the little separate patches of red and white blossoms that they had spread out on the flags to symbolise the living and the dead. Hundreds of candles glimmered in the obscurity of the cleared space where the Indians worship in their own way in the Christian churches, grouped in whispering semicircles round the candles, while their shamans passed from group to group, swinging incense-burners and muttering magical formulas. The Indians were dressed in the frozen fashions of the early sixteenth century: the striped breeches of Castilian peasants, the habits of the first few Franciscans who had scaled the heights to reach their villages, the cod-pieces of Alvarado's ferocious soldiery. They had left their babies hidden in the old people's care in the mountain caves, still remembering the days before the conquest, when at this season the rain god had taken the children for his annual sacrifice. These Indians were still surrounded by a world of magic and illusion, living characters in a Grimm's fairytale of our day in which the whites they see when they come down to the towns are enchanters and werewolves, who can kill with a glance, but are themselves immortal.

We went out into the sunshine again. A meteorite shower of parakeets fell screeching across the patch of sky stretched over the plaza. Soldiers, shrunken away in their American uniforms, were fishing in space with their rifles over the blood-red balustrade of the town hall, which was also their barracks. The green bell in the cathedral tower clanked five times, and the sleepers on the stone benches stirred a little in the vast shade of their sombreros. Calmo woke up an ice-cream vendor, bought a cornet, then said, 'I cannot eat it. The hot for my teeth is too great.' When speaking English he found special difficulty in distinguishing between opposites such as heat and cold.

We sat down in the car to decide what to do with the evening. The sleepiness of the place was beginning to paralyse us. Nothing stirred, but the vultures were waving their scarves of shadow over the flower beds. Calmo said, 'Yesterday a market-day, tomorrow a procession; so that today we have no prospect but an early night. There is really nothing to do.' As he spoke, a man came riding into the plaza on a tall, bony horse. The man looked like an Englishman on his way to a fancy-dress ball: he was lean, pink-cheeked, mildly aloof of expression, and his improbable costume of black leather with silver facings had clearly been hired out too often and was on the loose side for its present wearer. He was carrying a bundle of what looked like yard-brooms wrapped up in coloured paper. Calmo explained that these would be rockets for use in the next day's celebrations. The clip-clop of the hooves died away, and the silence came down like a drop-curtain. Huehuetenango was a place of apathetic beauty, built out of the ruin of a devastated Indian city. There was a sadness, a sense of forgotten tragedy in the air; and here it seemed that silence was a part of the natural condition. As Calmo had so often said, 'We Indians are a reserved people. Even in our fiestas. Our joys and our weepings are hidden away inside: for us only, you understand – not for the world.'

There was a notice over the hotel door that said, 'Distinction, Atmosphere and Sympathy'. The atmosphere was all-pervasive. The garden had been turned into a floral jungle encircled by borders of Pepsi-Cola bottles stuck neck-down in the earth. Quite ordinary flowers like stocks and hollyhocks were throttling each other in a savage struggle for living space, and hummingbirds like monstrous bees zoomed about the agonised sea of blossom. Goldfish bowls containing roses hideously pickled in preserving fluid, stood on every table-top. The bedroom towels were embroidered with the words, 'Sleep My Beloved'.

Food in this hotel was *American Plan* – words which have now been accepted into the Spanish vocabulary of Central America. They no longer refer to the system of charging for accommodation inclusive of meals, but describe a special kind of food itself – the hygienic but emasculated fare supposed to be preferred by American visitors, and now generally adopted on the strength of what are believed to be its medicinal and semi-magical properties. This time *American Plan* meant tinned soup, spaghetti, boiled beef and Californian peaches. The whole loaf of bread and a half-pound of butter of a generation ago had wasted away to two slices of toast and a pat of margarine. The milk was the product of Contented Cows, served in the original tin as a guarantee of the absence of dangerous freshness. We got through the boring ritual of dinner as soon as we could. The other guests – businessmen drawn from the elite ten percent of pure white stock – were still inclined to congratulate one another on the downfall of the last government, which had not been liked in commercial circles. 'A minimum wage. And why not? – I'd be the first. But when all's said and done, friends, what happens when you give an Indian more than forty cents for a day's work? You know as well as I do. He doesn't show up the next day – that's all. They've got to be educated up to it.'

After dinner I resigned myself to an early evening, and went to bed under a religious picture consisting of an eye that projected rays in every direction, and beneath it the question: 'What is a moment of pleasure weighed in the scales against an eternity of punishment?' I had hardly dozed off when I was awakened by an explosion. I got up and opened the window. The street had filled up with people who were all going in the same direction and chattering excitedly. A siren wailed and a motorcycle policeman went deafeningly past, snaking in and out of the crowd. There was another explosion and, as this was the homeland of revolutions, it was natural to assume that one had started. I dressed and went out into the courtyard, where the hotel boy was throwing a bayonet at an anatomical chart given away with a Mexican journal devoted to home medicine. The boy said that so far as he knew there had been no *pronunciamento*, and the bangs were probably someone celebrating his saint's day. I then remembered the lean horseman.

As the tumult showed no signs of abating, I walked down to the plaza, which had filled up with blank-faced Indians moving slowly round in an anti-clockwise direction as if stirred up by some gigantic invisible spoon. There were frequent scuffles and outcries as young men singled out girls from the promenading groups, breaking coloured eggs on their heads, and rubbing the contents well into the thick black hair. The eggs were being sold by the basketful all over the plaza, and they turned out to have been emptied, refilled with some brittle, wafer-like substance, repaired and then painted. When a girl sometimes returned the compliment, the favoured gallant stopped to bow, and said: 'Muchas gracias.'

Calmo, whom I soon ran into, his jacket pockets bulging with eggs, said it looked as if there was going to be a fiesta after all. He couldn't think why. There was really no excuse for it. The fashionable town-Indians, most of them shopkeepers, had turned out in all their finery, headed by the 'Queen of Huehuetenango' herself – a splendidly flouncy creature with ribbon-entwined pigtails down to her thighs, who was said to draw her revenues from a *maison de rendezvous* possessing radioactive baths. There was a sedate sprinkling of whites, hatted and begloved for the occasion.

Merchants had put up their stalls and were offering sugar skulls, holy pictures, plastic space-guns, and a remedy for heartsickness which is a speciality of Huehuetenango and tastes like inferior port. We found the lean horseman launching his rockets in military fashion from a wooden rack-like contraption. They were aimed so as to hiss as alarmingly low as possible over the heads of the crowds, showering them with sparks; sometimes they cleared the building opposite and sometimes they didn't. Other enthusiasts were discharging *mortaretes*, miniature flying bombs, which leaped two or three hundred feet straight up into the air before exploding with an ear-stunning crack. The motorcycle policeman on his scarlet Harley-Davidson with wide-open exhaust, and eight front and six rear lights, came weaving and bellowing round the plaza at intervals of about a minute, while a travelling movie-show was using part of the cathedral's baroque façade as the screen for a venerable Mexican film called *Ay mi Jalisco* featuring a great deal of gunplay.

A curious hollow structure looking like a cupola sliced in half had been built on the top of the town hall, and about this time powerful lights came on in its interior and nine sad-faced men in dark suits entered it by an invisible door, carrying what looked like several grand pianos. A moment later these pieces of furniture had been placed end to end to form an enormous marimba, under an illuminated sign that said 'Musica Civica'. A cosmic voice coughed electrically and then announced that in response to the esteemed public's many requests the municipal orchestra would have pleasure in rendering a selection of notable composers' works. Eighteen hammers then came down on the keys with a responding opening flourish, and the giant marimba raced into an athletic version of 'If You Were the Only Girl in the World'.

Calmo and I took refuge from the torrent of sound in a tavern called The Little Chain of Gold. It was a place of great charm containing a shrine and a newly installed jukebox in addition to the usual accessories. The main room was decorated with beautiful calendars given away by Guatemalan bus companies, and a couple of propaganda pictures of mutilated corpses put out by the new government after the last revolution. The Little Chain advertised the excellence of its 'hotsdoogs'. Most of its customers were *preparados*, Indians who had done military service and had rejected their tribal costumes in favour of brightly coloured imitations of American army uniforms. Some of them added a slightly sinister touch to their gay ensembles of reds and blues by covering the lower part of their faces with black cloths, a harmless freak of fashion which I was told had originated in a desire to breathe in as little dust as possible when footslogging along the country roads.

Calmo said that the main difference between a preparado and a tribal Indian was that the preparado, who had acquired a civilised taste for whisky, couldn't afford to get drunk so often as an uncivilised drinker of aguardiente.

We drank the aguardiente. It smelt of ether and had a fierce laboratory flavour. Every time the door opened the marimba music pressed on our eardrums. Calmo made an attempt to detain one of the serving girls. 'Don't go away, little treasure, and I'll bring you some flowers from the gardens in the plaza, whatever they fine me.' He received so baleful a stare for his pains that he dropped the girl's hand as if she had bitten him. At last the hour of civic music ran out. From where we sat, we saw that the Mexican outlaws had ceased to gallop across the cathedral wall. The crowds had thinned into groups of stubborn drunkards. Calmo was becoming uneasy. 'In my opinion it is better to go. These people are very peace-loving, but when they become drunk they sometimes assassinate each other in places like this. Not for malicious reasons, understand me, but as the result of wagers or to demonstrate the accuracy of their aim with the various firearms they possess.'

We paid our bill and had just got up when the door was flung open and three of the toughest-looking desperadoes I had ever seen reeled in. These were no shrinking Indians, but hard-muscled *ladinos*, half-breeds who carried in their faces all the Indian's capacity for resentment but none of his fear. They wore machetes as big as naval cutlasses in their belts. For a moment they blocked the doorway eyeing the company with suspicion and distaste, then one of them spotted the jukebox, which was still a rarity in this part of the world. His expression softened and he made for our table putting each foot down carefully as if afraid of blundering into quicksands. He bowed. 'Forgive me for addressing you, sir, but are you familiar with the method of manipulating the machine over there?'

I said I was.

'Perhaps then you could inform me whether the selection of discs includes a marimba?'

I went over to the jukebox. These ladinos, I thought, would still be living the frontier life of the last century; a breed of tough, illiterate outcasts, picking up a livelihood as best they could, smugglers and gunmen if pushed to it, ready, as it seemed from the frequent newspaper reports, to hack each other – or the lonely traveller – to pieces for a few dollars, and yet with it a tremendous, almost deadly punctiliousness in ordinary matters of social intercourse. I studied the typewritten list in Spanish. There were several marimbas. The ladino looked relieved. He conferred in an undertone with the other two fugitives from justice, came back, bowed again, and handed me a Guatemalan ten-cent piece. 'If you could induce the machine to play "Mortal Sin" for us, we should be much indebted.'

I returned the ladino five cents change, found a US nickel – which is fairly common currency in Guatemala – and put it in the slot, while the three ladinos edged forward, studiously casual but eager to watch the reptilian mechanical gropings by which their choice was singled out and manoeuvred into the playing position. 'Pecado Mortal' turned out to be a rollicking *son* – a kind of paso doble – executed with the desperate energy of which the sad music-makers of Central America are so prodigal.

Calmo and I were halfway through the door when I felt a tap on the shoulder. The principal bandit was insisting that we join him for a drink. 'Otherwise, my friends and I would feel hurt, gentlemen.' He laid bare his teeth in a thin, bitter smile. We went back and sat down again. While he was getting the drinks Calmo said, 'In the education of our people the most important thing taught after religion is *urbanidad* – good manners. Even those who have no schooling are taught this. I do not think that we should risk offending these men by showing a desire to leave before they do.' A moment later our bandit was back with double aguardientes and a palmful of salt for us to lick in the proper manner, between gulps. The music stopped, and his face clouded with disappointment. Behind him a lieutenant loomed, swaying slightly, eyes narrowed like a Mongolian sage peering into the depths of a crystal, mouth tightened by the way life had gone. He was holding a coin. 'Might I trouble you to perform the same service for me, sir?' he asked politely.

It turned out that the second *mestizo* wanted to hear 'Mortal Sin' again. 'It is remarkable,' he said, 'and most inspiring. I do not think it can be bettered.' The three tough hombres moved away uncertainly towards the jukebox again, simple wonderment struggling beneath the native caution of their expressions. The needle crackled in the ruined grooves, and we heard the over-familiar overture of ear-splitting chords. Someone found the volume control and turned it up fully. Every object in the room was united in a tingling vibration. The second bandit drew his machete with the smooth, practised flourish of a Japanese swordsman, and scooped the cork out of a fresh bottle of aguardiente with a twist of its point. Two more members of the band stood waiting, coins in hand.

'Mortal Sin' had been played five times, and we were still chained by the polite usage of Central America to our chairs, still gulping down aguardiente and licking the salt off our palms, when it suddenly occurred to me that it was unreasonable that an electric train should be rumbling through a subway immediately beneath us in Huehuetenango. I got up, grinning politely at our hosts, and, balancing the liquid in my glass, went to the door. The lamps in the plaza jogged about like spots in front of my eyes, and then, coming through the muffled din from The Little Chain of Gold, I heard a noise like very heavy furniture being moved about in uncarpeted rooms somewhere in space. The world shifted slightly, softened, rippled, and there was an aerial tinkling of shattered glass. I felt a brief unreasoning stab of the kind of panic that comes when in a nightmare one suddenly begins a fall into endless darkness. Aguardiente from my glass splashed on my hand, and at that moment all the lights went out and the music stopped with a defeated growl. The door of The Little Chain opened and Calmo and one of the ladinos burst through it into the sudden crisp stillness and the moonlight. Calmo had taken the ladino by the forearm and the shoulder – 'And so my friend we go now to buy candles. Patience – we shall soon return.'

'But in the absence of electricity,' the ladino grumbled sadly, 'the machine no longer functions.'

'Perhaps they will restore the light quickly,' Calmo said.

'In that case we shall play the machine again. We will spend the whole night drinking and playing the machine.' The ladino waved in salutation and fell back through the doorway of The Little Chain.

We moved off quickly under the petrified foliage of the plaza. Nothing stirred. The world was solid under our feet again. A coyote barked several times sounding as if it were in the next street, while a distant clock chimed sweetly an incorrect hour.

'A quiet evening,' I remarked. 'With just one small earthquake thrown in.'

'A tremor, not an earthquake,' Calmo said. 'An earthquake must last at least half a minute. This was a shaking of secondary importance.'

There was a pause while he translated his next sentence into English. He then said: 'Sometimes earthquakes may endure for a minute, or even two minutes. In that case it is funny...No, not funny, I mean very serious.'

The Snakes of Cocullo

So much of the traditional life witnessed by Lewis has vanished, but remarkably this outlandish festival, which he visited in 1989, seems to have thrived. You can see photographs and read a description by googling 'abruzzissimo ancient snake festival'. A most atmospheric video of the festival can also be found by googling 'Nick Hankins cocullo, Italy Magazine'. In the accompanying brief article, the author reveals that the Toothache Bell, mentioned by Lewis in the article, is still now firmly in use, despite having been banished by the Church.

Edward Burman, a writer living at L'Aquila, had been asked by a friend to show Lewis around. Burman and Lewis subsequently became friends, though Burman told Lewis's biographer that his first impression was of 'a dull old man bent over the bar'. He changed his mind over the subsequent days. He observed Lewis 'talking to shepherds, old ladies, and priests. I watched him coax information from stubborn interviewees with the most innocent-sounding questions'.

First published in the *Independent*, October 21, 1989

COCULLO SITS ON A HILLTOP IN ABRUZZO, on a level with Rome, but under the Apennines. This is a land of dark, lumpy mountains, empty roads threading through the valleys, and nothing in the silent fields to attract even carrion birds. An occasional village is crammed on to the top of a steep rock pinnacle, some half-empty, some wholly deserted. Fragments of the old southern customs survive where there are still inhabitants. *Guaratrici* (female healers) and *prefiche* (professional mourners) still serve local needs. Makers of amulets and those who concoct love philtres carry on a semi-clandestine trade. This is the part of Italy where magic and Christian faith are hardly separable. The 'Catholicism of the people' exists alongside the authorised version of the religion.

It is very much a family affair. The saints are still rewarded for their successful intercession in village matters, blamed for failures, and carried in procession to pay one another courtesy visits. Some houses have protective formulae against the Evil Eye carved in the stone over their doors.

The fame of Cocullo lies in the survival here of a snake-cult dating from pre-Roman worship of Angizia, goddess of agriculture and snake-charming in the ancient Marsican culture. On 19 March every year, the young men who have inherited or acquired the intuitions and skills required in the capture and domination of snakes go into the surrounding mountains for the ritual of the annual hunt. The snakes taken at this time are brought back to the village where – treated with a certain indulgence, even affection – they are prepared to adopt the principal role in the festival held on the first Thursday of May, when the procession of the *serpari* (snake-men) takes place. Before this they receive the blessing of the Church, and are then 'offered' to San Domenico. The saint – also a snake-charmer in his time – arrived in Cocullo in AD996, to take over officially from the goddess. The ritual, however, seems to have continued much as before.

The pagan goings-on at Cocullo seem to have been practically unnoticed even in most of Italy until a visit paid to the village in 1909 at the time of its snake festival by a Mr W. H. Woodward, who thereafter gave an account of his adventures in the *Manchester Guardian*. Mr Woodward arrived as the proceedings were about to start. Making straight for the church, he was surprised to find, a half-hour before High Mass was due to be celebrated, that a number of shepherds were kneeling at the altar rail, each with several huge white wolfhounds held on a leash, their muzzles resting on the rail. The dogs were there to be blessed and at the same time to be 'reverently' touched by a relic left by San Domenico when taking his departure from the village. This took the form of a shoe from his mule. The shoe was employed as a talisman against rabies. An even more cherished gift made by the saint, Mr Woodward was told, was a tooth he had wrenched from his jaw on the moment of parting. Ever since, sufferers from toothache in Cocullo had been able to cure themselves by kissing the relic, attaching a cord to the troublesome tooth and then using this to pull the special toothache bell in the church.

There were more surprises in store for Mr Woodward. Next day the procession took place, and he was startled by the emergence from the church of the image of San Domenico entwined with numerous snakes, and followed by members of the clergy, each carrying a serpent. It seems likely that vipers were present among the snakes carried at this time, for Mr Woodward goes on, 'The crowd hails him with prayers and invocations. Despite the seeming peril, hands are put forward to touch the saint... The venerable priest under the canopy carried his votive serpent with no sense of horror as being an evil thing, but rather with a caressing friendliness...'

This was all too much for the pillars of the established Church in Italy. The report was brought to the notice of the Bishop of L'Aquila, who had jurisdiction over the region: he went into scandalised action. It was deemed inadvisable to attempt the total suppression of a ceremony that had been going on for more than two thousand years, so the procession was allowed to stay, but other 'barbarous superstitions' practised in Cocullo were banned. Thenceforward no animal was to be taken into the church, and the toothache bell was banished. Also abolished was the custom by which earth dug from a cave in which San Domenico, a hermit, had preferred to shelter was sprinkled round the village houses to keep the snakes away. Some years passed before cautious backsliding was reported. The dogs excluded from church instead received the magic touch of the mule's talismanic shoe within sight of the saint's image through the open door. The shoe had now been renamed 'the branding iron', and every sheep leaving Cocullo to join the winter migration to the plains of Apulia was branded in the same way. A decade passed before the toothache bell was smuggled back and put into service again, and now, once more, people sneaked by night into the cave under the church to scrape up the soil considered still to be saturated with spiritual radioactivity from the saint's body. It is rumoured that, even as recently as 1986, one or more snakes found their way into the church itself after the May procession, under the pretence of a competition in which they were judged for colour and size. Prizes were distributed.

It was a sharp, clear morning in Cocullo on 4 May of this year. The first days of spring had breathed a little snow over the mountain tops but enormous violets showed in the road verges all along the steep climb up to the village. The houses were twisted turban-fashion through the upthrust of rocks, a little unbalanced in their grouping by a basilica that gave the place something of a Greek appearance. Under this was the cave where San Domenico had once sheltered and performed his miracles, but the building had been partly destroyed by the earthquake of 1981, since when the cult had moved its centre to the Romanesque parish church at the top of the hill. This occupied most of one side of the small square, which otherwise contained a handsome bandstand, a deeply cavernous bar and a few substantial houses with balconies.

The *serpari* pacing below were not as expected. As keen protagonists in an ancient magico-religious rite there should have been something about them, a certain non-conformity of appearance and manner setting them apart, but they were no different from any gathering of young men with time on their hands in a small-town square, except for the imposing presence of the snakes, some of which were surprisingly large. Big snakes roped round their necks hung down to the knees of well-pressed trousers, and small ones were worn, coiled like bracelets, round the wrists. The snakes, while alert and watchful, were sluggish in their movements. The snake-men stroked them gently and in response the snakes lifted cautious, swaying heads to study them with brilliant eyes, and thrust flickering tongues through mosaic lips.

The snake-men were willing and even eager for their snakes to be handled by strangers, although they were clearly nervous that they might be damaged. A young man in a blue suit, who appeared to exercise some special authority, was quite prepared to talk about his involvement with the *festa*. Alberto Lanzara had been born into a shepherd family – all the men of Cocullo had begun life by following the flocks – but now he was a technician in the Fucino Telecommunications Centre, Telespaziale, off the next exit but one from Cocullo on the autostrada. The easy accessibility of this area of snake-fetishism to the temple of high technology made it possible for him to pop backwards and forwards in his Fiat to supervise the collection of the snakes and their care. It was an occupation providing perfect relaxation from the mental effort demanded by his work. This, he said, had been an excellent year for the snakes. The weather (sun following rain) had favoured their appearance on 19 March – the traditional date on which, if disposed to collaborate in the ritual, they could be expected to emerge, wriggling from the earth. Thereafter they were kept in roomy earthenware jars, fed and tended until the day of the ceremony, and immediately afterwards returned to the place where they had been found, in the knowledge that they would present themselves and await recapture the following year. No poisonous snakes were now employed in the ritual, although one of the preferred varieties, the *colubri dalle quattro linee*, was the largest found in Europe. Lanzara understood that in a previous era his predecessors charmed snakes with the music of flutes, and by spitting upon them, but nothing of that kind still continued.

The arrival at about this time of the pilgrims marked another stage in the proceedings. Parties walked painfully across the mountains from local villages, or were brought by bus from the towns of Frosinone or Sora, where descendants of people who had emigrated from Cocullo had established pockets of the cult. The first contingent, comprising elderly persons of both sexes, struggled up the hill into sight, lighted candles clenched in hands, heads bent – humbled by this moment of the great day of their year. A soft mewing of hymns was smothered suddenly under the vast wheeze and groan of their bagpipes.

Firecrackers exploded overhead and, passing under a beflagged arch, the pilgrims found the *serpari* awaiting them, and reached out to be refreshed by the touch of the snakes.

With this influx the mood in the square quickened. New faces had appeared; faces sculpted with the long agony of field labour, and the faces of the middle class imprinted with city calculation and stress. The crowd filled all the open spaces, crammed into the church under the firmament of candles, and into the tunnel of a bar where there were thimblefuls of raw spirits on offer with ritual bread baked in the form of coiled serpents.

Midday approached, and with it the climax of the ceremony. A woman, who had been crouched behind the toothache bell with plastic bagfuls of earth from the old cave, packed up and departed; girls dressed in the marvellous uniforms of antiquity dismantled a pyramid of sweet cakes for distribution among the snake-men. The saint's image on its platform appeared at the church door, and with this the crowd shouted all at once – a sound that surprised like a clap of thunder out of clear sky. This was the moment of the blessing of the snakes and their 'offering' to San Domenico, which took the form of dressing the image with their writhing shapes until every part, head, body, arms, pastoral staff and metal aureole, squirmed with stealthy serpentine movement.

The procession began, seen from the square as a slow, twisting advance through the static mass of the crowd. The top half of the image, jerking forward foot by foot, was occasionally blocked from view by children hoisted upon paternal shoulders, upheld arms holding cameras, or the magnificent hats of the police who were clearing the way.

Mixed in with locals and pilgrims were a small number of young men clearly from the outside world, some of whom I had seen arrive in big cars with Roman numberplates, which they parked at the bottom of the hill. As late arrivals they found themselves at the back of the crowd, and now, suddenly, they formed themselves into a wedge and surged forward with such determination that they were able to reach up and fondle the snakes as they passed. And it was clearly all-important for them to be able to do so.

In this, for me, lay the surprise of the day. These men were not here for a tourist spectacle, or for an excuse for a few hours' escape from the world of banks. They were as much participants in the ceremony as the local peasants and shopkeepers, or the pilgrims who were now walking backwards at the head of the procession in order to be able to keep their eyes fixed upon those of the saint. Whatever the credences involved, they shared in them.

With that, the procession passed out of the square and the show was over. Many of those with cars now took themselves off to Villalago, a matter of ten miles away on the Lago di Scanno, where San Domenico had taken refuge uncomfortably for a year or two in a tiny cave over the lake. Like most of the heroes of religion he was a poor man, who had difficulty in feeding his mule, let alone himself. He appears also in the legends as an animal-lover, who objected to what he called fat laymen fishing in the lake for mere sport, and was apt to turn their catch into inedible scorpions and toads. In the close-knit family atmosphere of local religion he is spoken of as a well-liked relative, recently deceased.

A singular fact emerges about his cave, for in living memory (and as mentioned by W. H. Woodward) young children were taken there to bite through the necks of captive snakes. But why? The practices of magic, which so often present a reverse image of logic as we see it, can be strange indeed. The serpent, associated in the Bible with temptation, the Devil and banishment from Eden, appears in the mythology of old Europe in the benign form of Aesculapius, god of medicine and healing.

The cult seems always to have been strongest in the Cocullo area, where a universal medical cure has been compounded throughout history. This *teriarca*, which could be taken either in liquid form or applied as a salve, contained thirty ingredients, one being crushed vipers' heads, and it was on sale by a pharmacist in Rome – who obtained it in Abruzzo – as recently as ten years ago.

In a shepherd community, such as Cocullo, snakebite was once a common cause of premature death. Yet, in the festival, snakes are treated with affection, even a kind of reverence. The emphasis is strongly on propitiation rather than retribution. Why, then, this killing of snakes by the children at Villalago? The probable answer is that at this point, and even in this place, the ancient ceremony in the goddess's honour would have reached its climax with the sacrifice of the snakes.

Whenever such sacrifices – either of humans or animals – were performed, it was normal for the victim to be accorded the most solicitous treatment until the culminating act. In this case a bonus lay in the hope that in death the snake would transmit to the child some of those qualities, particularly sagacity, for which it was renowned.

Fighting the Mafia

Lewis was already an expert on the Mafia when he attended this trial in 1968. His insights and his connections with Sicily were helped by marriage to his first wife, whose elegant father was – at the least – strongly connected to the organisation. In his twenties Lewis's father-in-law had been arrested for a serious crime in Catania, escaped and was smuggled into the United States (possibly in a coffin with air-holes). Once there, he joined the Unione Siciliana, which was gradually infiltrated by the mob.

Over the years Lewis made repeated journeys to Sicily, and his reports on the Mafia received the remarkable accolade of being serialised in six consecutive articles in the New Yorker. *These articles were later published as a book,* The Honoured Society *(1964), which also included material the New Yorker had entirely excluded – about the Mafia's activities in the United States.*

In Tristram Powell's brilliant film about Lewis (google 'journeyman arena Norman Lewis') you can see Lewis at another astonishing mass trial in 1985, this time of 252 caged defendants, who were accused of being members of the Camorra.

First published in the *Independent*, July 22, 1989

MY EXPERIENCE OF A SICILIAN MAFIA TRIAL was extraordinary. It had attracted some international interest, having been widely advertised as designed to lay once and for all the hideous Mafia ghost. Others followed, each more spectacular than the last, with the accused men caged like animals in specially built courtrooms, and the judges escorted by armoured cars from their homes to the Palace of Justice. All foundered in boredom and disbelief, and the secret government of the Mafia continued as before.

I attended this trial in 1968 on behalf of a London newspaper, and as they had asked for photographs, I went to a friend, on *L'Ora* of Palermo, who promised to produce a photographer. 'He's a low-grade man of Respect,' my friend said. 'You may not be inspired by his photography, but he knows how to handle the judge – which is what really counts.'

Next day I went down to the courthouse where this man awaited me in surroundings that differed hardly in the matter of noise and excitement from a marketplace. Lo Buono was small and dynamic and bursting at the seams with a kind of genial cynicism. He carried an immense old-fashioned camera with tripod, and at the moment of entering the courtroom where the nineteenth day of the trial was about to begin, we were halted by an usher. His manner was exceedingly deferential. 'Will you be taking photographs today, Signor Lo Buono?' he asked. 'That is my intention,' Lo Buono replied, and the man smiled and bowed. At that moment we had been standing under a notice which announced that photography was forbidden under pain of the severest sanctions.

The court was in session with the wives and children of the accused men seated in the front row of the public benches. They were strikingly middle class in appearance, the women dressed meticulously as if for a first communion service in church. It was quiet in this room after the clamour of the antechambers and the air was heavy with a church-like odour of hassocks and varnished wood. What might have been a vestry door at the back opened and the eight prisoners filed in, led by a carabiniere with a gun. They were attached to a long chain, from which a second carabiniere freed them as soon as they had been seated in a double row in the dock, although they were still manacled at the wrists. All the prisoners wore immaculate sports clothes with open-necked shirts, and boasted impressive suntans, despite the fact that some had spent a year or two on remand in that notoriously sunless prison, the Ucciardone.

At this point Lo Buono opened up his camera, walked over to the dock and took a series of photographs of the accused men, none of whom gave evidence of noticing his presence. Next he photographed the judge, who acknowledged what might have been a routine courtesy with the slightest of smiles.

Now came the most extraordinary episode of the morning. One of the carabinieri, key in hand, went down the rows of prisoners releasing each man's left hand from the small chain attaching it to his right wrist. With this, while the judge and miscellaneous court officials turned their attention to other matters, the women and children got up, left their seats and made their way in orderly fashion over to the dock where moving scenes of family reunion were enacted. While the judge wrote in a book, the two carabinieri joined each other for a whispered chat. Hugs and kisses were exchanged, and the prisoners groping in their pockets produced sweets for the children and small gift-wrapped packages that might have contained perfume for the wives. I jerked Lo Buono's sleeve and gestured in the direction of this scene; he shook his head. 'Impossible,' he said. 'No one can take that picture.'

It was the last day of a trial in which once again the Italian state had revealed itself incapable of inflicting defeat upon the Mafia opponent. The Anti-Mafia Commission had been in operation for five years (it was to struggle on for another eleven), but had achieved nothing. In all trials that had taken place it was now assumed that the verdict would be 'not guilty'. Many were abandoned when witnesses for the prosecution retracted their evidence, went into hiding, fled the country or even committed suicide. One or two who had recklessly stuck to their guns could expect to be dealt with in exemplary fashion, such as the prosecution witness in the case of the mafioso monks of Mazzarino, who was found half-dead with a hand cut off. The contention of the counsel for the defence was that the victim had carried out the amputation himself.

Since it had become pointless to call witnesses for the prosecution, a new strategy had been adopted in the case I was attending. Instead of trying prisoners for specific crimes they were known to have committed, the charge was of 'association to commit crime', in which the prosecution promised to furnish evidence requiring no corroboration by testimony in court. The FBI had worked with Italian police agencies, offering proof that three of the prisoners – the possessors of dual US and Italian nationality – were heads of Cosa Nostra 'families'. The report was that they had gathered in Palermo at a time when the drug connection between Corsica and the States had come under police attack, with the intention of transferring the European base of the traffic from Corsica to Sicily.

It was hard to believe that the dignified, even benign-looking men in the dock could be overlords of the world of international crime, nor did their records, according to the defence, lend credibility to this point of view. Several were known for their association with leading personalities of the Church. Two had sons training for the priesthood. Another had paid for the building of an orphanage out of his own pocket. Vicenzo Martinez, the very pattern of a Sicilian gentleman, was a war hero who had lost an arm in action and been decorated for bravery (his citation was read out in court). He reminded one newspaperman of an 'aloof and splendid Coriolanus'. John Bonventre, one of the alleged Cosa Nostra chieftains (he had studied for holy orders in his youth), was head of a charity organisation promoting the welfare of Italian immigrants in the States. He treated the court with exaggerated respect, dropping a 'Your Excellency' into his interchanges with the judge sometimes twice in a sentence. He was also inclined to moralise. At one point the judge commented on the anxiety revealed in an intercepted letter sent by Bonventre to an American mafioso.

Bonventre: 'Your Excellency, I was worried about the times we live in. You can't pick up a paper without reading about some terrible thing ... I always say a quiet conscience is a man's dearest possession, Your Excellency, I'm sure you would agree with me.'

Judge: 'Not quite so much philosophy.'

With that, on 25 June 1968, the trial ended with the clearance of all the accused for lack of proof, and their subsequent release. That afternoon I saw an enraged Colonel Giuseppe Russo of the carabinieri, who had been sent from Milan to conduct an all-out war on the Mafia and to see to it that the accused men were not able to slip through the fingers of the police. He was a man of the north, handsome, clean-cut, frank, and with a take-it-or-leave-it manner. His office betrayed a liking for military order of a severe kind, with handcuffs – made to his own design, he confided – used as a paperweight on his desk. Contempt for the Sicilian environment in which he now found himself oozed from him. He referred to the Mafia as 'an oriental conspiracy fostered by local interests'. 'I have been sent here to finish it off,' he said, 'and that I propose to do.'

'The Mafia feeds on respect,' he continued. 'They have been able to convince people that they are all-powerful, and our first step is to destroy this legend. This must be done by public humiliation. Take the case of Coppola (one of the three Cosa Nostra chiefs). The custom here is to carry out arrests at night. I sent two men for him at the time of day when the neighbours would be about to see what happened. They chained him up and dragged him away. He lost face. For him things will never be quite the same.'

I called on Boris Giuliano, chief of the Pubblica Sicurezza of Palermo. His collaboration in the recent case was described by Russo in an offhand and grudging manner. Giuliano was a younger man than Russo – lively in southern style. It turned out that he spoke fluent idiomatic English with a London accent and vocabulary, picked up, he cheerfully informed me, while working illegally as a waiter in various Soho restaurants. He was clearly happy to talk to an Englishman. Giuliano was even franker than Russo and it was soon evident that he was no more impressed with Russo, the new arrival from Milan, than the colonel had been with him. He was particularly horrified at the account of Coppola's arrest. 'Russo's signed his own death warrant,' he said. 'They'll let things ride for a bit. Give the dust time to settle, then take him out. I give him five years.'

'What would you have done?'

'I'd have turned up in a plain car, been very polite, and given him a half-hour to get his things together. If there'd have been a woman around, I'd have bowed and apologised for the intrusion, and any kids in sight would have had a pat on the head. I have to do what I'm told. This man only has to lift a finger to have you snuffed out.'

Boris's predictions turned out to be correct. The blundering and impetuous Colonel Giuseppe Russo had gone blindly to the attack of an opponent he did not understand. He had arrested suspects by the hundred, nearly all of whom were released through lack of evidence, thus surrounding himself with implacable enemies who were prepared to bide their time. This came in July 1977 when an expected phone call summoned the colonel to a mysterious rendezvous. He was heard to say to the subordinate he summoned, 'This is the breakthrough', before the two men dashed off. The bodies of both men, riddled with bullets, were discovered some days later in a remote part of the island.

During my talk with Boris, I mentioned that there was some talk that at last there was a possibility of smashing the mafia. 'You can't,' he replied. 'At best you can contain it, which is at least something. It's the way of life down here, and to some extent we're all in it. Supposing you run a charity organisation, and somebody came to you with an offer of tens of millions of lire to build an orphanage. Would you ask him where the money came from? Do you believe the bishop is going to investigate any suspiciously large contribution to Church funds? They call that buying in. You heard about the project to build a marina? This is extortion money being laundered through the banks. When you have a national bank working with the Mafia, what are we supposed to do?'

There was a moment of confusion in the office. He was called away and came back saying he had to go out but would like to see me again. I told him I would be staying for a couple more days, and gave him the name of my hotel, and it was left that he'd give me a ring and pop over if he could.

Next morning he turned up at my hotel and we went down to the bar for a coffee. Sicilians who live at home in semi-darkness behind shuttered windows appreciate the maximum of light when they go out to relax. This place, despite the hour, was ablaze with the glitter of chandeliers. It delighted in Empire-style gilt furniture, which Sicilians also like. The waiters wore uniformed in scarlet and gold braid, like hussars, and the service was fast and good.

Boris Giuliano made it clear that he approved. 'And what do you think of it?' he asked.

'No complaints,' I said.

'In Milan they tell you to empty the boot of your car and leave it unlocked. None of that nonsense here. Nobody will touch a thing. For what they give it's cheap, too. They know how to buy at the right prices. This is the best hotel on the island. Know who owns it?'

'I can only guess.'

'By staying here you're contributing to mob funds. Want to take your trade elsewhere?'

'No,' I said. 'There's such a thing as carrying your principles too far.'

He laughed. 'Well, now you see how it is. You're one of us now.'

He had a couple of hours to spare, he said, and wanted to show me the sights. It came up that I'd been in Palermo before. When? he wanted to know. Five years before, I told him. 'You won't recognise it,' he said. 'Nobody could.'

We drove up to the end of the Maqueda for a view of the new Palermo, rolling through the low hills in a great red tide of bricks into and over the grey city. 'Fifty construction companies are working out there,' Boris said, 'with a Man of Respect on every board. They have the planning department in their pockets.'

'So they're taking over the city?'

'Yes, but if they didn't the Roman banks would, and the money would be siphoned off to Rome. This is a very complicated situation. The Mafia puts down petty crime and it makes work. It's found jobs for thirty thousand so far. Kids who used to live on bread and olive oil now eat meat. The inescapable fact is it has its uses.'

'This used to be a beautiful town,' I commented.

'Where the developers have been kept out, it still is,' he said.

We turned back and parked in an alleyway behind the Quattro Canti, still the noblest of road intersections, and set out into the back streets winding eventually down to the sea.

For the moment, the men changing Palermo were too absorbed in the new city to spare time and energy for the transformation that would sooner or later follow here. A cupola left by the Arabs lay cracked like an eggshell in a forgotten garden, where water running in a marble conduit showed through a filigree of leaves. Lizards darted in and out of the cracks in a Norman wall, and someone strummed on a mandolin against the fading rumpus of traffic. We found a bar down by the fish market into which legless ex-soldiers, face downwards on boards, dragged themselves by their hands to be fed by fishermen with the contents of sea-urchins. Nowhere could the heartlessness and the compassion of the Mediterranean have been more bitingly presented. Our presence went unnoticed: no one would listen in.

'All the books tell you the same thing,' Boris said. 'We're in the unique position of an island that's been invaded and conquered by foreigners six times in succession. Every fresh batch of foreigners changed the laws, which meant laws ceased to exist. We've been slaves to six masters. They left us with nothing. The Mafia had to exist. It defended us, fought for us, then conquered us. Now we've been conquered a seventh time. We got rid of the others, but we'll never get rid of this lot, because they speak the language. They're not foreigners. They're us.'

Sicilians are lonely. The sense of isolation from which so many of them suffer sets them apart from the other races of Europe. It is a trait manifesting itself in a number of ways. There are no country houses, and no small villages on the island. People live in towns, where, much as they may be inclined to keep their own company, they are comforted by the sound of voices, the sight of traffic, of people in the streets. Nevertheless they are lovers of the countryside from which they feel themselves debarred, and enjoy nothing more than to celebrate a festa by driving out into the grandeur of their empty landscape for a picnic and the collection of wild flowers, to which they are addicted. The problem then arises where to pull off the road. Drivers cruise along on the lookout for a pleasant spot, but also for company. Within minutes of a driver parking his car, he may expect to be joined by another. Each party, pretending not to be aware of the other, will get on with the business of lighting a barbecue fire and fetching water from a nearby stream. No greeting passes. Sometimes, when no second car arrives, those who have chosen the picnicking site will pack up and move on elsewhere.

Boris Giuliano was yet another lonely Sicilian ruined in the matter of his capacity to resist loneliness and isolation by the conviviality of the years spent in London, then returned to a society where reticence and secrecy were the norm. He loved the excuse to speak English which, I suspect, may have induced him more than once to put aside his work, and spend an hour or so with me in a remote corner, where he would be allowed to feed on memories of Soho's Greek Street.

I left Sicily and thereafter we exchanged sporadic correspondence from which I learned of journeys made to the United States. It had become clear that the Mafia had moved in to control the traffic in heroin; the secret laboratories now produced one-quarter of the world's supply. It was no longer possible for a policeman to continue to stand on the sidelines and talk about containment. On one occasion when he made a brief stopover in London I met him for an hour or so at the airport. He was on his way to Washington to confer with the FBI, and mentioned that a previous visit was concerned with the assassination of President Kennedy. He had proffered the theory, later accepted by many Americans, that contrary to the findings of the Warren Commission, this had been organised by the Mafia. His contentions suggested the plot of a novel which I subsequently wrote about the assassination.

In the early part of 1979, Boris's letters stopped, due I supposed to the increasing pressure of his work. In July of that year, two years after Russo's murder, he visited Marseilles and Milan in the course, as later revealed, of investigations into the allocation of spheres of influence in the narcotics trade between the United States and Sicilian Mafia. It has been surmised that high government officials of both countries found themselves compromised as a result of these investigations.

On 21 July he was back in Palermo, and at exactly 8 a.m., as usual, called in for a coffee at the Bar Lux, a few yards from where he lived. He stood at the counter to drink it, then chatted for a moment with several regulars before turning to go. At that moment there were about twenty customers, only one of whom could be later traced by the police to give an account of what happened next. This undoubtedly reluctant witness said 'I noticed a man who was trembling. He was white in the face. He must be ill, I thought. My first impulse was to offer to help. When the Signor Giuliano went towards the door, the man followed him. He drew a pistol and shot him three times in the neck. The Commisario fell face downwards, and the man then fired four more bullets into his back.'

Boris Giuliano was accorded in death the extraordinary civic accolade of *Cadavere Eccelente*, with which only six (including Russo) had been honoured in the decade, and was carried to the grave in a hearse drawn by twelve horses. By subsequent accounts of his career, he may have come as near as any single man to breaking the stranglehold of the Honoured Society.

The Bandits of Orgosolo

In the summer of 1966 Lewis visited Sardinia, partly because he was fascinated that bandits could still thrive in Europe. He asked the British consul to help find him a local guide for the bandit area. This turned out to be a wise precaution, considering that not long afterwards, a Dutch journalist travelling alone in the area was shot and seriously wounded.

Banditry persisted in Sardinia for a long time after Lewis's departure. In one famous incident in 1985, four bandits and a police officer were killed in a firefight; but today, apart from some bouts of kidnapping, banditry has almost vanished.

Orgosolo is now famous for its murals, a custom that probably began three years after Lewis's visit, following a protest against a NATO base that would have been built on land used by shepherds. In this small town, there are now more than a hundred extraordinary murals, some of them protests, some in the style of Picasso, some are images of rustic life. Many of the murals can be seen if you google 'Vagabundler.com Orgosolo'.

First published in the *Daily Telegraph*, November 18, 1966

TOWARDS THE END OF OCTOBER 1962, Edmund and Vera Townley, a middle-aged British couple on holiday from Kenya, who were making their way back to England by easy stages, arrived in Sardinia. Edmund Townley was employed by an import-export firm in Nakuru as well as possessing a half interest in an apiary which was doing well. He was also a notable jack-of-all-trades, who had been farmer, miner, and road-engineer in turn, as well as a bit of an amateur detective. The Townleys were regarded as a quiet couple, who didn't go out much, happy in their home life. They were a good-looking pair, and Vera had once been almost beautiful with strong, classic features. Edmund has been described by those close to him as being capable sometimes of aggressiveness, and he was outstandingly devoid of physical fear.

While the life of the Kenya highlands mostly suited them very well, they were both uneasy about the prospects for European settlers in independent Kenya. In Edmund Townley's case, there was some special additional motive for nervousness. He had been actively involved in the Mau Mau emergency, both officially as a screening officer, and as a private individual organising his own information network which had been responsible for the capture and death of several terrorists. Now he had learned that his name was on the ultra-nationalists' blacklist. This holiday, therefore, was to serve a double purpose. On the way home, the Townleys decided to visit the Mediterranean, and in particular Italy, with the idea perhaps eventually of buying land there for their retirement. Like so many Britons in their situation who have passed the active years of their lives under the African sun, they found it hard to believe that they could reconcile themselves to the climate of their native land.

The Townleys had always enjoyed pioneering, and now they were on the look-out for a place where they could start from scratch once again, clear a piece of land and start a beekeeping project. They had nearly settled for the Canary Islands, but as Mrs Townley spoke fluent Italian, it seemed more sensible to settle in a place where this could be put to use. Sardinia seemed the next best choice, and here, at least, there would be no language problem while, from their first enquiries, all the other natural advantages they hoped to find in their new homeland seemed to be present.

Sardinia, indeed, had a great deal to offer. In spite of the Aga Khan's luxurious settlement near Olbia in the north, the country was largely undiscovered by tourists, and land prices had not begun to soar as they had elsewhere in the Mediterranean. So far the coast hadn't been disfigured by chaotic development projects, as for example had the Costa del Sol in Spain. The cost of living was substantially less than on the Italian mainland, the beaches were the finest in the Mediterranean, domestic help was cheap and plentiful, the people charming and hospitable, and the towns clean – some of them built round a core of noble architecture. This was a land, in fact, possessing all the warmth and geniality of Italy, minus slums, smells and noise. If anything more was asked of it, this was an archaeologist's paradise, littered with dolmens, prehistoric 'giants' tombs', *nuraghi* (only one out of some seven thousand scientifically excavated), Punic cemeteries, shrines to the gods of Carthage and Rome, and rocks bearing mysterious inscriptions.

The Townleys planned to spend two weeks touring Sardinia, and, having flown from Rome to Sassari, they hired a Fiat car and set out for the interior. Ten days later, they arrived in Nuoro which is roughly in the centre of the island. Here they were in the foothills of the Barbagia – some of the wildest and least-known mountains in Europe. Sardinia is an island, but it is also a country in its own right, and it is big enough – one hundred and seventy miles in length from north to south – to possess real rivers and impressive mountain scenery. The sense of confinement, and in the end the claustrophobia of the small island, doesn't exist in Sardinia. Looking southwards from the window of their hotel room the Townleys might have imagined themselves confronted once again by a vast African horizon, although not so much the Africa of their own green highlands of Kenya as the Africa to the far north of their home on the barren frontiers of Ethiopia.

Nuoro has many attractions for the discerning tourist. It has stood apart from this century, a leisurely introspective town built in a graceful but haphazard fashion on the lower slopes of the sugar-loaf mountain of Ortobene. It is supremely Sardinian, and women in from the country still walk its streets in the bold flamboyant folk-costumes in a style inherited from the Middle Ages. Official brochures claim the view from the top of Ortobene to be the most striking in Europe. In fact, one looks out across a wide valley at an awe-inspiring recession of granite plateaux and peaks: a glittering hallucinatory whiteness where the sun striking the hard rock-surfaces counterfeits glaciers and snowfields. These are the mountains of the Barbagia – the word is from the same root as 'barbarian'. They are only 5,000 feet high, but almost as remote to humanity as the Himalayan peaks, and they are the last refuge of some of Europe's rarest animals, including a species of pygmy wild boar, as well as the indigenous home of the moufflon, elsewhere extinct. *Insani Montes* – the dangerous mountains – Diodorus of Sicily called them in the atlas he made of Sardinia in the first century BC. There has been no time in recorded history when outlaws have not roamed the Barbagia, and they are still as inaccessible to the prudent as they were when the Carthaginian, the Roman, and the Aragonese generals set up their outposts on the farther side of this valley of Nuoro, and went no further.

The Townleys stayed the night at the Jolly Hotel, one of an Italian chain set up throughout the country to relieve the asperities of tourism in such provincial towns. In the morning, they told the receptionist that they would be keeping their room again that night, but as it was a fine warm day, had decided to go for a drive in the country. They asked for and were given packed lunches.

Leaving Nuoro, they followed the main highway for five miles in the direction of Orosei, and then turned off into the narrow, winding and deserted road that leads to Orgosolo. Whether they knew it or not, the unknowing British tourists were now entering a most remarkable area. After two or three miles, the road passes through Oliena. Next comes Orgosolo where, barred by the mountain of Supramonte, the road loops away to the right to join the main Nuoro-Cagliari Road five miles further on. The population of these small sad towns and of the mountains behind them are mostly the descendants of Nomadic hunters that peopled Sardinia in prehistory. Franco Cagnetta, the Italian social historian, had written of Orgosolo, 'Here life is essentially unchanged after thousands of years; one is at the centre – all the more astonishing because the centre itself does not realise it – of a civilisation that is infinitely retarded; that has inexplicably survived. This is the most archaic community in the whole of Italy – perhaps of the Mediterranean basin.' As the people, so the landscape which has in part formed them. The mountain of Supramonte, which blocks the horizon south of Orgosolo like some flat-topped iceberg, is the bed of the sea thrown up by a cataclysm of 100 million years ago; its surface strewn with rocks gouged by the wind into fantastic shapes. The mountain has been hollowed out by vanished rivers, there are fissures a half-mile deep, as well as vast unexplored caves, primeval forests of chestnut and oak, and the ruins of nuraghic villages visited only by armed shepherds. The visitor to these parts from the outside world is warned not to leave the road, for this is the traditional stronghold of the bandit and of the vendetta.

Orgosolo, too, with its aroma of immeasurable antiquity has something to detain the traveller. In the greyness and the ugliness of its streets, one still sees figures from pre-history: old men in the stocking caps of the bronzes of the nuragic period of 1000 BC; an occasional shepherd carrying the triple reedpipes depicted in the same bronzes. Sixty years ago, the whole town was built of *fughiles*, the most ancient of stone habitations consisting of a single circular windowless room (in which it was impossible to stand upright) and a hole in the roof to let out the smoke of the fire burning in the centre of the floor. A few *fughiles* still exist even if they are disguised these days behind the façades of normal houses. Students of folklore find Orgosolo of extreme interest; at the high spot of any celebration is the apparition of the *mamutones*, the ancestral spirits, in sheepskins and tragic masks carved in wood, transporting the onlooker into an eerie animistic world that lurks here in the shadows behind a perfunctory stage-setting of Christianity.

The invisible life of the community is equally singular. Nothing but lip-service has ever been offered to the State, and the only laws respected are the ancient customs codified in the *Carta de Logu* of 1388, on the eve of the extinction of a thousand years of Sardinian independence. Never since the overthrow of the rule of the Sardinian judges by the Kingdom of Aragon has the presence of authority possessed legal validity in the eyes of Orgosolo, which initiated nearly 600 years ago perhaps the longest resistance-movement in human history. Within the provisions of the famous *Carta* are laid down in minute detail the rules for the conduct of the vendetta. Orgosolo's only building of significance is the Church of San Leonardo with the famous churchyard and its row after row of small wooden crosses marking the graves of men who have met tragic deaths. It has been stated in the Italian press that of a population of 4,500 over 500 men have died through the vendetta since the war alone. By local standards, none of these killings have been crimes: at the most, they are the malefic links in a chain of cause and effect, the payment of debts of blood, the almost mechanical retributions decreed by a revengeful Stone Age Jehovah.

The spot chosen by the Townleys for their picnic-lunch was a tiny triangular field half a mile from the outskirts of the town. Looking down from Orgosolo, it is the only green and pleasant place to be seen in any direction among the browns of the lean, sun-scorched earth. The owner of this small oasis of wildflowers and grass died mysteriously leaving no heirs to cultivate it. By chance, it offered an unusual and obvious amenity in the form of a small, oblong, flat-topped rock. It was the only place for a picnic along the whole of the road they had come, and the Townleys, having pulled their small car into the roadside, climbed the low bank into the field, unpacked their luncheon boxes, and set out the contents on top of the rock. This would have been about mid-day, and the Townleys had their lunch and were perhaps given a little time, too, for relaxing in the pleasant sunshine before they were interrupted by the appearance of a stranger.

Two days later, on October 30th, the London *Times* published a short description of the discovery of their bodies by shepherds. They had been riddled by bullets but, said *The Times*, 'the motive for the killing is not clear. Wrist watches and other objects of value were left untouched. The theory is that they may have come across a band of outlaws, who impulsively shot them and fled.'

Other reports were more erratic and fanciful and, in the case of the *Daily Telegraph*, even self-contradictory: '... they were preparing for a picnic ... the couple must have stumbled on a bandit hideout, and the bandits in the dusk mistook them for approaching police and opened fire'. And a few lines further on: 'According to a reconstruction of the crime, the attack took place (by the roadside) shortly after the passing of the regular bus in the early afternoon'. The *Daily Telegraph* has the couple killed by a large-calibre shotgun which it erroneously describes as 'the customary weapon of Sardinia', but next day it found that Mr Townley's own pistol had been used. This report ends by sketching in the Orgosolo background, where some years before more than twelve people were killed 'one by one after announcing the next victim's name in crude paint on the white walls of the churchyard'.

But even *The Times* theory was not a tenable one. The Sardinian outlaw is rarely a pathological criminal, but almost inevitably a man who considers himself an unfortunate victim of circumstances with as clearly defined a moral code as the man who has not been obliged to take to the mountains. Under pressure of hunger, he will commit acts of banditry, but the Robin Hood image is always there at the back of his mind. He robs with a certain flair, never molests a woman, never takes from the poor. If obliged to kill, such a man does not act impulsively but after extreme premeditation. Never in these mountains had a bandit been known to kill a foreigner. Seventy-five years previously, in fact, when two Frenchmen had been abducted by bandits who had believed them to be Italians, they were released as soon as the mistake was realised and sent back to Nuoro laden with propitiatory gifts. All Sardinia was aghast at this meaningless tragedy of the Townleys, and, in due course, other and even less profitable theories were produced in an attempt to explain the inexplicable.

One of these was an attempted sexual assault on Mrs Townley, but it had to be dismissed as more than unlikely, for not only was the lady fifty years of age, but there were no signs whatever of any interference or struggle. A local newspaper improbably suggested that the Townleys' death might have been the result of a suicide pact and printed an interview with a fellow guest at their hotel who, although he understood no English, claimed to have overheard them quarrelling. But if this was so, how was the fact to be explained that the Townleys had been killed with an Italian weapon which was never found?

By the middle of November, more sinister allegations were appearing in the Italian press. The suggestion now was that the apparently motiveless killing of the Townleys had been an act of terrorism designed to disrupt the nascent Sardinian tourist industry, and to discourage such projects as the Aga Khan's development on the Costa Smeralda. On 15th November, *Le Ore* announced:

> Sardinian banditry is perhaps divided into two camps; one committed to impeding the country's tourist development in which numerous financial interests are involved, and in particular those of the Aga Khan Karim; the other to protecting the operation. Naturally, the bandits of the first group are in the pay of obscure personalities and the most reactionary cliques, who are interested in keeping the island in backwardness and misery. In this light, the murder of the English couple would appear not as a stupid and meaningless crime, but an act of purposeful intimidation intended to terrify the future clients of Sardinian Tourism.

A far-fetched solution, perhaps, but not without its germ of possibility. In our own days, vestigial feudalisms, which have survived centuries of opposition by political reformers and well-meaning governments, have collapsed and died when exposed to a tourist boom lasting hardly more than a decade. In 1950, in the South of Spain, where the conditions in those days were roughly comparable to the most distressed areas in Sardinia at the present time, the Andalusian field worker was paid fifteen pesetas a day, and when laid off between sowing and harvest, sometimes lived on such things as roots, frogs and snails. The same man, transformed into an unskilled labourer on a holiday building site on the Costa del Sol or in Majorca, is now paid three hundred pesetas a day. His daughter, rescued from the expiring feudalism of the South to become a chambermaid, earns almost exactly the same in one week as in a year in service in one of the great houses of Andalusia. Despite all the increases in the cost of living, the advance in the standard of living of the labouring class is huge.

The object lesson is not lost on the landowners of such countries as Sardinia where tourism remains an infant in swaddling clothes. Labourers employed on the Aga Khan's project near Olbia receive at least ten times the pay of a peasant on an estate, and work an eight-hour day as opposed to anything up to fifteen hours demanded from his workers by one of the landowners of the old school. A shepherd, who has been tempted from his skilled and ancient trade to become a waiter in one of the Costa Smeralda's luxurious and expensive hotels, may occasionally expect to receive in a single tip as much as he could have earned in a week trudging over the desolate mountains after his sheep. Insidiously and indirectly, everywhere, a tourist boom destroys privilege and imposes its own democracy whatever the form of the regime. It begins by mopping up the pool of unemployed upon which feudalism depends, and in the end entices away the workers that remain, thus depriving the feudalist of the labour he needs to carry on. He cannot possibly feel anything but hatred for the interlopers of the tourist trade, and if he is strong, unscrupulous and bred in a tradition of rapid authoritarian action, he may be ready to fight back – and with whatever weapon he can.

But the theory of the Black Hand of the diehard feudalist and devilish manipulations behind the scenes – however tempting to the Latin sense of the dramatic – had to be abandoned. A succession of macabre happenings, following immediately on the heels of the Townley killings, were considered and eventually related to the murder of the two British visitors. Now, too, the suicide-pact rumour, so eagerly seized upon by the Sardinians following the story that the Townleys had been shot by Edmund Townley's own gun, had to be relinquished.

On November 2nd, the *Daily Telegraph* said:

The bullet-riddled bodies of two notorious bandits were found today under a bush less than two miles from the spot where a British couple, Mr and Mrs Townley, were killed last Sunday.

One of them, Salvatore Mattu, twenty-three, was said to have killed a policeman when he was only nineteen. There was a price of £600 on his head. The other, Giovanni Mesina, forty, was released from gaol a short time ago.

Police investigating the deaths of Mr and Mrs Townley are inclined to admit that some connection exists between the two bandits' execution and the motiveless murder of the British couple. The most likely theory is that other bandits executed the two for having killed foreigners.

In addition, the other bandits hoped that by 'having done justice' for the murder of the British couple the duel between bandits and the police will again return to its previous comparatively leisurely course.

The use here of the word 'executed' is significant, and it is correct enough in the case of Mattu, although a much less appropriate definition of the manner in which Mesina met his death. What this report does not make clear is that the discovery of Mesina's body followed that of Mattu. The story of Mattu's supposed murder of a policeman at the age of nineteen cannot be confirmed. At the time of the Townley incident, Mattu was a fugitive from justice suspected of the kidnapping and murder of a rich landowner two years previously. Mesina, having come out of prison, had married and settled down in the town. Like forty percent of the employable men of Orgosolo, he was without a job.

Four days after the Townleys met their deaths, Mattu's body was found – in accordance with local tradition, 'by a young and innocent child'. He had been shot to death and the corpse was displayed in what might be described as ceremonial fashion. Like a princeling of Ancient Egypt prepared for his journey to the underworld, his weapons and portable possessions had been placed at his side. This signified to a student of the mores of Orgosolo that he had indeed been executed following sentence by a secret court of the heads of the clan-families of the town. Someone let drop the fact that among the objects found with the body had been a pair of binoculars belonging to Edmund Townley. The natural assumption was that the crime for which Mattu had been judged, sentenced and executed in such short order was the Townleys' murder. But before the amateur investigators had had time to prise more information out of their contacts in Orgosolo, the situation was complicated by the discovery of Mesina's corpse. There was nothing this time to suggest a formal execution. Mesina had simply been murdered by a burst of fire from a sub-machinegun, and his body flung contemptuously face downwards on the ground. But the fact that both Mattu and Mesina had been found in the same place in some way linked their deaths together. It was known that Mattu and Mesina were sworn enemies.

Meanwhile, police investigations had dragged to a standstill defeated, as usual, by Sardinian *Omertà* – the silence, the wilful ignorance, the honourable non-cooperation with the law, which is the normal citizen's defence against what is seen as the inhumanity and the essential 'foreignness' of Italian justice. Several hundred carabinieri and members of the Public Security force abetted by a helicopter wandered aimlessly and forlornly about the rocky trackless wilderness of the mountain of Supramonte, lost themselves, broke their limbs when they fell into crevices, chased after shepherds who behaved like deaf mutes whenever they were cornered, and shone their torches into the blackness of caverns as big as cathedrals in which a thousand bandits could have hidden themselves and never been found. In Sicily, a politician was quoted as saying, 'a pity we can't lend them our Mafia'.

The middle-class Sicilian and Sardinian attitude towards both the Italian police forces are practically identical. It is one of amused contempt. This quip was made in recognition of the well-known fact that, after the Italian police (aided by an army division) had battled ineffectively against the thirty bands of outlaws infesting Sicily after the end of the last war, these bandits were liquidated in a manner of months, when the job was unofficially confided to the Mafia. But then there has never been a Mafia in Sardinia, where this famous and ferocious secret society is temperamentally as alien as it would be, say in Holland or the West Riding of Yorkshire. At the time of the Townley tragedy, there were ten outlaws on the mountain of Supramonte alone, despite the fact that a recent intensive police drive had resulted in forty of the citizens of Orgosolo being in prison serving life sentences.

As all these men would be normally covered by the blanket description of bandits, it is necessary to consider and attempt a definition of the word. Few of the forty lifers from Orgosolo dispersed about the island's maximum-security prisons would, in fact, have committed any crime by local standards. They would have been no more than the executors of long-standing feuds, quarrels passed down from generation to generation, faded and hardly identifiable hatreds taken over as a matter of social obligation by sons, grandsons and great-grandsons of the original disputants. Dimly reflected are the customs and the *lex talonis* of the nomadic hunters and migrant pastoral peoples of immense antiquity, of Homeric Greece and the Near East of the Old Testament, people without settled property forced to push on endlessly in the search for sparse pastures or follow the movement of hunted animals into territories claimed by other tribes. These conditions still persist in the Barbagia in Sardinia, and they have been fostered by poverty, isolation, the remoteness of the central government, and by tradition. In such fossilised societies, the power normally confided to the State in a more advanced civilisation is stubbornly retained by the family – the Mediterranean 'compound family' of the anthropologist, which may number anything up to 100 members – a tribe in miniature composed of the 'senior father', then sons and grandsons, and all the wives (the women leave their parents to live in the patriarchal home). The family inhabits a single house, or a number of adjacent houses, is ruled by a Council presided over by the 'senior father', and holds all family property in the way of buildings and livestock in common. The compound family's strength lies in its utter self-sufficiency and its single will; its weakness is its horrifying memory. It is found in remote corners of Sicily and Corsica, as well as in Sardinia. Even in the tiny Spanish island of Ibiza – hardly visible on a small-scale Mediterranean map – a few such archaic family amalgams still exist, living in fortified towers, *atalayas*, which are architecturally related to the *nuraghi* of Sardinia, dominated by formidable patriarchs who a generation or two ago would have seen to it that their commands were enforced, if necessary by knife or gun.

A life-sentence for carrying out the obligations of the vendetta carries no social stigma in Orgosolo – in fact, quite the reverse: it is seen more as a kind of sombre accolade, the admission to the exclusive club of men who have not hesitated to sacrifice themselves on the altar of honour. In gaol, the *ergastolano* with an honour killing or two to his credit is usually a model prisoner, entitled to an Olympian aloofness, treated with respect by prisoners and warders alike, often addressed by prison officers using the polite form *lei* where a swindling millionaire must content himself with the familiar and contemptuous *tu*. From *Enquiry on Orgosolo* by Franeo Cagnetta: 'An ex-prisoner who comes back from serving a life-sentence finds a wife who has awaited him in perfect fidelity, and who has brought up and educated his children. Thereafter he occupies in the community a position of special respect.'

Cagnetta was referring here to the vendetta killer, but another and far more numerous class of outlaw must be included in the general category of bandits. This is a purely Sardinian speciality, the *dogau* (a semi-bandit), who is the creation of a catastrophic error on the part of the police. The police's mistake was the employment of secret informers, and action was taken without any check being made on the veracity of their reports. This created an immediate vested interest in banditry. The informer – paid a small lump sum for every arrest – denounced all and sundry, and a single anonymous accusation was enough to secure a man's arrest and imprisonment perhaps for years while awaiting trial. Worse still, acts of banditry were encouraged and organised by informers, who took a share of the loot before selling the participants to the police. The logical outcome was that when a man – however impeccable his previous record – had reason to believe that someone had denounced him to the police, he took to the mountains. After that, it was usually only a matter of time before hunger drove him to become a bandit in reality.

These borderline outlaws – suspects who cannot be charged with any specific crime – still exist off and on by the hundreds in the mountains of central Sardinia, and they form a pool of tough and embittered humanity – samurai of our times, who are available for employment in any kind of dangerous or illegal activity. Last year, when possibly a record number of animals were stolen in Sardinia – no official figure is available because less than half the losses are reported – a rich man, the descendant of one of the original Italian 'Colonists', for whom the Sardinian underdog cherishes an inherited detestation, complained to a magistrate that the police couldn't protect his property. 'Why come to me?' was the magistrate's astonishing reply. 'Surely there are plenty of *dogaus* about on the look out for a job? Take on a few like everybody else does. You'll find you'll have no more trouble.' Cagnetta believes that practically every male citizen of Orgosolo has been a *dogau* at least once in his life.

The ill-fated Mattu and Mesina – the one presumed to have been sentenced and executed by a secret court, the other whose death still remains a mystery – had both started their careers as *dogaus*. After that, they kept the wolf from the door by staging a few unimportant hold-ups. These happen by the hundred in the remoter parts of Sardinia, are often carried out in an apologetic fashion with the nearest approach to courtesy possible when one man is holding a tommy-gun pointed at the chest of another, no one is hurt, and the bandit gets away with the equivalent of a few dollars. From modest hold-ups of this kind, however – almost enforced charities –

Mattu and Mesina graduated to kidnapping, in this way passing the point of no return. Public opinion in Orgosolo would still have remained sympathetic to them as the victims of circumstances, and when either man came into town to visit a parent or relation, he would have been given food and shelter, protected from the eye of the informer, and smuggled back after the visit to the safety of the labyrinths of Mount Supramonte. Even if Mattu or Mesina had found themselves compelled to kill a policeman while avoiding capture, it is unlikely that Orgosolo would have held it against them. But the purposeless murder of a foreigner would be regarded with horror, as a stain on the honour of the community and, therefore, to be dealt with implacably.

Orgosolo was once more in the limelight. Newspapers delved into history for accounts of the exploits of the innumerable bandit chieftains it had produced. The masterpieces of the illustrations of the old *Tribuna Illustrata* were rediscovered and reproduced, and they offered a highly seasoned choice of scenes full of heroic action, savagery and anguish. A favourite was the attack on the Cagliari-Ussassai train carried out in 1922 by a force of about a hundred bandits. Another was the last stand of Salvatore Pau. It shows the bandit perched on one leg – the other having been broken by a bullet – waving his pistol defiantly while smoke spurts simultaneously from the muzzles of the rifles of the dozen soldiers who have him cornered. A third records the return of troops from a successful anti-bandit operation – the bandits roped to the backs of horses, while the commanding officer receives the congratulations of a government official, who receives him with raised bowler hat. There was also a highly imaginative reconstruction of the marriage in 1929 of the bandit chief, Onorato Succo, celebrated in the Church of San Leonardo in Orgosolo after the regular publication of the banns, and followed by a banquet at which members of the local police force and government functionaries were entertained.

This was all very well, but most Italians took the view that such things belonged to the romantic past, and they were startled to learn that despite half a dozen full scale military expeditions since the turn of the century, the bandits were still there. A few serious commentators stopped to enquire into the anomaly that at a time when the nation's industrial growth-rate was the highest in Europe, areas existed in Sardinia where a tourist's life was not safe and where the social pattern typified by the tradition of the vendetta hardly differed from the description of such authors as Diodorus and Strabo.

Isolation, poverty and neglect no doubt explained the situation in part, but there were other areas – in Calabria and in much of Sicily – even more poverty-stricken, if slightly less isolated. This being so, the determining factor must be sought in history.

The Mediterranean islands, Sardinia included, became the colonies of the ancient world as soon as the continental nations developed into sea powers. The colonisation was brutal, wasteful and forthright; analogous to the destruction by the Spanish conquistadors of the Caribbean Indians, and their replacement with plantation slave-labour brought over from overseas. In the late Bronze Age, the indigenous civilisation of Sardinia had reached a degree of prosperity evidenced by its trade in luxury articles with Egypt, Phoenicia and Etruria, and it was protected by 6,500 fortresses – the *nuraghi*, some of them most ingenious and complex in construction, and large enough to shelter a whole village in times of danger. By the end of the sixth century, the Carthaginians had occupied all but the mountainous areas of the island, dismantled the *nuraghi* and massacred all the natives within reach. These they replaced by Libyan slaves, who laboured and died to produce wheat to feed the Carthaginian armies. Diodorus noted that a few Sardinians managed to survive by building underground shelters, and these could only be caught by waiting until they were forced to come out to visit their religious shrines – a problem the Romans surmounted by importing police dogs to chase the refugees from their hideouts.

New conquests brought new masters. After the Romans, came the Byzantines, then, following a thousand years of independence, the Aragonese – bringing with them their own brand of torpid feudalism – then the Piedmontese and the Italians. Only the mountains of central and eastern Sardinia were never occupied and settled – visited at the most by punitive expeditions – and here in the Barbagia with its spiritual capital, Orgosolo, a little of the old independent Sardinia, however degenerate, survived intact.

All the men, who farmed in the plains, made roads, and built towns, were strangers and enemies to the untouched mountain peoples, and the Italians, who took over Sardinia in 1848, were as foreign as their predecessors. With the Italian arrival, the Sardinians got their first taste of capitalism and found it even more bitter to their taste than the anachronistic feudalism to which they were accustomed. First of all, the common grazing lands were expropriated, sold by auction, and acquired at knock-down prices by the church. After that, a law was passed permitting any Italian colonist to claim as much land as he could afford to build a wall round. Italian timber companies then came on the scene and cut down the forests – a project which was completed in ten years. As a result, the climate changed and a number of small rivers dried up. The equivalent of the Elizabethan 'Statute of Sturdy Beggars' was enacted to deal with the vagrant dispossessed, who soon filled the prisons. 'The time has come to ask ourselves if, in the past, we have not wasted scruple,' Cavour said. '... Treat them as the English treat the Irish.' The Italian police and troops, who poured into the island, did their best to comply with true Britannic ferocity.

Until 1848, the people of Orgosolo had lived as semi-nomadic shepherds, pasturing their animals on the mountains in the summer and moving down to the valleys in winter. Now they found access to the old grazing grounds denied to them, and much of the hill pastures had withered away. Some starved, but many turned to banditry. The acts of legalised robbery, by which they were deprived of much of their livelihood, have never been forgotten and are the unfailing justification for the criminality of these days. When a bandit kidnaps a rich man and holds him to ransom, he says, 'They robbed us, didn't they? I'm only getting back a bit of what belongs to me.'

The loss of the traditional grazing lands meant that the law of the survival of the fittest was applied with mathematical exactitude in central Sardinia. In good years, the herdsmen of Orgosolo contrived to pull in their belts and carry on somehow, living on their sheeps' milk curds and the occasional animal rustled from the flocks of the rich Italian settlers. But after a bad winter, the choice was rebellion or death from hunger. The best pastures were claimed by other villages further down the valleys, and over and over again they were forcibly seized after terrific pitched battles. These were the notorious *bardanas*, raids organised for the extermination of competitors for meagre food supplies, and every man of Orgosolo expected to be conscripted by the shadow council of village elders for such an expedition at least once during his lifetime. All the nearby villages, Oliena, Marmojada, Formi, Desulo, Arzano, and Locoe, were the targets of these assaults. Locoe had to be abandoned for a number of years, and in an attack on Tortoli in 1894, every male inhabitant was either killed or wounded. On more than one occasion armed bands from Orgosolo had fought their way right across Sardinia to Oristano on the coast to capture the salt needed for their cheese-making.

The men of Orgosolo went to the *bardana* to clear the way for their flocks, on foot, on horseback, even on bicycles; a small, compact and infallibly victorious horde reflecting in microcosm the sorties of famished Asiatic nomads into the plains of Eastern Europe of the Middle Ages. Orgosolo, the tiny microcosm of a State and conscientiously ignoring the State that contained it, saw these local conflicts as necessary, justifiable and as patriotic as the wars waged by the nation in pursuit of its larger interests. Reprisals by police or army were never more than temporarily effective. One might as well have tried to debar the bedouins of Arabia from their oases.

Parallel with these wars in miniature went the endless internal struggle of the poor against the rich – against the descendants of the Italian colonists, who had established themselves in the fertile valleys, and the rich native families, who had slowly accumulated grazing land which they now let to the disinherited at extortionate rentals. The shepherd steals with perfect conscience from the rich proprietor and, year by year, the number of cattle thefts increases. There is little trace of the Christian ethic in these mountains where Christianity gropingly established itself only by the early seventeenth century. The impoverished, self-sufficient, semi-nomadic shepherd simply cannot understand the philosophy of non-resistance, of turning the other cheek, of laying up for oneself treasures in heaven. The very word 'good' has its own special meaning in Orgosolo. A good man in the Orgosolo version is one who never puts up with an injustice, and his opposite ranks with a cuckold in the scale of public contempt. Meekness and submission belong to the code of the man who has allowed himself to be disarmed.

A final factor – one that is not completely detached from the subconscious – completes the picture of the shepherd-warrior mentality. Wealth, possessions and the *strength* they imply are magical substances, wholly good. The unconcealed ambition of every shepherd is to possess flocks and land, and to lead the life of a rich man. To be rich is to be virtuous and if in seizing the possessions of others more virtue is acquired, then the act is sanctified by the end.

Both Giovanni Mesina and Salvatore Mattu were the sons of families that had never quite got on their feet, that had always lacked the mystic attribute of possessions, and could do no more than reproduce generations of landless, hired shepherds, who still in the year 1966 must be on watch all day or all night over their master's sheep for a payment in kind of twenty sheep per year.

Both men had been *dogaus* many times. In a criminal or semi-criminal capacity, Mesina – who was by this time on the threshold of middle age – had made something of a reputation for himself. He was bold and intelligent, the stuff of which the founders of the powerful families of Orgosolo are made, and he could stand alone. In Sicily, he would probably have been a fairly influential member of the Mafia.

Mattu, in his early twenties and having still to win his spurs, had become a junior member of a band by October 1962. In 1960, both men had been denounced to the police as implicated in the murder of Pierino Cresta, who had been kidnapped and then killed as a matter of principle when the ransom was not forthcoming. Giovanni Mesina has been arrested, but Mattu had managed to escape and had been in hiding ever since on Supramonte. After two years, Mesina was released but found himself under a cloud back in his home town, where it was rumoured that he had secured his release by putting the blame for the Cresta killing on the missing Mattu.

By a strange, almost exhibitionist quirk in the occult mind of Orgosolo, journalists are warmly received and spoken to with surprising frankness. The friendships formed by journalists in this most reticent of towns show them glimpses of a secret life that is completely denied to the police in their isolation from the community. Thus it was that although the Townleys' killing remained officially a mystery, by the end of the month Italian newspapers were publishing an account, pieced together from inside information, of what happened at Orgosolo shortly after noon on 28th October, and the succession of bloody events of the week that followed.

Mattu, it is accepted, had challenged Mesina to a duel to be fought in traditional style, 'at high noon', and the place chosen for this encounter was none other than the green field where the Townleys had stopped for their picnic.

It was Mattu who arrived first, and found the English couple resting after their picnic. He kept out of sight and waited for some time, and then, as there were no signs of Mesina, decided to return to his mountain hideout. At that moment, he noticed a pair of binoculars left by Edmund Townley on the ground beside the rock that had been used as a table, and although Mattu is supposed to have protested, when hauled before the secret tribunal which sentenced him, that he had no intention of robbing the Englishman, he said that he felt justified in his present emergency in helping himself to the binoculars.

He must have appeared a fearsome figure to the Townleys as he came into the open, unkempt from his two years of hiding in the mountains, and armed with pistol, as well as a sub-machinegun, hand-grenades, and a dagger. What precisely happens next? In the light of what we know of Townley's character, his aggressiveness, was he imprudent – even contemptuous – in his handling of this desperate man? The accepted facts are that Townley, realising that Mattu was a bandit, put his hand in his pocket, perhaps to offer him money, and Mattu, mistaking the movement for an attempt to draw a gun, drew his own pistol and shot him dead. He then destroyed Mrs Townley – out of pure compassion, as he claimed, 'so as not to leave a widow'. The cynics of Orgosolo presume that he simply decided to eliminate an eyewitness.

At this moment, Mesina, attracted by the sound of shooting, came on the scene and immediately realised what had happened. He hid until Mattu had gone off, taking Townley's binoculars with him, then scrambled down the hillside for a closer inspection of the bodies. He is alleged to have said that he realised that there could be no question of a duel now because he would have considered himself dishonoured to fight with a common murderer. Instead, he went back to Orgosolo and told all he had seen – not, of course, to the police but to the shadow authorities of the village. Mattu was promptly caught, interrogated, sentenced and executed. But in the light of what followed, one has the feeling that Mesina had few sympathisers in Orgosolo – a man the whole town believed had allowed himself to be broken by the police.

It appears too that Mattu had left powerful friends, who were not disposed to allow the matter to rest as it was. He had been an associate of the then celebrated Muscau band, and a mutual allegiance may even have been established by an archaic blood-mingling ceremony. However much the Townley murders undoubtedly scandalised the bandits on Supramonte, there were loyalties that could not be dissolved, and consequently only one course to be followed. Within hours of the traditional exhibition of Mattu's body, Mesina had been spirited away from the house where he lived with the wife that he had married only twenty days before, and was never seen again alive.

The stern obligations of the vendetta now fell upon the Mesina family, and ritual vengeance was entrusted to twenty-year-old Graziano, a young man of saturnine good looks and acute intelligence, as he frequently demonstrated in his subsequent trial. At the time of his brother's death, he was being held by the police in gaol under investigation over a charge of sheep-stealing, and on being told what had happened, he carried out the first of his many gaol-breaks.

The Mesinas had once before being involved in a vendetta. This was the celebrated Great Quarrel – a small war *à outrance*, lasting from 1903 until 1917, conducted between a number of allied families over a disputed inheritance. In the course of this, although the Mesinas survived, some families lost all their males – including any children over the age of thirteen. One faction was headed by the town's priest, the 'senior father' of a group of a half-dozen families, a sinister cleric whose popularity in Orgosolo can be measured by the fact that he was accustomed to celebrate Mass with an armed policeman standing on each side. Father Diego Cossu, a rich man, and an efficient deployer of the power inherent in his position, hit on the ingenious idea of buying the complicity of the police to have his opponents – the Mesinas included – declared bandits. This was effective in terms of short-term policy – perfectly honest citizens, who happened to be the father's opponents and who decamped in terror, being promptly shot down by the police. In the long run, the plan failed simply because the many fugitives the police failed to imprison, or kill, were transformed into real desperados. The Great Quarrel produced several novel situations in the history of outlawry and the vendetta. There were occasions when the police masqueraded as bandits and murdered along with the bandits, and others when bandits dressed up in borrowed uniforms and passed themselves off as police. Some rich sports even formed a band as a diversion from the boredom of hunting and cards – Robin Hoods in reverse, who armed and disguised themselves to rob the poor.

This was the golden age of Orgosolo's special contribution to the arts, the funeral lament; and Banneda Corraine, most famous of Orgosolo's professional mourners, found her vocation when at the beginning of the vendetta her brother died in ambush. She was eighteen years of age and the most beautiful girl in Orgosolo. She sang,

> Oh, my brother Carmine, flesh of platinum and porcelain
> Where is Carmine, tinkle of precious metals, glimmer of gold?

Her laments have the passion and the imagery of the poetry of Garcia Lorca, perhaps occasionally even the *Song of Songs*, and are still sung at Orgosolo funeral wakes. Some of Banneda's hyperbolic laments seem unsuitable. Someone described at his death as 'The Golden-Eyed Flower' was a bald, middle-aged man of repellent ugliness, who had committed twenty cold-blooded murders, and had made no attempt to stop a lieutenant from strangling two thirteen-year-old children of the enemy clan. When asked why he hadn't shot them, instead of strangling, he replied, 'To save an effusion of innocent blood.'

There has always been a class of professional peacemakers in Orgosolo, whose office it is to settle the differences of warring families when things appear to be getting out of hand. This they usually do by arranging a marriage between suitable members of the opposing parties. Such a traditional marriage of convenience was once attempted in the Great Quarrel, but was quashed by the macabre Father Cossu who objected that as all the families were related, the laws of consanguinity would be endangered by the proposed solution. However, with the near-extermination of family after family, peace had to come in the end and when it did, it was the relief of all Italy. The petty slaughter in Sardinia had been an unpleasant distraction to a nation in arms, absorbed with the patriotic holocausts of the First World War. Seven bandits, who had been in prison six years awaiting trial, were pronounced innocent and released as a gesture of goodwill on the authorities' side. A celebratory banquet at Orgosolo followed at which the survivors were reconciled. It was graced by the presence of the Prefect of Sassari, the Bishop of Nuoro, a member of parliament, numerous police officers and the richest of the local landed gentry. Civil dignitaries and men who had committed multiple homicide with huge prices still on their heads, embraced and got drunk together. Mesina's grandfather knelt with the rest to receive the bishop's blessing, but nothing is recorded of his action in the vendetta. He was one of the small fry who had served without distinction or notice in the band led by the Golden-Eyed Flower.

On November 3rd, 1962, the day after Giovanni Mesina's death, the body was taken to the Mesina house for the lying-in-state. Here, with all the members of the family present, the professional mourners entered the room and began their dirge. Only in the last verse, after a recital of the virtues of the dead bandit, of his strength, his charity, his courage, and his manly beauty, came the moment so long awaited: the shrieking denunciation by the leading wailing-woman of the name of the man held responsible for his death. However much this may have been common knowledge beforehand, the mourners would have kept up a ritual pretence of their ignorance of the killer's identity until this moment. But now, against the sobbing of the women of the household and the shrieks of the corps of wailers, the calls for vengeance were heard. This is the moment when the death sentence of the family council is entrusted to the member or members of the family most fitted to carry it out.

In a large clan well-supplied with vigorous males, the execution will take the classic form – a purely Sardinian variation on the theme of the vendetta in which the honour and responsibility is shared by several volunteers. These, because justice should be seen to be done, choose a public place to approach their victim, draw him aside, whisper his sentence to him, and shoot him down. Instantly, the streets empty, passers-by slip into obscure alleyways and disappear. Doors and windows close. Nobody has seen or heard anything. The town averts its head and acquiesces in its muteness in what has happened. But a small family like the Mesinas must cut its coat according to its cloth. The only suitable male – if one exists at all –

may have emigrated, or he may even be in gaol. When the news of his brother's death reached Graziano Mesina, he was in prison, held as a suspect. By feigning madness, he had himself transferred to the prison infirmary, and from this he easily escaped and made for the region of Orgosolo.

For ten days he scoured the bandit hideouts, the caves and grottoes on Supramonte searching for the men who had killed his brother. Failing to find them, he decided to enter the town itself and arrived there on November 13th just after dark. He was incited by others, the prosecution said at his trial, to do what he did.

He was seen by a great number of people that evening as he walked up the narrow, badly lit main street. His appearance must have been dramatic indeed, for despite the presence of a strong body of police in the town, he was armed to the teeth including the inevitable grenades and sub-machine gun, and it was evident to the bystanders from his 'iron face', as they described it, that he was about to accomplish a 'mission of honour'.

Mesina went into the town's principal bar almost opposite the town hall, which is hardly larger than a cell and furnished with a few shelves carrying bottles of wine and cognac, an enormous refrigerator and three low tables with even lower bench-seats about nine inches high. Antonio, the bar's proprietor, was refilling a row of the tiny wine glasses used in Orgosolo. 'As he came in our eyes met, and I knew what he had come for,' he says. Mesina said nothing. He simply gestured with his machine-gun, and the patrons quietly left their tables and lined up against the wall. Among them was Giovanni Muscau, twenty-two-year-old brother of Giuseppe Muscau. Mesina believed Giuseppe to have been Mattu's friend and protector and to have ordered the killing of his brother, and so – as Giuseppe could not be reached – he had decided to make do with Giovanni. Graziano beckoned to Giovanni Muscau to leave the men standing against the wall, shoved him against the bar with the barrel of his gun, and then fired two bursts into his chest. Muscau slid to the ground and Mesina gave him a final burst as he lay there.

Now Mesina turned to leave and the incredible happened. The custom of Orgosolo absolutely forbids interference in a vendetta by outsiders, and even recommends an onlooker, who believes a vendetta killing to be about to take place, to throw himself face downwards on the ground, to avoid seeing, and therefore being capable of identifying the assailant. The deed is done; the women draw their black veils over their faces, the men slip away into the shadows, the executioners pocket their weapons and disappear.

In this case, to the astonishment of all Sardinia, what happened was that as Mesina turned to leave the bar, someone picked up a bottle and struck him on the head from behind. He fell to the ground, stunned, and was then overpowered and handed over to the carabinieri. This was a break with the past indeed and the notables of the town are said to have shaken their heads in consternation at what was regarded as evidence of the moral corruption of their young men. Terrible reprisals were predicted but, so far, the Mesina faction seems to have been content to bide its time. Memories are long in vendetta country, and it is nothing for a man to nurse his private vengeance for ten years or more – even to appear to have become reconciled to his enemy – while he awaits the right time and place for the settlement of the score.

A few days after my arrival in Sardinia, Graziano Mesina stood up in the iron cage, in which he had been kept like an animal in the courthouse of Cagliari, to hear sentence passed upon him. This conclusion of the sanguinary episode in the bar at Orgosolo had been long deferred because, in the meantime, Mesina had broken Italian records by escaping from five different gaols and one prison hospital. He had never used the slightest violence in these evasions, nor had he attempted to avoid re-arrest. Throughout the trial, he had shown no more than the mildest curiosity in what was going on, 'the master' – as one report put it – 'of a sphinx-like imperturbability'. When asked why he had killed the innocent young Muscau, Mesina considered the question for a moment and said, 'It was his brother Giuseppe I was after. I thought that by killing Giovanni I might tempt him down from the mountains to settle accounts with me.'

Present to hear this admission along with a great contingent from Orgosolo, the women with their black veils drawn half across their faces, the men in stiff dark suits kept for trials and funerals, was none other than the famous Giuseppe himself. Giuseppe Muscau had been captured and put on trial for banditry a year or so before, and as happens in about two such cases out of three, he had been acquitted for lack of sufficient proof. He is now unofficially the town's leading citizen, described as the possessor of great dignity and charm as well as something of a poet, and the highest honour Orgosolo can confer upon a visitor is to arrange for a presentation to the great man.

Giuseppe's demeanour on this occasion remained stolidly unrevealing, matching in every way in correctness by the standards of Orgosolo that of the protagonist in the dock. Both men were in the eye of a critical public. If things didn't change, presumably it would one day be Giuseppe's sacred duty, or that of his son, to kill Mesina – but it would be many, many years before that day could arrive. Only once Mesina was stirred from his apparent indifference. This happened when the Public Prosecutor suggested in his final speech that Mesina had killed a helpless and unarmed lad because he had been afraid to confront this smallish, mild-looking, middle-aged man sitting there with bowed head and clasped hands in the body of the court. Mesina smiled.

The defence's only hope was to extricate him from the ultimate calamity of a life sentence, and the strategy applied was an uphill struggle to create sympathy for a man who clearly hadn't had much of a chance in Orgosolo's battle for survival. 'The negative circumstances of his childhood', as the defence counsel called them, were enumerated. Graziano Mesina had been orphaned at the age of twelve, and then a few years later, the family suffered 'moral and economic disintegration' as a result of the arrest of the three adult brothers who were kept two years in prison on suspicion of murder before it was decided to release them as innocent. Graziano had had to support his mother and sisters through the long months of misery and near starvation. Then came the Townley affair, and the eldest brother's death. 'Don't think of Graziano Mesina as a cold-blooded murderer,' his counsel pleaded. 'He's just an impulsive headstrong boy, incapable of premeditation.' He gave a few instances of Mesina's typically impulsive actions such as tearing down the sheepfold of a man who had killed his dog, and then, perhaps to demonstrate that his client was essentially reasonable, recalled Mesina's protest earlier in the trial: 'After all, the younger brother was in the bar as well, and I might have finished him off too while I was about it. But there was no question of that. One was enough.'

A psychologist's report was read out in court, which described Mesina as legally sane and of above-average intelligence, although egocentric and remarkable for his 'moral coldness'. Sentence of twenty-six years was then passed, and the judge added that only a consideration for the special social climate, of which the prisoner was a product, had prevented him from sending him to gaol for life. Emotional scenes are not unusual in Italian courts at moments like this, but here Orgosolo dominated in its taciturn acceptance of victory or defeat. The sombre men and women in the public gallery got up and filed away in silence. No one looked again in the direction of the prisoner still standing motionless and expressionless, hands clasped behind his back in the cage, waiting for the chains to be fastened on him.

On the cross-country journey to Orgosolo, one need only leave the main coastal road at Cagliari to experience an immediate transition from a familiar to an alien civilisation. In a matter of minutes, the Bruegel-like world of the laborious peasant bent over his crops, is left behind, and one finds oneself enclosed without warning in noble and arid landscape, devoid of humanity. In this hard air, details of rock, tree and ruin are painted with gothic exactitude; rusted ferrous earth is relieved with the greyish green of oaks; sun-flayed mountains lie all along the horizon; there are no isolated houses, no small villages; an occasional town like Santu Lussurgiu is the site of an ancient nomad encampment built where precious water gushes miraculously from a rock. Besides the flinty chatter of wheatears and the occasional screaming of an eagle, there is an omnipresent sound that is at once gay and sinister. This is the lively discord of bells – all of different tones – as a flock of goats goes by. They come through the dark bloodily red trunks of the cork-oaks at a quick, stealthy trot moving as fast as a man can walk. One knows that the shepherd is there too slipping from tree to tree, or out of sight over the lip of a ravine, or behind the rocks; never coming into view. The sensation is an uncomfortable one remembering that there is nothing of the meekness of the shepherd of Christian parable in this man, that he is a cruel, hungry dreamer with a gun, and that in this austere, archaic world where human life counts for so little, the shepherd is often separated by a hair's breadth from the bandit.

Santu Lussurgiu is a cheerful-looking village unusual for the fact that most of the houses have flower-gardens, and, here on a hillside in the pinewoods, the government has built a tourist hotel with a swimming pool and children's playground, and picnicking areas with fountains and waterfalls among the trees. Nightingales were singing in all the bushes when I was there. I was the only guest staying at this hotel, which has some fifty rooms, and the barman was also waiter, chambermaid and receptionist.

'Things aren't so bad as they were,' he said 'Two years ago, we had hold-ups almost every day, but last year it calmed down a bit. We're keeping our fingers crossed.... Dangerous to go out alone? Not really, but it's a smart idea not to carry too much money on you if you go out for a stroll.'

The day's newspaper from Nuoro was open in front of me and there was Santu Lussurgiu itself in a headline: 'Like the Wild West,' judge says. I read on. 'In this part of the world, life seems more and more to imitate the standard Western movie, a continual real-life battle between outlaws and the sheriff and his men – and all we ordinary citizens can do is to look on.'

In the year 1966, in fact, there are estimated to be a hundred bandits at large in Sardinia, about ten of them regarded as particularly dangerous. The majority are centred in the province of Nuoro where the Questore (the chief of the Public Security Police) recently said: 'At nine o'clock, people shut themselves in their houses. Outside you'll find only police and soldiers. All traffic stops at night. If there's a car on the road, you can safely say that it belongs to a bandit.'

Through the window, I could see the barman's children picking wild narcissi at the edge of the wood. Santu Lussurgiu looked as peaceful at that moment as a garden-suburb of London.

'The other day they put the pressure on a neighbour of ours, Francesco Atseni,' the barman said. 'You can see his house from here. Told him to hand over five million lire – or else. He went straight to the police, and while he was about it, bought himself a new rifle. What good did it do him? They got him all the same, and not only him but his shepherd Salvatore. Waited outside the house one night and machine-gunned the pair of them. We've learned our lesson now.... I expect you've heard of the famous Antonio Michele Flores of Orgosolo. He used to operate round here until the police killed him last year. He was only twenty-five when he died, and he'd been a bandit since he was fifteen. I saw him once or twice. Good-looking kid, but his eyes scared me stiff.'

I brought up the fact that Orgosolo was thirty-five miles from Santu Lussurgiu, but the barman said that this was no problem for the special kind of bandit Orgosolo produced. They'd been known to cross the country and carry out a raid as far afield as Oristano, fifty miles from their base. When I told him I was going to Orgosolo next day, he was astonished. Not one Sardinian in a thousand has ever visited Sardinia's most famous town.

Sedilo was two villages, one *nuraghe*, and a cavalcade of gypsy horsemen further on. It looked deserted, and in fact half its population happened to be in Cagliari at that moment for the trial of Pepino Pes (sometimes known as the bandit of the decade) who had been born there. Pes, a lover in the grand manner, as well as a mere killer, with some facial resemblance to young Ramon Navarro, was alleged at the trial to have paid forty-thousand lire (fifty-five dollars) for a killing, when too busy to attend to the job himself. He had many friends in Sedilo still, and one of them had written that day to each of the judges of the Supreme Court threatening them with death. 'Not perhaps the best possible part of the world to be in for the next week or two,' the British resident in Cagliari had said of this region. 'Always a fair amount of highway robbery when a big bandit trial's going on. These people's families need money for their defence counsels. They're punctilious about paying for their legal advice.' To avoid discouraging me, the lady then added, 'Mind you, the chances of being held up aren't terribly high. Say, one in ten at the most.'

The last stop before Orgosolo was Oliena. This town has stood a dozen times in the path of the erupting Orgosolo horde, and as a result has a makeshift and haphazard frontier character. Carlo, the guide I had picked up in Nuoro, was a native of Oliena, and he pointed out a local Alamo where, in the days of his grandfather, a last-ditch battle had taken place between townsmen and invaders. Now cautious and exploratory friendships were beginning to link the two communities. The two wars had exercised a liberalising influence, and the fierce endogamic rule of Orgosolo had been relaxed to permit one or two outside marriages. Carlo was very proud to have friends in Orgosolo.

Oliena seemed to believe that tourism would eventually appear like some fairy prince to rescue it with a kiss from the servitude and drudgery of the present, and as an act of faith, and quite astonishingly, a roadhouse had been built on the outskirts of the town, overlooking a natural curiosity, a deep, onyx river gurgling out of an unexplored cavern in the mountainside. Thirty or forty tables were laid in the dining room, in a vaguely Hawaiian ambience, neat little waitresses with pretty identical Sardinian faces stood by, and the menu offered *porchetto* (roast sucking pig), but no guests arrived. The only customers at the bar were police, and armed shepherds in velveteen who stacked their repeater rifles in the corner before ordering their drinks. A rich farmer of the neighbourhood, Antonio Listia, had been carried off from his home by four armed men on the previous day, and as the mechanism for paying the ransom had broken down, his life was feared for. The shepherds were members of the search party.

Hereafter my journey followed a Sardinian no-man's-land, a deserted landscape composed of the cautious greens of spring, but dramatised with a bold infusion of red – the red washed walls of a refuge for road-menders; the sanguinary red of newly flayed oak trunks; the bluish bruised-red of the Sardinian prickly pear, which grows here everywhere and is quite unlike the prickly pear in other Mediterranean countries. Supramonte rose up over the horizon, silhouetting the green hills against the skull-whiteness of its rock.

A last curve in the road revealed Orgosolo clinging to a hillside, drab as the outskirts of some mean industrial town. Greystone unfinished houses stood among old middens of building materials. In a moment, a dejected street began, hardly wide enough for two cars to pass. Then an arched doorway under the sign *Municipio*, through which a man could be seen hunched over a desk in a dim bare room, announced that we were in the administrative heart of the town. Black hairy pigs cantered up and down the street, and a sharp penetrating odour of animals hung on the air.

A few yards up the street from the decrepit Town Hall was Tara's barber's shop reopened some years back under new ownership. Tara himself was under one of the wooden crosses in the churchyard. He had been suspected of informing to the police, and after his assassination, his body had been exposed with the corners of the mouth carved to the ears (the punishment prescribed for the false witness in the ancient *Carta de Logu*).

The atmosphere of this town was furtive. Although architecturally it was at first sight quite formless – a jumble of mean, dissonant buildings – one soon defined premeditation in this anarchy. Houses were built at more than one level on steeply sloping, zig-zagging alleyways having in many cases, I learned, multiple exits, escape routes to interconnecting cellars, concealed passages and rooftops. It was a town designed to shelter the fugitive; a labyrinth behind blind walls and barred windows, where a sick or wounded outlaw unable to face the life on Supramonte could take refuge for weeks and months at a time. Paska Devaddis, Orgosolo's only female bandit, who died of tuberculosis in 1914 after a short life full of trouble, never once left the town.

But one extraordinary circumstance separated Orgosolo from any other town I have ever been in, with the exception of a Welsh mining village: the people sang. Groups of sombre Goya-esque figures gathered outside the small taverns waiting for room at the tables inside, and the sound of music and of splendid male-voice choirs poured out into the street. By now, Carlo had found his friends, and after a whispered discussion as to the suitability of such a visit, I was conducted to the bar where Graziano Mesina had killed the young Muscau and shown the bullet holes left by the fusillade.

In this bar, too, they were singing *sos tenores*, seated at dwarfs' tables, one man leading with the first verse, and then three more joining in with the chorus, their hands cupped over their mouths to form a kind of resonant chamber; and in this way imparting to the voice a harsh, nasal quality recalling the sound of bagpipes. Musicologists say that *sos tenores* are African or Asian, rather than European, in their musical affinities, but beyond this they appear to know little about them. They are very beautiful, strange and exceedingly melancholic. All the songs I heard that afternoon were on the themes of parting, sorrow and death, and a typical one began: 'Let me live on in hope. Don't tell me my days are numbered.'

The women of Orgosolo were hardly to be seen – black-shrouded creatures that flitted from doorway to doorway with no more than a brief Islamic revelation of the eyes. The menfolk both in appearance and manner demolished another preconception. One would have expected the reputation of Orgosolo to have been reflected in at least some hint of fierceness or facial brutality. But nothing could have been less true. On all sides, one saw faces of great sensitivity and refinement – often not the faces of our time, but rather the heroes of Attic and Etruscan pottery: long straight noses with the slightly recurved nostrils of the Bedouins, bold Iranian eyes, and mouths often of feminine softness.

And the last thing they wanted to talk about was bloodshed and banditry. One of them, Orgosolo's most celebrated singer, a twenty-four-year-old unemployed shepherd, said, 'What's so important is to understand the social background of these things.' Giuseppe Muscau was a heroic figure as a man of action, that was agreed (I was invited to meet him when he got back to Orgosolo), but higher still in the social scale were the troubadours – *sos poetas*. The high spot of their year was the August *festa* when the professional singers and poets came down from the mountains to take part in the contests in improvised verse – marathon calypsos lasting as much as twelve or fifteen hours on a subject announced by the judges without previous warning. Last year's subject, to which several thousand verses had been dedicated, was 'space travel'.

Later that evening, the time came for the pilgrimage to the small green field by the roadside below the town. Three shepherds went with us, including the singer, Salvatore, and we climbed the bank and stood among the grass and the flowers, and the men took off their caps. An old man leading a goat down the road stopped to cross himself.

Salvatore said, 'The people put up two crosses, and the children used to bring flowers every day. In the end, the authorities had the crosses taken away. I suppose they wanted to do their best to forget the thing.'

He added, 'All Sardinia turned its back on us. In Nuoro, they sell the cheeses we make in Orgosolo, and the shops sent them back. You might say that we ourselves were stunned. People kept asking each other how such a thing could have happened? Understand me, a man gets killed for some reason, and then his relations get together to even the score. That's the custom. But this thing didn't have any meaning. We felt as if there was a debt to be paid, and it was hanging over us. It threw everything out of balance. We are not criminals. We are an oppressed people.'

The time had come to return. The men covered themselves. Orgosolo was ahead, a dark jagged silhouette against the fateful shape of Supramonte, glowing benignly in the evening light.

A shepherd said, 'Something has changed since then. People showed a little of what they felt when they caught Mesina and handed him over to the police. We were sick of the whole thing. The papers said that the fellows who turned Mesina in weren't long for this world, but you can see for yourself, nothing has happened to them at all. There's been a change of heart. Even the police feel differently about us these days. They leave us in peace.'

Salvatore said, 'One of the *poetas* composed a beautiful lament for the two strangers. You must come back in August if you can, and you will hear him sing it.'

Return to Naples

Norman Lewis was an Intelligence Officer in Naples during the Second World War, and his experiences at that time were recounted in his masterpiece Naples '44.

This visit took place more than three decades later. In view of those wartime experiences, it is interesting to learn that the intervening years hadn't, he felt, destroyed the essential character of the city he loved so much.

In an interview with Lewis's biographer Julian Evans, Ken Griffiths, the photographer on this trip, described Lewis's methods and lack of assertiveness. 'We just walked around. Other writers used to say things like, "We are going to do this, we're going to do that" ... But Norman never said that. He just said "Oh I've got an idea, this might interest you."'

Norman's affection for Naples may have made him a bit optimistic on the subject of Neapolitan crime. Soon after he departed, Griffiths and his assistant were briefly kidnapped by their armed driver. Without much difficulty they managed to escape.

First published in *The Sunday Times*, April 6, 1980

O F ALL THE GREAT CITIES Naples has suffered least at the hands of that destroyer of human monuments, the dark angel of Development. Pliny himself, who once stood on a headland there to watch the great eruption of Vesuvius 'shaped like a many-branching tree' in the moment of the obliteration of Pompeii, would have little difficulty in picking out the features of our times. Nor would

Nelson and his Emma, who chose roughly the same viewpoint to watch the eruption of their day – nor, certainly, Casanova looking down from his gambling house over the layered roofs and the soft-yellow walls of volcanic *tufa* which hoard and dispense the special Naples sunshine. Hardly a stone of Santa Lucia has been disturbed (except by air-bombardment) since its celebration in the ballad of the nineties. When the traveller of the last century was adjured to 'See Naples and Die', it was notwithstanding the competition offered by so many glittering rivals. How much more valid and enticing is the invitation, now that so many of these rivals have withdrawn into their shells of concrete.

Naples is a once-capital city, glutted with the palaces and churches of the Kingdom of the Two Sicilies. Seen from the heights above it, it is a golden honeycomb of buildings curved into a sea which, beyond a bordering of intense pollution, is as brilliant and translucent as any in the world. It is built on ancient lava fields, and has been threatened by numerous eruptions – only one of which, in 1855, came near to engulfing the city: it was saved by the miraculous intervention of a statue of its patron saint, San Gennaro, taken to the Maddaloni Bridge, where it spread its marble arms to halt the passage of the lava. The history of Naples abounds with similar marvels, all of them believed and recorded by responsible citizens of the day: a plague of mermaids, figures in Giotto's frescoes in the Castel dell'Ovo – so marvellously drawn that they were actually seen to move – and, more recently, the prodigies performed by Padre Pio, the flying monk, who flew from an outer suburb to the rescue of Italian pilots shot down in combat with Allied planes. He bore them safely to earth in his arms. Dependably in March of every year the dried blood of San Gennaro liquefies in its ampoule in the Cathedral – the most hallucinatory of spectacles surviving from the Middle Ages.

It is characteristic of Naples, described by Scarfoglio as 'the only oriental city having no resident European quarter', that one of its kings, Ferdinand I, should not only have delighted to play the hurdy-gurdy but have commissioned Haydn to compose six nocturnes for performance on the instrument. He was the ruler of a people infatuated with music, and there is music still, everywhere in the Neapolitan air. There can be few more poetic experiences in the local manner than to visit the Parco della Rimembranza, where the young of the city go to make love in their cars, then clamber down the cliff to a point where, even though the departing fishing boats are out of sight, they can be tracked by the trail of their mandolin music.

Naples has been taken by a long succession of foreign conquerors, the cruellest of them being Lord Nelson, who collaborated in the fearsome slaughter of the city's liberals; and possibly the most corrupt being the Allies in the last war, who virtually handed over civic control to the American gangster Vito Genovese, in the guise of adviser to the Allied Military Government – an experience from which the city has never wholly recovered. A continuing resistance to so many alien conquerors has sharpened the native capacity for self-defence, and, since most of the laws were made by their oppressors, they have a tendency to mistrust law in general. Neapolitans are gregarious and spirited with a frank devotion to the pleasures of the table and the bed. In the last war, Naples was almost certainly the only city in the theatre of warlike operations where civilian employees of our armed forces could apply for transport facilities to their homes at noon, to enable them to fulfil their marital obligations.

Those cities, such as Naples, which remain wonderfully unchanged, have usually survived not because developers have recognised their charms, but because, for one reason or another, the developers see no hope of a return on their money.

Indeed the economy of Naples is chronically ailing and slides from one crisis to another. It is generally accepted that an expanded tourist industry could be its salvation, but the tourists do not come. Last year *Il Mattino* listed some of the reasons why Naples fails to entice foreigners to break their journey on their way to Sorrento or Amalfi, spending a night or two in its half-empty hotels. Sorrowfully the newspaper admitted that Naples had become the hometown of petty criminality. In the past twelve months, 77,290 minor crimes had been reported, but in only 1,300 cases had an arrest been made, or the criminals even been identified. During this period 29,000 cars had been stolen in the city – possibly a world record taking into account the number of vehicles registered. The Vespa-mounted *scippatori*, the Black Knights of the alleyways, buzzing in and out of the crowds in search of camera or handbag to snatch, had become so commonplace a sight as hardly to evoke interest or comment.

From a glance at the newspaper's statistics it seemed, too, that an evening meal out was to be recommended neither to the native citizen nor to the visitor to Naples, since fourteen leading restaurants had been raided by bandits in the past twelve months. It was the kind of experience most of us would want to avoid, but a Neapolitan friend involved in a hold-up had been stimulated rather than alarmed. He had been invited to Da Pina's for a christening celebration. A nice party, he said. The best of everything, with the wine flowing like water. But about halfway through the proceedings three hooded men carrying sawn-off shotguns had walked in and ordered the guests to lie down on the floor. He was impressed with their courtesy, their correct use of the language, and by the way they addressed their victims using the polite *lei* rather than the familiar *tu*. All in all, it was a bit of an adventure, he said, and well worth the trifling four pounds or so he had been obliged to part with. His only fear had been that by some incredible mischance the police might show up and start a battle, as they had done at Lombardi's Pizzeria last June, when fifteen customers were wounded.

But many of the coups pulled off by the organised gangs, the Camorra, which imitate the Mafia of Sicily, are theatrical rather than violent. Three robbers who succeeded in sealing off Parker's Hotel from the outside world, and who took two hours to ransack it from top to bottom, prepared and consumed a leisurely meal before departing. The recent capture of the Ischia ferry-boat was another episode that might have been modelled on a film; having robbed the passengers with the now familiar show of civility and regret, the bandits leapt to the deck of a following motor launch, then waving farewells and blowing kisses to the girls before vanishing into the night.

If one has an affection for such movies as *The French Connection* this is an environment not wholly without its own brand of attraction. What in its way can be more pleasant than to draw a chair out on to the balcony of a room in the Hotel Excelsior overlooking the exquisite small harbour of Santa Lucia, and there, glass in hand and without the slightest risk to one's safety and comfort, play the part of an extra in such a film? The view is of the majestic fortress of the Castel dell'Ovo, dominating a port scene by a naïve painter: simple fishermen's houses that have become restaurants, painted boats, tiny, foreshortened maritime figures, going nowhere in particular, a quayside stacked with the pleasant litter of the sea.

There is less innocence in the prospect than at first meets the eye, because a corner of the port has been taken over by a fleet of some forty large motor launches, painted the darkest of marine blues, devoid of all trappings, and having about them an air of sinister functionalism. From time to time one starts up with a tremendous chuckle of twin 230 Mercury engines. It then manoeuvres in swaggering fashion round the other boats, then heads for the horizon, trailing a wake like a destroyer.

This is the fleet of the best-organised and most successful *contrabandisti* in southern Italy. In these launches (which give the impression of having been specially designed for the trade) cigarettes are smuggled, and who knows what else gets picked up in the incessant rendezvous with ships that steamed out from the ports of Tunisia. Smuggling is hardly the word to describe these operations, all stages of which, though taking place in Italian waters, are on open display. The boats come and go throughout the day, unload their cargoes without concealment and cut a few jubilant capers in the harbour before tying up. There are no signs of the law in the harbour area, and motorcycle policemen passing through Santa Lucia do so hurriedly with eyes averted. Understandings have clearly been reached at high levels. Customs launches lack the speed to catch the *contrabandisti* at sea, and rarely dare to enter the port. Occasional disagreements among the smugglers themselves can, however, be explosive: a week before arrival at my hotel, guests had a ringside seat at a brief battle, followed by a spectacular incineration of boats.

Smuggling is viewed by Neapolitans with tacit approval, if not with enthusiasm, and the benefits of the direct trade with North Africa are immediately visible to the man in the street. There is hardly a pavement without a small boy seated at a table to offer Marlboro cigarettes, made in Tunis (only the government health-warning is missing), at less than 500 lire as opposed to the 800 lire charged in the shops. The authorities seem to regard the traffic as hardly more than an inevitable evil. 'I refuse to admit that this is a crime,' said Maurizio Valenzi the Communist Mayor of Naples. 'For me it is an illegal solution.' The Mayor shared the frequently voiced Neapolitan view that his city is the victim of a calumnious outside world. 'If you are looking for a crime on a big scale, go to Rome or Milan,' he said. 'The worst thing that can happen to you here is to have your pocket picked. Nobody gets violently robbed in Naples and we treat women with respect. Whoever heard of a Neapolitan being pulled in for knocking a child about? Even the Red Brigade don't operate here.'

Valenzi is as Neapolitan as Brezhnev is Muscovite; he is lively of expression and gesture, a distinguished painter, and a first-rate oratorical performer in a city in which few politicians can survive without the knack of rhetoric and powerful voice. His appearance recalls the views of matters of dress held by Togliatti, the leader of the Italian communist party for nearly forty years: 'What pleases me is to see a comrade dressed in a good double-breasted suit – if possible, dark blue.' Valenzi is wholly congruous in the rococo furnishings, the marble and the glitter of the Naples Town Hall. He is fired by local patriotism, impatient of criticisms of his city, and particularly saddened by those contained in a book by a Communist author, Maria Antoinetta Macciocchi, who had been a parliamentary candidate for one of the poor quarters of the city. 'She wasn't much liked here,' the Mayor said.

Macciocchi had mentioned that the rat population of central Naples was seven million. Many of these, she said, were shared out in the *bassi*, those claustrophobic dwellings consisting of a single room that line the streets of the old town, in which as many as fifteen members of a family may live as best they can with no windows, with the street doors shut at night, with no running water and with a closet that is behind a curtain. The Mayor, who showed a partiality for euphemism, shied away from the word *bassi*, but agreed that nearly seventy thousand families live in 'unhygienic houses'. 'The municipality,' he said, 'has plans to do something.'

'Our submerged economy' was Valenzi's description for the child-labour which exists in Naples to an extent found nowhere else in the Western world. There is no way of calculating the number of children from the age of eight upwards employed in cafés, bars, or the innumerable sweatshops tucked away in the narrow streets; but there are certainly tens of thousands of them. It would appear to be another 'illegal solution'. Naples has the highest birth-rate in Italy – twice the national average – and it is an everyday accomplishment for a woman to have borne ten children by the age of thirty-five and to have completed a brood of fifteen or sixteen by the time she ceases to reproduce. Such families are a source of complacency rather than despair. One is assured that they testify to a woman's sexual attraction and her husband's virility. More importantly, perhaps, they represent an insurance policy against economic disaster. When up to five or six children contribute small regular sums to the budget a family is not only more affluent but also securer than a less numerous one in the trap of chronic unemployment.

These are the facts of Neapolitan life against which Mayor Valenzi struggles like Canute against the waves. If the child in proletarian Naples is an economic weapon in the family armoury, it follows as a consequence that such central areas of the city as the Vicaria district have the highest population density in Europe – possibly in the world – with up to three people occupying every two square metres. But if overcrowding, and its damaging effect on public health, are the most pressing problems that face the Mayor and his council, it is the terrific anachronism of child labour with its whiff of early Victorian England that gives the city a bad name.

Therefore gestures have to be made, and from time to time the police are ordered into action to close down all establishments employing child labour, and to punish their owners with exemplary fines. What follows is economic disaster for all involved – sometimes desperate impoverishment for the families thrown back on the providing power of the father who, statistically speaking, can expect to spend a third of his life unemployed. At this point, the exploiters and the exploited only too often join forces in protest, and their votes are lost to whatever party is held responsible for their plight.

How is this situation to be tackled? How can any political party hope to put an end to the Neapolitan tradition of the large family which engenders so much poverty? Schooling in Italy is compulsory up to fourteen years of age, but the school inspectors are as helpless as the politicians. The little courtyards tucked away everywhere in Naples are full of small boys aged upwards of eight years who work ten or twelve hours a day, for as little as £2 a week, stitching and gluing shoes. In the family atmosphere that pervades even the workshop, it would be hard to find a happier-looking, more intelligent collection of children. But none of them will ever read or write.

Naples sharpens the stranger's wits and teaches him to look after himself. The lesson is not a difficult one to learn, and in a matter of hours, days at most, amusement is apt to take over from indignation. One exchanges laughter with the agreeable young man who offers a perfect imitation of a Seiko watch that only ticks for a minute or two after it is wound up; or, on taking a taxi, one points without severity to the meter inevitably still registering the fare clocked up by the last passenger. There are basic precautions to be taken: passports and valuables are automatically committed to the hotel's safe, and only enough money carried to meet immediate requirements. When parking a car it is not a bad idea to secure the steering wheel with a chain and padlock. These things attended to, one can relax and join in the local games.

Our own arrival in Naples was on the second day of the ancient and popular feast of Santa Maria del Carmine, held in the streets adjacent to the old church at the far end of the port. Del Carmine is the parish church of one of a number of districts, once virtually separate villages. Each has its history, traditions, customs – and often the enfeebled remnant of a once-powerful ruling family. And such was the spirit of rivalry between one district and another that fifty years ago intermarriage between districts was rare.

The church possesses a picture of a 'black' Virgin, held responsible for many cures, in particular of epileptics and lepers and of those afflicted with all kinds of pox. The most unusual and attractive feature of the *festa* is the 'burning' of the church tower, by the setting alight of bales of straw fastened to its walls with the intention of cleansing it, and thus the district itself, of evil spirits during the forthcoming year. It was a disappointment to learn that the tower was not to be 'burned' on this occasion: repairs to its structure had been found necessary, and the scaffolding was already in place. The cancellation of the ceremony had cast a certain gloom in the neighbourhood, which depends largely on fishing, and which therefore feared that catches might be affected.

The Corso Garibaldi, a wide if dishevelled street running past the church, was filled by early evening with a holiday crowd. Here all the familiar ingredients of a Neapolitan *festa* were assembled: the stalls with tooth-cracking nougat, solid cakes, and cheroots; the shooting booths; the intimidating display of strange shellfish; balloons and holy pictures – and black-market cigarettes.

In Naples the cult of the enormous Japanese motorcycle has arrived, and they were here in fearsome concentration, roaring through whatever space they found among the press of human bodies. We saw one elfin girl mounted on a Kawasaki 'King Kong' hyper-bike. Children are not overprotected in Naples. The minimum age for a Vespa rider seemed to be twelve or thirteen; and crash helmets are taboo.

These are the occasions when, in holiday mood, Neapolitans resolutely suspend normal belief. A professional 'uncle from Rome' was there, aloof and immaculate in his dark suit, ready to hire himself to any family wishing to impress its guests on an occasion such as a christening, wedding or funeral. *Magliari* – confidence tricksters who flock to all such *festas* – were there in numbers, instantly recognisable, even to an outsider, by the apparatus of their trade.

Superior hoaxers convincingly present themselves as rejected suitor, thus offering the 'silver' service bought for the wedding that they claim will no longer take place. *Magliari* in truck-drivers' overalls and with oil on their fingers flog trashy radios and defective tape-recorders 'off the back of a van'. A local boy in burnous and headcloths, skin yellowed by several layers of instant-tan, hawks vile carpets which, he claims, have been brought over from Tunisia with the cigarettes. How do Neapolitans – those masters of guile – allow themselves to be taken in?

Until two years ago the seller of Acqua Ferrata would have been here. This most esteemed and expensive of curative waters, nauseatingly flavoured with iron, was drawn from a hole in the ground somewhere in Santa Lucia and offered – exactly as illustrated in the Pompeii frescoes – in containers shaped like a woman's breast. Since then, following a typhoid scare, the health department has stepped in and Acqua Ferrata is at an end – temporarily, perhaps – to be replaced with a poorish substitute: fresh lemonade animated with bicarbonate of soda.

One figure alone from the remote past had survived at del Carmine: the *pazzariello*, the joker of antiquity, also shown in the Pompeii frescoes. Once he drove out devils, and as recently as the time of the last war no new business could be opened before a *pazzariello* had been summoned to lash out with his stick at every corner of a building, where a devil might have concealed himself. The office was an honoured one, hereditary and indispensable, too, in a city where even now people cross the road in the Via Carducci to avoid passing too close to a building notoriously under the influence of the Evil Eye. But now the magic power of the *pazzariello* has drained away; the one we saw, doing his best to dodge the motorcyclists as he capered about in the Corso Garibaldi, was there to advertise a fish restaurant.

Our visit to del Carmine provided a mild adventure. Among the exhibition of holy pictures, most of them crude versions of the celebrated ikon on display in the church, we noted one of a strikingly different kind; a portrait of a somewhat stolid-looking middle-aged man, stiff in formal suit: *Il Santo Dottore Moscati*. It turned out that the holy doctor was a GP of the district, recently deceased and newly canonised by popular acclamation (without reference to Vatican or Church) as a result of a number of miraculous cures he had effected.

The display with its popular new saint seemed to call for a photograph, but the elderly lady in charge fought shy of the camera. She retreated in haste, shielding her face with one of the ikons and leaving her husband to conduct any further investigations. The old man made no objection to being photographed after we had agreed to buy a picture of Dr Moscati. Since the light was already failing, the camera was set up on a tripod and preparations put in hand. Immediately a crowd began to collect, drawn away from the competing attractions of a shooting booth and a nearby church. A Neapolitan friend who had guided us to the *festa* became concerned, feeling that we were attracting too much attention and were too vulnerable, surrounded by photographic gear, to a passing *scippatore*. But the crowd was co-operative and affable; a new face and an excuse to exchange a smile with a foreigner. People were actually jostling each other and manoeuvring to be included in the photograph, so that, realising that we were among friends, all warnings were ignored, and the photography went ahead.

A moment later there was a sudden chill in the atmosphere and the smiles faded. A grim-faced, gesturing man had pushed himself to the front to demand payment of 50,000 lire – about £30. His story was that he was acting for the owner of the pictures; but it was to be supposed that he was an enforcer of the protection gangs said to levy a toll on most Neapolitan business enterprises. We decided to resist the extortion: there were four of us, and we were certain that we had the crowd on our side. And the situation was saved when the helpful old man had the courage to announce that he had never seen the presumed gangster before in his life. With this, the unwelcome stranger went off, and the emergency was at an end.

In 1943-44 I spent a year in Naples, arriving a day or so after its capture from the Germans, when the city lay devastated by the hurricane of war. The scene was apocalyptic. Ruins were piled high in every street, and people had to camp out like Bedouins in a wilderness of brick, on the verge of starvation.

It was Naples' calvary of fire and destitution; the days of reproach through which it finally emerged so astonishingly unmarred. The Neapolitans' salvation was their fortitude, their incapacity for despair. Perhaps, too, there was a kind of austerity in their make-up which was unsuspected in southerners – a readiness to make do with little, and a lack of affinity with the frantic consumerism already beginning to dominate Western European society.

Revisiting Naples, more than thirty years later, I saw it as a city that has achieved its own kind of emotional stability, being content to drift with no eye to the future, and highly resistant to change. Economically it has stagnated where the industrial North with its separate identity and ideals has pushed further and further ahead. Nearly half the Neapolitan workforce is unemployed and under-employed, but Neapolitans help each other. The income per capita is only a third of that in Milan: but, for me, Naples would always be the better place to live. The sensation of continuity – that here in Naples one was recapturing the vanished past – was reinforced by a visit to a famous shore-side restaurant, unaltered in any way in its furnishings and atmosphere from the days in 1944 when it had been full of Allied officers and the barons of the black market. The house troubadours, facsimiles of their fathers, attended the guests as ever to strum the everlasting *Torna Sorrento* on their mandolins. The same algae-spotted showcase with its display of octopus and crabs was there still and so, too, was the old man hunched behind the antique cash register with its bell chiming like the cathedral's angelus.

All the rituals had been preserved. Fish were still presented with hooks hanging from their mouths to suggest they had been cut in that very instant from the line; and what used to be known as the 'show-fish', a majestic bass or *merou*, passed on a lordly salver from table to table to cries of admiration from diners who should have known that, whatever they ordered, it would not be this that they would eat. At the proper moment the visitors' book was produced – but here, at least, there had been changes. All the great, blustering Fascist names had been weeded out thirty-five years before, as well as any names of subsequent American generals. Enduring fame now belonged only to such as Axel Munthe and the very local girl, Sophia Loren, surely well on her way to popular canonisation. Neapolitans had thrust the politicians and the soldiers out of their memory.

Close to us a number of tables had been pushed together to accommodate a family, later identified as a man of fifty, his wife, two teenage children, a son in his twenties and the daughter-in-law, their three children, the host's elder brother, his widowed sister, and the grandfather, who was placed out of respect at the top of the table and to whom the show-fish was first presented for his nod of approval. In Naples there are no babysitters: the family takes its pleasures and suffers its tribulations as a unit, and the aged are excluded from none of its experiences.

With the exception of the eldest son, in his *moda inglese* pinstriped suit, and his stylish wife, the family seemed less than affluent; yet it was clear that a small fortune was being spent on this meal. By the time coffee came I found myself chatting to the head of the house. He had just been released from hospital – hence the celebration. The family went out on the town two or three times a year, he said, 'whenever an excuse can be found'. In this way any savings must vanish.

It was the kind of household based on a three-roomed flat – the young couple and their children would live separately – with nothing on hire-purchase, the minimum of furniture and a kitchen of the old-fashioned kind with nothing electrical in it apart from the toaster.

The father went on to say that he had been employed as a mechanic in the Alfasud factory, then laid off. He added that while drawing what benefits he could, he managed to get his hands on a list of Alfasud buyers in the area and, by servicing their cars at cut price, had been able 'to keep the soup flowing'. His daughter went to school but took time off before Christmas to make figurines for Nativity cribs, which at that season were in great demand. If necessary, his wife could always turn her hand to sewing umbrellas for sale in the London stores. Should a financial emergency arise, the eldest son, who 'worked on the boats' – he nodded in the direction of the piratical launches in the harbour – could be counted on to pitch in. '*Si arrangia*,' he said: 'We get by somehow.'

Sicily

The lunch in this article happened during a Sicilian trip that Lewis made in 1984 with the photographer, Don McCullin, for an article that never got published.

Although the trip was no great success, it renewed his fascination with Sicily. To his friend the Sicilian journalist Marcello Cimino (who had helped him with The Honoured Society*) he wrote, 'You will be staggered, and probably horrified, to learn that I have decided to write a novel having a Sicilian background. This, I realise, is an impertinence, but it is a country which fascinates me endlessly, and the treatment will be sympathetic.'*

The novel was never written, but he eventually wrote a brief travel book, In Sicily *(2000), which was dedicated to Cimino, who had since died.*

The feast of the inexplicable 'Sainted Physicians', described at the end of the article, is still hugely popular.

First published in the *Independent*, January 13, 1990

SICILIANS LIVING IN THE tightly packed, traffic-jammed city of Palermo naturally do their best to escape, whenever the opportunity arises, into the country, or to one of the rare havens of peace still to be discovered by the sea.

Inland, a favourite excursion is to a small town with a relaxed and somewhat ecclesiastical atmosphere. Twelve churches – some superbly baroque – are crammed into the surroundings of the small square. In addition to its ample provision for the devout, the town possesses seven schools, an excellently equipped hospital, and various benevolent institutions. People drive more than thirty miles from Palermo just to stock up with exquisite bread baked in wood-fired ovens and to buy meat free from those adulterations and tamperings associated with city markets. The streets are clean, there is no petty crime – the last burglary took place three years ago. It is a relief to visitors from the city, where muggings happen in broad daylight, to find that here they can stroll in the streets by night in perfect confidence and security.

This is Corleone, made famous by the book and the film *The Godfather*, and generally accepted as being under the stern and watchful control of a man held in custody since 1974. In the maximum security prison of Termini Imerese he is currently serving a life sentence for multiple murders and for being 'promoter and organiser of a criminal organisation'. This surely is a phenomenon without parallel in the modern world and hardly in history. Many Palermitans seeking refuge from the bustle, clamour and insecurity of the city have decided to settle here, although once, back in the Forties, thirteen bodies of murdered men were recovered from the streets in as many days. It holds special attractions for families with children to be educated.

If schooling is of no consequence, townspeople in search of peace may opt for Ficuzza, five miles away, located in Arcadian surroundings under the portentous shape of the Rocca di Busambra, and at the edge of the Ficuzza wood, once the hunting preserve of the Bourbon King Ferdinand II, whose palatial baroque lodge is just round the corner of the single street. It was the ambition of the Sicilian friend who accompanied me on this trip, and who had taught philosophy for two years at Corleone, to buy a house here. Prices are high – for Ficuzza, too, is under uncontested, and therefore pacific, Mafia control. Cars are left unlocked in the street at night while the populace sleep quietly in their beds. Once a day or so, a policeman on a motorcycle may pass through without stopping. There is nothing for him to do.

The Corleone landscape is dramatic, even formidable, backed by harsh mountain shapes, perpetually misted and aloof; a proper setting for the atrocious deeds of the past. Forty years ago the traveller would have been careful to avoid the Ficuzza wood, for it was here that outlaws kept rustled cattle. In 1947 police investigated a deep crevice on the flat top of the Busambra mountain, discovering the remains of trade union leaders and peasant malcontents, dumped there by the feudal-friendly Mafia of those days. There are no isolated farmhouses hereabouts. Those who work on the land live in villages, often built in a circle for purposes of defence, their backs forming an unbroken wall.

This is bandit country, riddled with secret caves and hidden tunnels, for the bandits were here from the beginning of Sicilian history to the end of the Second World War. At that time thirty bands roamed these mountains, a pool of desperate men from which were recruited the private armies of the feudal landlords. In 1950, with the passing of the agrarian reform law, these things came to an end. The social conditions in which banditry had flourished ceased to exist, and the private armies had gone. Those bandits who remained at large were no more than an intolerable nuisance, and, with their usual efficiency, the 'Men of Honour' arranged for their extermination.

Many theorists see Sicily's history of banditry interwoven with that of the Mafia as a kind of continuing resistance to foreign occupations – six in all – which never permitted the creation of a stable state. As a matter of routine each incoming regime abolished all freedoms granted by its predecessor, cancelled the title deeds to ownership of land, changed all the laws to suit themselves. Thus, at intervals of approximately one and a half centuries, Sicilians found themselves reduced to pauperism by some new category of foreigners, governing inevitably through a corrupt and brutal police force. Since the state offered no protection, it fell to the individual to do what he could to defend himself, and his best recourse was to join forces with other victims of oppression in the organisation of underground action. Thus, according to the theory, the Mafia was born.

When in 1860 Garibaldi arrived on the scene to bring about Sicily's unification with Italy, it became clear that there was much to be done before the island could be governed from Rome. It was therefore decided as a temporary measure to make do with political remote control through 'families of respect' – which may have been in effect governing from behind the scenes for centuries before his arrival.

The surrender of land in 1950 to the once land-hungry peasantry saw the end of the old-fashioned rural Mafia, now that their function as guardian of feudalism had ceased to exist. Calogero Vizzini of Villalba, and Genco Russo of Mussomeli had shared enough power in 1943 virtually to hand over western Sicily to the invading American forces, with hardly the loss of a man. Their kind was now extinct, and with them had gone all the traditional godfathers, to be replaced by young men of quite exceptional ferocity, who began their conquest of the towns.

In 1950, at the stroke of a pen, a Sicilian lifestyle came to an end, and in the countryside the change was instant and profound. The old system here had been based upon a vast reserve of labour; now almost overnight the labour market collapsed. There was no one to sow, tend or reap the crops. Agricultural wages doubled, then increased five-fold, but there were still no takers. A few field workers busied themselves with the tiny patches allocated in the reform, but most of them either moved into the towns or went abroad – and they would never be back.

We had been invited to lunch at the *casa padronale* of Caltavuturo, south of Cefalù, where the land reform had cost our host all but 500 hectares of his original 2,000, leaving the estate from his point of view no longer a viable proposition. This reverse had been accepted with dignity, and good grace. The small fortress provided at least an excellent backdrop for the entertainment of his friends at weekends.

Life in the *casa padronale*, set in the empty magnificence of what might have been a Highland glen, clung to what it could of the style of the past. Power had gone, but its persistent wraith lingered on. A high wall with massive gates enclosed a courtyard in which, when we arrived, a baker was busy at his oven, and servants, formally hatted in 18th-century style, cooked an assortment of meats over a great brazier. The servants, led by the major-domo, came forward to shake the hands they would have once kissed. The baroness awaited us at the head of a marble staircase leading to the *piano nobile*, then presented us to the guests, all of them speaking perfect English, learnt in all probability from Anglo-Saxon governesses. Courtesy titles had been firmly retained.

The talk was of English literature of the 19th century, of a croquet lawn it was hoped could be created, and the possibility of introducing foxhunting in this moorland and scrub environment, so unfavourable, one would have supposed, to the sport. Although the people of the estate had received their 1,500 hectares, they had all left, and the heather spread a coverlet over the once-cultivated fields.

Lunch had been based upon a recipe chosen from a selection of British glossy magazines on display, the one medieval touch being that the bread was presented to each guest in turn to be respectfully touched. Strong Sicilian white wine was provided from the vineyard of the Conte Tasca D'Almerita, present for the occasion, who announced himself as grandson of Lucio Tasca, deviser of the plan in the late Forties for the 'tactical utilisation' of the bandits into a Separatist army. This, it was hoped, would detach Sicily from Italian sovereignty and offer it to the United States. Paradoxically, the grandson, as he told me, had been captured and held to ransom for some months in a cave by Salvatore Giuliano, most famous bandit of them all. 'He was extremely polite,' the count said, 'and never failed to address me by my title.' A memory caused him to wince. 'The food', he said, 'was monotonous.'

Caltavuturo was a quiet place, perhaps a little dull, but in the past excitements had been frequent. A feature of the house was a tower with two storeys. Three embrasures were provided in each room, through which rifles could be pointed at attacking outlaws, who had never succeeded in scaling the wall. Our charming hostess pointed out the six thrushes' nests, one per embrasure, each having five eggs. The thrushes had become house mascots, inordinately tame. Sometimes in the bad old days, when an attack was imminent, the nests had had to be removed, but this was done with great care, and as soon as the danger was at an end they were replaced. Usually the birds returned.

Sicily, apart from the coastal strip in which its principal towns are located, is fast emptying of its people. The autostradas, unrolling their ribbons of concrete across the island from north to south and east to west, are largely devoid of traffic. The monks drift away from the isolated monasteries, and the great feudal houses have lost all purpose. Fields that once produced Europe's highest yield of wheat are now submerged in gigantic thistles. Only shepherds inhabit this landscape, and if one makes a roadside stop they come scurrying down the mountain slopes to the car, desperate for a moment of relief from their loneliness. Like magicians they draw the newborn lambs from their sleeves, and unburden themselves of pent-up words. 'Don't go away,' they say. 'Why the hurry? Let's talk about something.'

Nostalgia still drags at those who have turned their backs on the scenes of their childhood and emigrated to the towns, and at holiday time they swarm out into the country to pay their respects at the shrines beckoning in the background of their lives. For the ex-villagers who have moved into Palermo, the most powerful of such magnets is the great temple of Segesta, and, making their pilgrimage by bus in spring, they deck themselves with red poppies in tribute, it is to be supposed, to the watchful spirits of the place.

The temple awaits them at the top of a steep slope, a ravine at its back. It is colossal and perfect although never completed; seemingly part of the present, since it is untouched by ruin. Here the mystery of antiquity is complete, for nothing remains but contoured fields concealing its foundations, and a theatre on a hilltop a mile away, from which it appears as a bright new child's toy. It stands in a wide encirclement of mountains, facing an escarpment of white rock, a black cliff launching its falcons over the valley, and, to the south, the dingy pile of Monte Grande. At the end of the day when the crowds have gone, this supreme monument dominates a place of supreme loneliness.

Back in the grim industrial suburbs after their brief escape to the country, the new townsmen and women, subjected to a turbulent and frequently violent environment, continue stoutly to defend village values. In Branchaccio, where the police barracks have been blown up on two occasions, and where last year eight men were killed in a shotgun massacre, female deportment remains a key concern. Therefore lengths of cloth are stretched along balconies to impede the view of feminine legs, and some doors have been modified to resemble those of a stable. These enable housewives to conceal the lower part of their bodies while chatting with neighbours, or buying from passing street traders.

Chaos – the word is hardly ever out of Sicilian mouths – reigns in places such as this, subjected to a divided Mafia, engaged continually in mutual slaughter over the division of the spoils. In nearby Bagheria the death roll among contending factions amounted to fifteen in the past twelve months. More important to many onlookers is the demolition of this enchanting seaside town – once a showpiece of baroque architecture – by illicit Mafia property development. Bagheria, favoured resort of the 18th-century nobility – who threw money to the wind in construction of fanciful palaces – has been buried under concrete. Only the eccentric and exuberant Palazzo Palagonia, with sixty-two ceramic monsters ranged along its surrounding wall, remains intact.

In the opinion of many Sicilian experts, the Mafia, with its close and fatal involvement in politics and high finance, can't be defeated in the foreseeable future. Current tactical problems in the struggle arise from an internecine war resulting in the destruction of strong bosses, leaving power vacuums to be filled. A case in point is the tragic history in recent years of the town of Alcamo, about forty miles from Palermo, following the sudden death of eight out of nine members of the Rimi family, an outstandingly successful firm supplying one sixth of the US consumption of heroin, besides being a major clandestine exporter of arms to the Middle East. Their elimination following orders from the prison cell in Termini Imerese was a catastrophe for the citizenry, who lived comfortably more or less as the people of Corleone do now – under nominal overlords who ran their profitable affairs and left them in peace. Anarchy followed their departure. Those who sought to replace them lacked the capital to take over the Rimi businesses and, in order to raise this, have devised a system of protection rackets from which there is no way of escape. An innovatory technique has compelled the banks to hand over their files, and on incomes thus revealed a percentage is levied. Resistants quickly change their minds when their cars or houses go up in flames. Farmers are brought to heel by the loss of valuable agricultural machinery. Since the police no longer count in a situation like this, the unfortunate people of Alcamo can only pray for a return of the old Rimi-style stability.

Organised crime has now spread to most towns in the island. Nevertheless, bright spots remain amidst the encircling gloom, and Mondello, a pretty seaside town 15 miles to the west of Palermo, is one of these. It is saturated with calm, family pleasures. Villas with absurd turrets and fake-antique fountains spouting water from grotesque mouths line a promenade along which carriages dawdle under the spiky shade of palms. Taped Neapolitan music wails in the cafés where customers sit through the day demolishing sculpted ice-cream; corpulent fathers, their trousers rolled up, net tiny fish in the shallows.

Soon after our arrival here on an evening in early autumn, a wedding party came on the scene. The theatrical setting of the Mondello waterfront is much favoured for the ritual photography following the church ceremony, staged in surroundings such as this, as my friend put it, 'to commemorate the event in the public eye'. The quay was instantly transformed into a stage upon which the bride glided on her father's arm. A corps of photographers were at work in the background, moving a Rolls-Royce here and a boat there, in preparation for an instant of extreme luminosity following sunset, when the division vanishes between sea and sky, and a lively refulgence touches every cheek. The wedding group formed and, as if at the touch of a single switch, the lights came on all round the bay. There was a soft, crowd-produced gasp of appreciation, the cameras flashed and the audience put down their ice-cream spoons and clapped.

Almost certainly among this gathering would have been Sicilians now living in the States who had flown over to take part in the feast of the 'Sainted Physicians', Cosima and Damiano, celebrated three miles down the coast at Sferracavallo. Their engagement is a strenuous one, for they join a group of about a hundred who carry the enormously heavy platform supporting the figures of the saints in a rapid, jogging promenade for hours on end up and down the streets of the village. In the course of this, as one devotee after another collapses from fatigue, another rushes to take his place. With every year that passes, the Sainted Physicians draw greater crowds, and the American contingent increases. What is extraordinary is that Cosima and Damiano have no history, and no one knows what this wild annual scamper through the streets is really all about.

If nothing else, it demonstrates the huge and often increasing strength of custom. A Bostonian participant, in Sferracavallo for two days, told a reporter: 'I suffer from depression. Most years I come over and do this, and that does the trick for a while. If I can't get there I phone in and listen to the music, which is better than nothing.' It is this stubborn traditionalism, this inextinguishable respect for the comfortable values of the past, that may provide a last-ditch defence in Sicily against the encroaching banalities of our age.

The Private Secretary

These events took place in the summer of 1945 when Lewis, still in the Intelligence Corps, was sent to Austria where, as part of his role, he was obliged to interrogate many Nazis.

When Lewis questions the 'secretary' in this article, he asks him whether he had been present at Salsk. This now seems even more prescient. Long after Lewis wrote this piece, investigations have shown that the Russian town of Salsk was the scene of horrors even worse than Lewis would have known about. Excavations in 2021 revealed that the Germans had massacred more than three thousand civilians in the town. Nearly all were the old, or women and children, even babies. Archaeologists found that the Nazis had saved on bullets: most of their victims were killed with a boot to the head, or crushing the skull against a rock face, or with a bayonet.

First published in *The Happy Ant Heap*
(Jonathan Cape, 1998)

THE SECRET PROPOSALS for the destiny of a liberated Austria at the end of the last war were no more impractical, even absurd, than any such army schemes. They were under discussion at Security Headquarters at Castellammare, southern Italy, and were listened to against a background of soft guffaws by section members with some experience of operations of the kind. Experts neutered by the unreality of a non-combatant war were sent out from London to speak of the many war criminals who had taken refuge in this basically pacific country, from which, at the moment of defeat, they would slip away across the Alps to Italy en route for South America. A plan devised to baffle them was explained. It involved the encirclement of the whole country by hundreds of miles of unscalable and impenetrable electrified fencing, linking radar-equipped strongpoints at twenty-mile intervals. Behind this fence the whole Austrian population, both military and civilian, would be confined, prior to investigation in the equivalent of a vast concentration camp, while Allied forces dealt with the expected pockets of armed resistance with firmness and precision.

Little in the history of warfare approached the ingenuity and the scale of this planned undertaking, yet when the talking was at an end, and we were finally despatched to the Italian frontier with Austria, our mood was one of profound scepticism. It was characteristic of this adventure that the Intelligence Corps sergeant with whom I had joined forces, a PhD in Hellenic studies, should be fluent in Greek of the time of Pericles, but spoke no German.

We arrived at the Brenner Pass three days after the cessation of hostilities. According to the plan, this should have been firmly closed, with a mixture of thirty-odd million Germans and Austrians penned in behind the fences awaiting our arrival. The stunning fact was that the pass was wide open and we squeezed our lorry into the roadside as a grey avalanche of humanity slid down through the valleys towards us. The fugitives were on foot or being carried in every conceivable conveyance from farm-carts with peasants wedged in among their cattle and mournful possessions as well as a circus steam-engine towing a truckful of performers, a few even still dressed for the ring, with a caged bear on a trolley. Many soldiers who had torn the distinguishing marks from their uniforms were mixed in with non-combatants of every kind and all ages. No one would ever know how many responsible for war crimes had been able to hide themselves away in this desperate snail's-pace exodus into the safety of Italy.

Some hours later, radiator spouting steam, we reached the top of the pass ready for the view of the first of the strongpoints, the electrified fence and the searchlights that would turn the Alpine night into day. Of these nothing was to be seen, and even the post with its notice proclaiming the frontier's existence had been dismantled and thrown to the roadside. Soon, with the frontier hardly two hours behind, we discovered in Austria an extraordinary normality. The war had never come as far as this and, with the first of the Russians still 200 miles away, Austria remained a fragment of the dismantled empire of the Hapsburgs. Country people free of wartime controls came and went as they had always done. It was a journey into the past, full of picture-postcard Tyrolean scenes: farmers in *Lederhosen* with wooden pitchforks, yoked oxen, and gigantic mountain dogs bounding to snap at our wheels.

On the second day we reached our destination, the small Alpine town of Engelsdorf. Here there were houses painted with flowers alongside angels in flight, and churches with enormously high steeples and clocks with little hammer-armed figures revolving to music as they struck out the hours. Carts, replacing the now-vanished cars, were drawn by oxen with splendidly carved horns. Women wearing billowing skirts and bonnets kept the streets clean, and scraped the ox-droppings into neat piles. We were halted by a girl herding geese, and later saw another shoeing a horse. From this small town all the men had been carried off to the war. Stopping, we listened to the silence, broken only by the tinkling of the ox-bells, which was to continue through day and night. Engelsdorf smelt of milk.

No hotels were open so we moved into a *Gasthaus* run by a Frau Pauli, possessor of the fairest skin, the blondest hair and, as a result of her endless labours, the hardest muscles of them all. She was assisted in her never-ending tasks by two girls, also blonde, beautiful and muscular. Sometimes they broke off from whatever they were doing to watch with bereaved eyes as a squad of newly arrived young British soldiers marched by. Asked what had happened to husbands or lovers, the answer was, 'Sir, the housepainter [Adolf Hitler] carried them away to Russland. We do not think we shall see them again.' Work, perhaps, saved them from repining, and for the abandoned women of Engelsdorf work was unending.

Two hundred yards from the *Gasthaus* at the end of the street three wonderfully polished cannons pointed down the open road to a checkpoint a couple of miles away on the boundary of the Russian zone of occupation. Germans still in uniforms rumpled from sleeping rough wandered aimlessly along the roads outside the town and we first ignored them, then came close to forgetting their presence. There were rumours of last-ditch stands being planned by Nazis who had not surrendered, and following investigations we concluded that they might be correct. Back in the imperial days, the milk barons of Engelsdorf had built vast semi-castles in which to enjoy the views over the mountains, lying in folds at the back of the town, and there were reports that some were occupied – and would be defended – by regrouped SS formations when the time seemed ripe. We in fact discovered a vast, decaying mansion with the SS still in possession, and we borrowed a half-company of infantry from the nearest Army HQ to deal with the crisis. What followed was something of an anti-climax, for having arranged for the surrounding of the supposed redoubt, my friends and I went in through the front door to find the SS occupants already lined up in the principal room, as if on parade in readiness for the offer of their surrender.

The takeover by our troops of this corner of south-east Austria provided hardly more excitement than the routines of civilian life. We learned of the commanding general's concern that the hunt for war criminals had produced scanty results and that he had spoken of his hope that the presence of trained security personnel might remedy the situation. By chance our arrival coincided with the capture of our only big fish to date, a Gestapo chieftain called Heinrich Poldau, whose mistake it had been to stay quietly at his house instead of mingling with the streams of displaced persons and ordinary soldiers who no longer attracted attention. The nature of Poldau's service committed him to a somewhat solitary existence and he had taken a small house out of town, where he had lived with an Austrian girl who had decamped as soon as the first reconnaissance car flying the Union Jack drove into Engelsdorf. Frau Pauli knew this girl, who had whispered to her of Poldau's secret activities. Now our hostess rushed to tell us of the prize lying within our grasp, and later that day we picked him up.

Meeting him, I realised that I had known almost exactly what to expect. This was a quiet man who concealed himself behind the façade of a small bourgeois life, a man who knew how to enter and leave a room without the occupants being aware of his presence. The house he had chosen to rent was a showpiece of domestic clutter. Trinkets of all kinds were pinned to the walls of his living room, including a pair of stuffed owls holding a stuffed mouse apiece in their claws. There was a shelf full of decorative pipes, and a faded print of the Redeemer, who, apart from his oriental garments, could have been a middle-class Austrian of the last century. This assemblage of objects if anything strengthened an underlying sensation of emptiness.

Poldau was a neat man in one of those rough-surface woollen jackets popular whatever the weather in places where mountains are at least in view. His hands were well looked after. He wore a plain ring and a plain watch. I was most taken by his face, which reminded me of one of those serious portraits carved in stone in the porch of a Gothic cathedral, depicting more calculation and less religiosity than those of its neighbours. I explained why I was there and he listened intently, nodding his head at the end of each sentence. Finally I told him he was under arrest and he bowed slightly and said, 'At your service.' I gathered from his accent that his English was good. I took him back to the *Gasthaus* where the girl called Fruli came down dust-covered from mending the roof, to serve him an evening meal. He was exceptionally polite to her, but she dumped the food down, turned her head and flounced off. After the meal I accompanied Poldau to the town jail – a medieval building entered through a portcullis – signed the register and left him to the jailer who, as usual in such institutions, appeared more than a little mad. Poldau's only reaction to his caperings was the slightest trace of a smile.

Back in the *Gasthaus* I sent off a hasty note by dispatch rider to the Staff Officer (Intelligence) at Klagenfurt, 'Poldau, Heinrich, believed Gestapo, held under arrestable categories. PIR follows.'

There was a reply within the hour. 'Good luck. Ascertain urgent priority whether Poldau with *Einsatz Gruppen*, Poland and Russia.'

The problem next day was where best to talk to this man. The prison environment is the one least likely to foster relaxed and possibly revealing conversation. Rough guidelines for dealing with such situations were offered in a leaflet passed out at Security Headquarters containing for me at the time the truly remarkable suggestion that the suspect under interrogation should be taken, if the surroundings were suitable, for a quiet walk. Bizarre as it sounded, it seemed in these circumstances not altogether a bad idea. As a precaution I borrowed a rifleman from the RA unit and then, collecting Poldau at the jail, told him we were going for a stroll in the country.

We drove up steeply into the foothills of the Glein Alps rising from the back of the town, and stopped for a last view of its pinnacles and golden roofs through the oaks spreading their foliage like a tinsel decoration in the summer light. I parked the car and told the soldier we were all going for a walk in the woods and he nodded, worked the bolt of his rifle and dropped a bullet into the breech. An extraordinary transformation took place in Poldau's manner and appearance. He breathed out, then laughed. 'Why is he afraid?' he asked. 'The war is over. No more war.'

'He's not the slightest bit afraid,' I told him. 'He has orders to carry out.'

Poldau laughed again, as I speculated on the change he had undergone. Now it was all over. The game was up. In some way he could now relax.

We set out on a narrow path through the woods, the soldier in the rear humming a monotonous tune.

'I have some questions to ask you,' I told him.

'Whatever you like,' he said. 'I think the time for secrets is at an end.'

'What is your rank in your organisation?'

'We do not speak of ranks. Only duties. I am a secretary.'

'Tell me how you were recruited.'

'That is simple. It was by accident. I was born in Görlitz near the Polish border, where most people spoke some Polish. When the talk of war began there was a call for Polish speakers. We were told only that the matter was confidential, having some connection with the police. It was of interest to me because the pay at the bank was very low, and because successful candidates would not be subject to military call-up.'

'So that was the Gestapo?'

'It was not a name we ever used. I was officially listed a member of the Statistics Bureau of the Security Police. After a few weeks' training I was sent to Poland.'

'Tell me something about your activities there.'

'Really it was no more exciting than the bank. It was office work as before, but I was sometimes employed to question Poles who were unable to speak German. With the end of the campaign I returned to records and statistics in Bremen. In the end I asked for an interview with the Chief Secretary and told him I was bored. He was sympathetic and said I could not be allowed to resign, but I might take six months' leave to join the army, after which I would return to the office.'

'So you got six months' leave from the Gestapo merely for asking? I find it hard to believe.'

'The Bureau was very flexible,' Poldau said – I thought with a touch of pride. 'I was encouraged by the Chief Secretary to take courses in army organisation and the Russian language,' he went on. 'After that my leave came through. I was under the minimum height for the Waffen SS, but the Chief organised things for me and I got in with the rank of *Standartenführer*, and was sent to the Eastern Front.'

'When you were in action in Russia were you a member of an Einsatzgruppen?' These were German death squads, responsible for mass murder, mostly by shooting.

'Of course,' Poldau said. 'All soldiers with my qualifications were automatically directed into such groups. Let me explain that their task was to ensure nothing could hold up the speed of our army's advance. Large areas of enemy territory were surrounded by our pincer movements in a matter of days and many prisoners taken without a fight. Thus thousands of these were left behind. Our regular troops could not speak their language, and *Einsatzgruppen* took over.'

'And many prisoners starved?'

'They starved because an army moving at high speed can carry provisions only for its soldiers. In a single day we might take twenty thousand prisoners and be compelled to leave them behind. Camps had been designed to hold them, but they could not be made ready in time. When two hundred thousand prisoners were taken and brought to Stalag VIIIB, we found this to be no more than a square of ground with no buildings of any kind. This was inevitable, because the roads were blocked with snow.'

'Were you at Salsk?' I asked.

'I was at Salsk, where not only the Russians but we ourselves came close to starving. It was ten days before food reached us, and then there was little of it. The Russians ate the bodies of their comrades who died from sickness or starvation. At first there were struggles over dividing up human meat, but then we permitted only doctors and butchers among the prisoners to do this, and order was restored.'

The rifleman, eager to listen to such gory details, had come closer and I waved him back.

'How many died in all?'

'In Salsk there is no way of knowing, because in some cases only the bones were left.'

'I mean in all the camps.'

'Figures were never given. Five million? Ten million? Perhaps more. I could talk to Russians but never understand their mentality. In battle they laughed at death. When we attacked our orders were to kill all wounded Russians left lying on the ground, because otherwise they would drag themselves to their knees as soon as we had gone through, and shoot us in the back.'

I was beginning at this point to question how much of this catalogue of horrors it was necessary to include in the report. A vague multitude of men had vanished from the earth as if through a monstrous conjuring trick. No more than a legend, eventually to be forgotten. The figures were guesswork. Those who had kept records had themselves been swept away, so there were to be no names on Russian memorials, no epitaphs to be inscribed. As García Lorca had written of a single brave man killed by a bull, 'a stinking silence settled down'.

We walked on and Poldau spoke eagerly, as solitary men sometimes do, of his childhood and background, and reverently of the skilful surgery of modern warfare as demonstrated in the blitzkrieg by which France had been overthrown in a matter of weeks, compared with those slogging years of trench warfare in the First World War. Only the snow had put an end to Germany's dream of carrying its eastern frontiers as far as the Urals. 'Not only Germany but Europe has been defeated,' he said, 'and Bolshevism remains intact.'

I returned him to prison, finished the report and took it to the Staff Officer (Intelligence), a classical scholar whose habit it was to slip admiring references to Caesar's campaigns into discussions of the military chaos of Austria. Major Stevens was a worrier also, distraught at that moment at the news that Russian Asiatic troops had broken into the town and were chasing all the women in sight. He ran through the report. 'My God,' he said. 'This man mustn't be allowed to slip through our fingers.'

'He won't. He comes into the arrestable categories. Pending further investigations, we can hold him as long as we like. But that's as far as it goes.'

'Even if he was involved in mass killings?'

'That has to be proved.'

'You say here he was at Salsk. Isn't that the camp where Russian Jewish prisoners were forcibly fed with excrement and drowned in urine?'

'Poldau denies that the Final Solution was ever employed in the *Stalags*. He admits the death total was high, but says that prisoners died of diseases, hunger and the cold.'

'What was this man doing in Engelsdorf?'

'He was sending back stories he'd made up about Austrian separatists so that they'd keep him here. He knew that Germany was finished and decided to lie low until it was all over. His plan was to stay here until things settled down, then change his identity and go home.'

'Could we use him?'

'In what way, sir?'

'It's been confirmed that war-criminal trials are to be held. This man was there. He's seen it all. He would make a sensational witness. Perhaps he could be sounded out?'

'In a way that's already been done. I believe he would agree to anything we propose.'

'With some sort of inducement no doubt?'

'It might help.'

'There's little we could offer. All these people will have to be let go in the end. Might be able to speed up the process, that's all. Any question of financial inducement has to be ruled out.'

'Money doesn't interest him. He's an abstemious sort of man. Strangely infantile. He had a collection of toy railway engines and is fond of animals. He mentioned he'd carried a white rat as a mascot through the Russian campaign. It died through eating unsuitable food. He likes painting.'

'What does he paint?'

'Sea views.'

'Is he married?'

'No. He is too devoted to his mother, he told me, to marry. The worst thing for him about the Eastern Front was that her letters took up to three months to come through.'

'In a way none of this surprises me,' Stevens said.

'It didn't surprise me either, sir.'

'Any thoughts as to what might help to bind him to our purpose?'

'So far as I'm concerned, discussions are complete and as soon as they're ready, he'll be sent to one of the camps in Germany. They're opening one near Bremen, where his mother lives. If it could be arranged for her to visit him there, I'm sure he would show his appreciation in any way he could.'

'Well, I imagine we could do something about that,' Major Stevens said.

A Few High-Lifes in Ghana

Lewis was commissioned by the New Yorker *to cover Ghana's Independence celebrations in March 1957.*

Before the actual celebrations he embarked briefly on what he described as a 'sightseeing trip'. Plenty of suitable destinations, both peaceful and alluring, were available, but Lewis ignored them all, characteristically choosing an area that was adventurous and risky.

Many years later in an interview with the Daily Telegraph, *he described his ideal destination: 'Generally speaking, it has to be a bit horrific. I like any place that has some atmosphere of suspense and danger.'*

First published in the *New Yorker*, November 15, 1957

THE IMPORTANT THING to bear in mind when visiting what was once the Gold Coast and now is Ghana is that the advice liberally proffered by the old Gold Coast hands in retirement will be designed to perpetuate a nostalgic legend. You are warned to prepare yourself for a scarcely tamed White Man's Grave, where you do not omit to take Sensible Precautions, stick to sundowners, keep your possessions in an ant-proof metal box, and wipe the mildew off your boots at regular intervals. I fell a victim to this propaganda to the extent of buying a pair of mosquito boots before I left London. They were made of soft, supple leather, fitted very tightly at the ankle, and they reached almost to the knee, beneath which they could be drawn tight with tapes. I put them on once only, in the privacy of a hotel bedroom, noting that worn with khaki drill shorts they made me look like some grotesque Caucasian dancer. After that I packed them away. It was a symbolic act. I had observed that in Accra, Europeans in these days seemed to make it a point of honour to go bareheaded in the noonday sun. The sundowners seemed to have gone out with solar topees. You popped into a bar and drank a pint of good German or Danish lager whenever you felt like it. It was hot in Accra, but not so hot as New York at its worst and – in the dry season – not so humid either, and down by the shore there was usually a cool breeze blowing in from the Atlantic.

Accra turned out to be a cheerful, vociferous town with an architectural bone-structure of old arcaded colonial buildings – some of them vaguely Dutch or Danish in style. The streets, in the English manner, had been cut in all directions. There were corrugated-iron shack warrens right in the heart of the town, a wide belt of garden suburbs, and isolated slabs of modern architecture looking like enormous units of sectional furniture. As usual, the English had thrown away the chance – always seized by the French in their colonies – of turning the seafront into a pleasant, tree-shaded promenade. The surf crashing on the beaches was out of sight behind the warehouses, put up at a time when trade was all. The merchants were content to put off their gracious living until they had made their pile on the Coast and could get away with it back to England before malaria or the Yellow Jack finished them off.

The streets were crowded with a slow-moving mass of humanity: the men in togas, the women in the Victorian- and Edwardian-style dresses originally introduced by the missionaries but now transformed by the barbaric gaiety of the material from which they are confected. The designs with which these cottons are printed demands some comment. They are produced in Manchester, in Brussels and in Paris, from African originals, and although they recall the most striking Indian saris, there is something fevered and apocalyptic in the vision behind the drawing itself, which seems to be purely African. The best results are supposed to be produced by West African artists as the result of dreams, and the artist may mix abstract symbols and careful realism in a single design, with a result that often has a drugged and demonic quality, like a descriptive passage from *The Palm-Wine Drinkard*. Dark blue birds flit through an ashen forest of petrified trees; silver horses with snake-entwined legs charge furiously into a sable sky; huge metallic insects glint among the lianas of a macabre jungle; the black bowmen of Lascaux pursue griffins, fire-birds and tigers over fields of gold, with autumn leaves the size of shields tumbling about their shoulders.

Amazingly an Englishman can be at home in this atmosphere, which somehow, in defiance of the genial African sun, the colour, and the seething vitality, succeeds in reproducing a little of the flavour of life in England. The African citizen in Ghana, for example, is reserved in his manner compared, say, with his counterpart in Dakar. It is hard to believe, in fact, that a century of independence will be long enough to expunge the essentially British odour of life in the Gold Coast: the cooking (Brown Windsor soup and steak-and-kidney pie), the class observances, the flannel dances, the tea parties, and the cricketing metaphors in the speech. Even the paint on the fences in a suburb of Accra is of a kind of sour apple green never found outside Britain and her dominions. The middle-class African of Accra, too, lives in home surroundings indistinguishable from those favoured by his equivalent in the London suburbs. There is the same affection for whimsy and humorous pretence: china ducks in flight up the wallpaper, Rin-Tin-Tin bookends, toby jugs, telephones disguised as dolls, poker-work mottoes, and jolly earthenware elves in the back garden.

It is generally believed that fraternisation between the whites and blacks is less complete in African territories colonised by the British than in those colonised by the French. This does not apply in West Africa. In Dakar the colour bar is officially non-existent, but it is exceptional to see an African in a good hotel or a fashionable restaurant. The reason one is given is that they do not feel comfortable in such surroundings. The natives of Accra are not overawed in this way, and there is a fairly proportional colour representation – say ten Africans to one European – in all public places of entertainment. It was in fact quite the normal thing that my first experience of Accra nightlife should be in the company of Africans.

This was in early March. Ghana was just about to receive its independence. There had been a week of celebrations, and the streets were awash with restless, slightly jaded revellers. My host was a minor political figure we will call Joseph, and he had brought with him his secretary, Corinne. We started the evening at the new Ambassadors Hotel, which is said to be one of the three best hotels in Africa. At this time there was no hope of staying there, as the Government of Ghana had filled it with foreign VIPs invited for the celebrations. We sat in the bar and admired the photomontage on the wall, which included a dancing scene from *Guys and Dolls* and the towers of the Kremlin. Behind the palms a pianist in tails was striking soft, rich chords on a grand piano. Joseph and Corinne ordered Pimms No.1, which was currently *de mode* in Accra. I noticed that the VIPs present included firebrands from British Guiana and Tunisia – temporarily tamed and transformed in glistening shark-skin – as well as a berobed African chief who wore ropes of beautiful ancient beads, and who waved genially and said 'Ta-ta' as we came in. There was no hope of a table for dinner at the Ambassadors, so we went on to another restaurant, and there, in the sombre and seedy surroundings of an English commercial hotel in a small Midland town, we made the best of a highly typical meal of fried liver, tomatoes and chips.

After that we visited a nightclub called A Weekend in Havana, outside which Joseph got into an altercation with a policeman over parking his car. There is often a fine Johnsonian rumble about such exchanges in Accra. 'You were attracted by the glamour of your profession. Now you must work,' was Joseph's parting shot. Later in the evening when a cabinet minister offered him an extremely stiff whisky he said, 'I am not, sir, a member of your staff, and am not, therefore, accustomed to more than singles.' Still later when an enormous Nigerian emir, gathering his robes about him, joined our table, Joseph remarked, 'I hear, sir, that your people reproduce at an alarming rate,' and the emir, who took this as a compliment, beamed with delight, and replied, 'You have been correctly informed.'

The nightclub turned out to be an open-air place, with the tables placed round a thick-leaved tree that gave off an odour of jasmine. A white dove-like bird circled continually overhead as if attracted by the powerful fluorescent lighting. When we arrived, the band was playing 'It's a Sin to Tell a Lie', and the dancers were gliding round in the stately Palais-de-Dance manner. About half those on the floor were in national costume and the rest in evening dress. The few European women to be seen were outclassed by the African women with their splendidly becoming gowns and their majestic carriage. Corinne sat happily commenting on the private lives of those present, and closing my eyes and listening to her remarks I could hardly believe I was not sitting in a similar night-spot in London, although Corinne's voice was somewhat richer and deeper than that of any conceivable English counterpart. 'Good heavens! Isn't that Doctor Kajomar with Mrs Chapman? Had you any idea that show was still going on, Joseph?' A girl swung past in the arms of her partner wearing one of the new slogan dresses with 'Justice and Liberty' printed large across her ample posterior. Corinne looking away as if pained said, 'Do you know – I do think people should draw the line somewhere!' Soon after this the band played a high-life – a dance of Gold Coast invention – which resembles a frenzied and individualistic samba. A party of British sailors from a naval vessel helped themselves to partners from the local girls and joined in this to the best of their ability and were much applauded by the Africans. The trouble about the British – Corinne had previously commented – was that they never let themselves go. Her ex-boss had been a Scotsman who had made her blood run cold by his habit of concealing his anger.

This gave me an opening to indulge in a favourite pastime, when in regions that are slipping, or have slipped, through the colonialist fingers – that of carrying out a post-mortem on the relationship between the two peoples involved, in the full prior knowledge that the findings – allowing for local variations – will be the same. Getting right down to bedrock objections, Joseph said, it amounted to the fact that the Englishman had never learned to stop complete strangers in the street, shake hands with them warmly, and ask them where they were going, and why. What was even worse, they had almost succeeded in breaking the people of the Gold Coast of such old-world African displays of good breeding, inculcated in all the bush-schools before the European came on the scene with his version of education and his insistence on the formal introduction. I knew this to be true. Only a few weeks spent in Africa – especially if not too much time is wasted in big towns where the real flavour of the country is subdued – are enough to convince one of the extreme and innate sociability of the African. Africans, as one discovers them in travelling in the villages of the interior, are never standoffish, rude or aggressive; always ready to receive the visitor in a courteous and dignified way. This is the tradition of the country, and even in such Europeanised areas as the Gold Coast, where a boy no longer spends four years or more under the strict discipline of the bush-school learning the ideals of manhood, a stern semi-Victorian training is usually carried out in the home, with what to me are excellent results.

It might even be reasonable to suggest that there are strong temperamental and emotional factors behind the façade of politics, which are in reality helping to strip the Briton of his colonies. Even in the Gold Coast, where the Englishman had learned to become a better mixer than his French neighbour, there remained a trace of that aloofness, that inability to get together with the African on a footing of absolute socially equality, which makes it so difficult for him to be loved as well as respected. Here, as in India and in Burma, the European clubs defended their exclusiveness to the last ditch. The Englishman was received socially by the educated African without any reserve whatever, but the African's civility was not fully returned, and there was an offensive flavour of patronage in this lack of reciprocity. The Accra Club admitted no African members or guests. At Cape Coast only two influential chiefs had ever succeeded in joining the white man's club. The Kumasi Club underlined its determination to hold out to the last by posting a notice which informed the members that 'due to the development of events' they would be permitted on and after Independence Day to introduce guests of 'any nationality' – although the names of such proposed guests had first to be submitted to the secretary. To these pinpricks, to which Africans submit cheerfully and without rancour, more serious wounds are added when they come to England as students and find that under some dishonest excuse – since the colour bar in England has no admitted existence – 85 percent of hotels and boarding houses will refuse to admit them.

No demonstration of the virtues of imperialism – the high-minded incorruptibility, and the like, of the white overlords – could quite compensate the new, nationalistic African for his being treated, whether overtly or not, as a member of an inferior race. Thus many Africans who have been hurt by the coolness of their reception in England have returned to Africa carrying the germ of a disease that is fairly new in that continent – anti-white racial feeling. Now the whites were on the point of surrendering their domination. The European clubs would open their doors to all. Dr Nkrumah's portrait would – despite the protest of the parliamentary opposition – replace that of Her Majesty Queen Elizabeth, on both stamps and coinage, and shortly the British Governor-General, Sir Charles Noble Arden-Clarke, would be asked to surrender his impressive apartments in Christiansborg Castle to Dr Nkrumah, and to retire to the modest accommodation previously prepared for Dr Nkrumah in the State House. But already, on the eve of independence, the newspaper editorials sounded a little less dizzy with success. There would be no colonial scapegoats about when things went wrong in the future. The Ghanaians would have only themselves to blame if the much-publicised corruption in their public men brought about their undoing as a nation, or if the disputes with the Ashanti and Togoland minorities were allowed to deteriorate until they exploded into civil war.

Many members of the newly freed colony regarded the victory of Kwame Nkrumah and his followers as the victory of an energetic political clique which had been able – sometimes by dubious means – to impose its will upon the politically lethargic general masses.

Such disgruntled opponents of the regime, who did not expect to participate in the fruits of the victory, were on the whole unhappy to see their white rulers depart, and there were refusals in several parts of the country to put out flags. That the English could pull out as they did with so little apparent reluctance, and so many protestations of goodwill all round, is due to the nature of their stake in the country, which does not in reality demand their physical presence. Ghana has been saved from the tragic situation of Algeria, and the almost equally unhappy situations of Kenya and South Africa, by the fact that it has never been considered suitable for European settlement. West Africa mostly had been protected from white ownership by malarial mosquitoes, an inexorable rainy season, and an absence of salubrious highlands where European farmers could have established themselves. When the demand for independence came, there was no reason not to accede to it. As things were, only British traders, technicians and colonial officials got a living from the country, and these would not be compelled to leave. The only conceivable losers might be certain African underlings with a preference for the devil they knew to the devil they didn't know and a suspicion they might be exchanging King Log for King Stork: these and the 300,000 small farmers of the Ashanti, who between them produce the cocoa that forms the country's wealth, and who in the long run – and at present with little political representation – must foot the bills run up by the politicians at Accra.

From the very beginning it has been commerce that has drawn the European to the Gold Coast, and from this commerce developed one of the gravest social cancers that have cursed the human race – the slave trade. With the exception of the Spanish, all the maritime European nations – Portuguese, Dutch, Danes, British, Swedes and Prussians – had at one time or another established strongholds in the Gold Coast, and squabbled among themselves over the rich loot in captives. The British proved to possess most staying power. In recognition of their straightforward and efficient business dealings, they finally secured a much sought-after contract for the supply of slaves to the Spanish colonies, held previously by the French and then the Dutch. In the end, over one-half of the total slave trade fell into British hands. It has been estimated that between 1680 and 1786, more than two million slaves were exported from the Guinea Coast, as it was then called. The wastage of life was tremendous. Livingstone believed that ten lives were lost for every slave successfully shipped, and even at sea the carnage continued. French ships' stores, for example, included corrosive sublimate, with which slaves were poisoned when the ship was becalmed in the Middle Passage and supplies ran low (the French defended the practice as being more humane than the British and Dutch one of simply tossing the starving slaves into the sea).

The slave trade was always utterly shameful, but contemporary accounts by those who took part in it are full of conscience-salving devices. Much was made of the slave's happy opportunity to be brought into contact with Christianity. Slavers piously presented themselves as the rescuers of prisoners taken in African wars who would otherwise have been slaughtered, making no mention of the fact that it was they, the slavers, who encouraged or even organised the wars. The slave merchants could be tender, too. 'I doubt not', says William Bosman in a letter written in 1700 from the castle of St George d'Elmina, 'but that this trade seems very barbaric to you, but since it is followed by mere necessity it must go on; but yet [in branding the slaves] we take all possible care that they are not burned too hard, especially the women, who are more tender than the men.' I have visited this castle and seen the rooms where the slaves were confined and where they were auctioned. What particularly struck me was the arrangement by which heads of families who had brought some dependant to be auctioned were admitted to watch the proceedings from a chamber overlooking the auction room, where they themselves would not be exposed to the reproachful gaze of their victim.

William Bosman, despite his name, was a Dutchman, a man of severe morality and regular habits, who much deplored the intemperance of his English trade rivals entrenched at Cape Coast

Castle some thirty miles away: '... The English never being better pleased than when the soldier spends his money on drink ... they take no care whether the soldier at pay-day saves gold enough to buy victuals, for it is sufficient if he have but spent it on Punch; by which excessive tippling and sorry feeding most of the Garrison look as if they were Hag-ridden.' The English, Bosman observed, were also much given to a plurality of wives, particularly the chief officers and governors of the castle, while two of the English company's agents had married about six of the local ladies apiece. This enterprise was the Royal African Company, promoted under a charter granted by Charles II. His Majesty was the principal shareholder in the venture, in which the whole of the royal family invested money. In spite of Bosman's poor opinion of the garrison a dignified protocol, as befitted a royal enterprise, was observed in all the company transactions. Slaves were branded, as a compliment to the Duke of York, the company's governor, with the letters 'D.Y.' – and the brand used was of sterling silver.

Denmark was the first European power to abolish the slave trade, by a royal order in 1792. The British followed in 1807, although a general European agreement was delayed for another twelve years by the French, who hoped in this way to gain time to be able to crush the rebellion in Haiti and restock the colony with fresh slaves. The century that followed saw the gradual adjustment of the Gold Coast to legitimate trade, based at first principally on the extraction of gold (the guinea was originally coined from the gold secured from the Gold Coast), and then the cocoa bean. The first cacao tree to be grown is supposed to have been brought from Fernando Po in the eighties of the last century, as the result of the enterprise of a native blacksmith, and each pod is said to have sold for one pound. By 1949 the Gold Coast was producing as much cocoa as all the rest of the world put together. It is now one of the richest areas in Africa, and its total revenue from all sources is about ten times that of the neighbouring republic of Liberia, which has never been under colonial domination. It is a curious illustration of the mentality of nationalism that the politically educated citizen of Ghana now tends to play down the importance of the slave trade in the history of his country. The subject when raised is likely to be changed or to be brushed aside as historically insignificant. The memory is clearly considered derogatory to the dignity of a modern nation.

The emergence of this modern nation could never have been delayed more than a few years, but the fact that the Gold Coast became Ghana in March 1957, and not perhaps twenty years later, is largely due to the energy and the tactics of its leader, Dr Kwame Nkrumah. Dr Nkrumah was born in 1909, said by some to be the son of a market woman, and by others, of an artisan. After a few years spent in the teaching profession he went to America and gained the degree of Bachelor of Sacred Theology at Lincoln University, Pennsylvania, which a few years later granted him an honorary doctorate. When he returned to Accra, Nkrumah took over the nationalist movement, founded a new party, the Convention People's Party, with its slogan 'SG' (Self Government) and 'Freedom' (the two syllables are pronounced in Ghana as two separate words). Nkrumah's tactics began with a boycott on European goods, and from this, rioting and looting developed. Two short prison sentences followed, both invaluable to the progress and propaganda of the CPP. Nkrumah was released from the second of them to become officially first 'Leader of Government Business' and then Prime Minister over a predominantly African team of ministers.

Democracy is liable to be transmuted by the old tribal tradition of government into a parody of what is understood by that word in the West. Political issues are decided not so much by party programmes – which are quite beyond the comprehension of village electors – as by the political personalities involved, and the crowds swarm to the support of the energetic and flamboyant leader. The enfranchisement of the black masses spells the end of the white man's domination – not because there is any solidarity in colour except an artificial one in the course of creation at this moment – but because the white man cannot compete with the African's knowledge of native psychology, and cannot in our time, even if he would, play on the African electors' hopes and fears with the deadly expertness of an ex-tribesman. As an illustration of what is happening all over those parts of Africa where the electoral system has been introduced, a party will often choose as its emblem an animal known for its sagacity and strength – say the elephant – while the opponents may decide on the lion. The election, in the unsophisticated countryside, now resolves itself into a contest between the merits of these two animals. The elephant followers will obviously be unsuccessful in districts where a herd may be running wild and trampling the crops, whereas lion supporters can have no hope of gaining ground in remote pastoral areas where lions still sometimes carry off livestock.

African political parties – and this applies not only to those of Ghana, but to the whole of West Africa – change their programmes and their affiliations in such a way that not even a trained student of politics can keep up with them. Their appeal to the mainly illiterate elector must then be simplified to the point of absurdity. The standard of political advancement of the village masses may be judged from the fact that when just before the election Nkrumah and his supporters carried out a perfectly normal animistic ceremony which consisted of formally asking the support of the spirits of the Kpeshie lagoon near Accra, the rumour became general that the Prime Minister had called on the gods to kill all who voted against him. Many electors, as a result of this, abstained from voting. Again in the Ashanti country, where Nkrumah isn't liked, his supporters had successfully spread the report that the Duchess of Kent when she arrived for the Ghana celebrations had actually crowned Nkrumah king of Ghana. Many people say that Dr Nkrumah would like to be not a mere prime minister but a real king – and not a king over Ghana alone, at that. French newspapers published in Dakar report that when, several years ago, he visited a celebrated witch-doctor in Kan-Kan, in the French Sudan, this was the prize foretold when the auguries were taken from the blood of a sacrificed chicken. After the independence celebrations Dr Nkrumah visited Kan-Kan again, but in the meanwhile the old witch-doctor had died, and, as his successor was not yet fully trained, no cock was sacrificed this time. A friend of mine who saw Dr Nkrumah on this occasion noted that he was carrying a copy of Machiavelli's *The Prince*.

The most frequent charge levelled against the CPP is that of corruption, and even to the casual observer it would seem that many Government functionaries live in a style remote from that made possible by their salaries. By the time I visited Ghana it was said that no man could expect to get on the shortlist for any Government appointment without a scale payment 'to the party funds', while a British senior police official who was staying on, admitted that the length of his service probably depended entirely on how soon it was before he received an order to turn a blind eye on the misdoings of someone in a high place. This general corruption in African politics is excused, even defended, by some observers, on the ground that it is strictly in line with the ancient tradition of the country. Every formal human contact necessitates its appropriate offering. When a man leaves on a journey all his friends make him a gift, however trivial, and when he returns he will be welcomed with another small offering. The successful conclusion, in the old days, of an initiatory stage in the bush-school was signalled by a shower of congratulatory presents. A girl expected to receive tributes of beads and cosmetics not only for her wedding, but when she was officially recognised as marriageable. No dispute could be brought for a chief's adjudication without a 'mark of respect' being offered by both parties in the case. One of the worst torments of African travel until very recently arose out of the necessity of 'dashing' every chief one visited on one's travels, and then there was the problem of disposing the livestock that were frequently 'dashed' in return. When I once paid a courtesy call on an important dignitary living in a remote part of the country where the old way of life was still followed, I was startled after we had shaken hands to be told by the chief that he could not receive me, 'without warning'. What he meant by this was that I had not given him time to find a suitable 'dash', and our meeting must therefore be considered as without official existence. It also meant that the two bottles of beer I had brought for him could not be decently handed over until I had gone. These are the usages of highly complicated civilisation; they are all-pervasive, and when – as at present – the old order breaks down and politicians take over from the chiefs, nothing is easier than the transition in almost imperceptible stages from the ceremonial gift to the outright bribe.

The official jollifications that took place in March 1957 in Accra, it should be stressed, celebrated in reality a situation virtually in existence since Nkrumah became Prime Minister in 1952, so that by the time I visited Ghana the country had been to all intents and purposes independent for several years. The formal takeover was accompanied by all the public junketing one would have expected, but as these were not particularly characteristic of West Africa, I took the opportunity two days before Independence Day to go on a sightseeing trip outside Accra. Hiring a taxi I drove to Ho, capital of Togoland, a hundred miles away. Although it had been feared that the Ashanti minority – many of them were opposed to union with Ghana – might cause trouble at this historic moment, it was in Togoland, in fact, where rioting was actually going on, and to which troops had been sent.

This particular day turned out to be a coolish one. We drove eastwards from Accra along a good asphalted road, shortly, as the road left the coast, entering the rainforest belt. Here opulent woods replaced the parched scrublands of the coastal areas. There were frequent giant anthills by the roadside, pinnacled like Rhine castles painted in the background of German old masters. I was disappointed to see no animals, no flowers except for a few meagre daisies growing in the verges, and no birds except for turtledoves and an occasional lean dishevelled-looking hornbill. The African native's access to firearms has brought about the virtual extermination of all edible animal species in the Gold Coast. It turned out indeed to be a great day in the driver's life, when later in the trip a hunter offered us a large cane rat – practically the only form of game obtainable in these days. The price asked for this animal was twenty-five shillings.

The driver beat the man down to fifteen shillings and told me that it was a bargain at that figure. We passed through nondescript villages plastered with advertisements for Ovaltine, Guinness, and Andrews Liver Salts. Africans it seems are easily persuaded to worry about their health. There was a decrepit shack of a restaurant called 'Ye Olde Chop Bar', and a drinking saloon called 'Honesty and Decency'. We met a great number of what are called 'mammy-lorries' coming down for the Accra celebrations. These trucks, which are owned by the world's most prosperous market women, are famous for their names, which – following the principle used in the tabloid headlines – usually attempt to crowd too much information or comment into too few words. The result is sometimes unintelligible to the outsider. We saw trucks with such names as 'Still Praying For Life', 'Trust No Future', 'Still As If', 'One Pound Balance', 'Look, People Like These', and 'As If They Love You'. These trucks are driven with abandon, and the wrecked and burnt-out shells litter the roadside. The African brand of driver's fatalism is even more irremediable than most, due to the fact that the African tends not to believe in the existence of inanimate matter. Trees and rocks are capable of locomotion, so that after an accident a driver – washing his hands of something so completely outside his control – may simply say: 'A tree ran into me.'

We crossed the new bridge over the Volta and immediately entered a new country. This had been German colonial territory until 1919, when the country had come under League of Nations control and been divided rather crudely and purposelessly between the British and the French. There were a few signs of jubilation in these villages. In the outskirts of Ho a shop still carried the title Buch Handlung, although it no longer sold books. At this point, we ran through the tail end of a rainstorm and the thick spicy odour of an old-fashioned grocer's reached us from the wet jungle.

I had a letter of introduction to Mr Mead, who had been formally known as Resident of Togoland but whose official title had now for some time been modified to Regional Officer. My arrival could not possibly have been worse timed. The situation at that moment in the surrounding villages was officially described as explosive, and Mr Mead, whose job it was to see that no explosion took place, had had no sleep for several nights. A minor upset had been caused by a tornado that had ripped through the edge of the town that afternoon, torn off some roofs, and put the town electricity supply out of action. Finally the Regional Officer's wife was in the last stage of a difficult pregnancy, and a car stood by, ready, in an emergency, to rush her to Lome, the capital of French Togoland where, Mr Mead said, the medical services were better developed than the local ones.

Mr Mead faced these difficulties with an Olympian calm. We dined splendidly on the vast polished veranda of the Residency, served by white-coated, whispering stewards who moved as stealthily as Indian stranglers. Almost certainly, having first discovered through the steward that I was travelling very light, Mr Mead had asked to be excused from dressing for dinner, and we ate in civilised, tieless comfort. Like all great administrators the Regional Officer seemed to admire and respect the customs of the people he ruled, although he thought that they were rather letting the side down in their violent and non-constitutional reaction to their integration with Ghana. My host was a master of magnificent understatement, and his only complaint arising from the vexations of the moment was that there was a shortage of bath water. About halfway through the meal a dispatch-rider arrived with an urgent message. The Regional Officer, after hurriedly excusing himself, departed for his headquarters for another sleepless night, carrying with him a copy of *À la recherche du temps perdu*.

Next day I set out to see something of Togoland. Protocol first required a visit to the paramount chief of Ho, but here a difficulty arose. A schism had taken place in the leadership of the Ewes of Togoland over the issue of their permanent incorporation with Ghana, and a very strong minority had asked to remain under British rule until such time as they could unite with their brothers in French Togoland to form a separate nation. When the division of Togoland had taken place in 1919 about 170,000 Ewes found themselves transferred to the British, and about 400,000 came under French control. The Ewes complained that they suffered by this change of masters. The British slice, in particular of the ex-German colony, they said, became no more than an unimportant appendage of the Gold Coast, and from 1919 onward no Ewe had much hope of self-advancement unless he left his native country – as great numbers did – and migrated to Accra. The two political factions dividing the country – those in favour of the CPP and union with Ghana, and their opponents who had lost the recent elections – now regarded each other with implacable hostility. What was perhaps the most extraordinary feature of the situation was that the original paramount chief who headed the apparently pro-British faction had come under Mr Mead's displeasure, and diplomatic relations had been broken off between him and the Residency. The British, in fact, officially supported a new pretender from the royal family – a member of the once revolutionary CPP (which still talked sometimes about breaking the chains of imperialism – although in these days with no really convincing show of acrimony).

It was a problem to know which chief to visit first, as it had been hinted to me that either might feel himself slighted if it came to his knowledge that I had placed him second on the list. In the end I decided to make it the dissident chief who was notorious for his readiness to take umbrage. I found him living in a small single-storey house. From the bareness of the furnishings and the absence of comfort, I got the impression that this chief – like so many minor potentates of West Africa – was a poor man. Chiefs are elected from a number of suitable candidates drawn from the royal family and I was told later that no Ewe candidate stood much chance of election unless he was the kind of man who got up early every morning and set off, hoe over his shoulder, to work on his farm.

Chief Togbe Hodo's reception was not a genial one. I found the chief in his courtyard, wearing his working clothes and seated on a piano stool. He was a man of about sixty. According to old-fashioned local usage he affected not to notice my entry, and appeared to be absorbed in his study of the faded coronation picture which provided the room's only decoration. A 'linguist' invited me to seat myself on a worn-out sofa and whispered that the chief would answer my questions when his council of 'wing-chiefs' arrived. He then fiddled with the knobs of a radio set until he found a station broadcasting hymns, and turned this up to a fair strength. The council of wing-chiefs, who had evidently been fetched from their work, soon trooped in, and seated themselves on a miscellany of chairs that had been placed round the courtyard. There were eight of them, and one of them wore a carpenter's apron and sat clutching a plane. This was my cue, as Mr Mead had warned me, to get up and shake hands with each chief in turn, starting with the man on my right and working my way round the circle. Speaking on behalf of the paramount chief, who now appeared to have noticed my presence for the first time, the linguist now said, 'You are welcome. Pray begin your questions.'

Formal palavers of this kind form a great part of African small-town life and any visitor from another country, however unimportant he may be, is expected to enter with good grace into the spirit of the thing. I don't remember what questions I asked, and these certainly only provided an excuse for the exposition by the council of their views on the burning theme of union with Ghana. Chief Togbe Hodo's Grey Eminence turned out to be a nonconformist minister, who had been hurriedly sent for. The Reverend Ametowobla had been at Edinburgh University, and he spoke with persuasion and grandiloquence in the soft accent of the Scottish capital. There had never been much hope for Togolanders, he said, since the Germans who wanted to make a show-colony of it had left. Now that they were to be delivered up to the mercies of the politicians of the Gold Coast, there would be none at all. One of the chiefs in attendance was old enough to remember what it was like under the Germans. Herzog the German governor had wanted to outstrip the Gold Coast and had set to work with tremendous energy to develop the country. There has been compulsory schooling for all, whereas in these days most of the population was illiterate. On the other hand the Germans had introduced forced labour. Chief Togbe Hodo made no contribution to this discussion except in his native tongue.

I believe that he understood English but would have considered it undignified to dispense with his interpreter on such occasions.

The other chief, Togbe Afede Asor, turned out to be young and agreeably expansive. Once again there was the business of waiting for the assembly of wing-chiefs before he could speak, but after this he brushed ceremony aside. We shook hands. I said, 'How do you do?' and the chief smiled widely and said, 'Okay'. After a brief discussion of local affairs, the chief asked if I had any objection to his performing a libation. This pagan custom is in wide use all over the Gold Coast, despite the most vehement protests from the Christian clergy and in particular from the Bishop of Accra. On national occasions it is carried out by Dr Nkrumah himself in exactly the same way as his counterpart in Europe might lay a wreath on a cenotaph. Dr Nkrumah when he makes a public libation uses the traditional Hollands Gin, but Chief Togbe Asor said that Black and White Whisky would in his opinion be just as acceptable to the ancestral spirits to whom the libation would be made, and he liked it better himself. We went into the chief's living-room, which was densely furnished in Victorian style, and there the chief poured about a teaspoonful of the whisky out onto the green linoleum, at the same time praying in a loud matter-of-fact voice for the success of any mission I happened to be on, and – as at that time he still supposed me to be a Government servant – promotion in my particular department. After that we completed the ceremony in the improved fashion by drinking a stiff whisky apiece ourselves. Chief Asor told me that he was a Catholic, and that among the Catholic flock in Togoland only chiefs were allowed to pour libations and possess more than one wife. As another chiefly privilege he had 'medicine' buried in his backyard to protect the household from malevolent spirits. When I left he invited me to come round next morning at six, when he would sacrifice a sheep in honour of the flag-raising ceremony of the new nation. He also presented me with a neatly written biographical note, reading as follows: 'Togbe Afede Asor II was born in June 1927 by Fia Afede XII of Ho Bankoe and Abla Dam of Taviefe. He was educated at the

Catholic Mission School from 1936-46. He was Assistant Secretary to the Asogli State Council from 1947-52. He was installed on 22nd February, 1952, on the ancient Asogli Stool of Ho. Togbe Asor II was the descendant of the great grandfather Asor I of Ho who led the Ewe emancipation from Notse 360 years ago. Hobbies: Table-tennis, Walking, Gardening.' The stool referred to here is the ancient West African symbol of kingship: the counterpart of the crown in Europe. It is kept under close guard by a functionary known as the Stool Father, whose power may almost equal that of the chief. The stool is considered to be impregnated with a magical essence, which in the old days was 'fed' or revived, by the blood of human sacrifices, and although it is too small and too sacred to be sat upon, a chief may be held in contact with it in the seated position from time to time, to allow him to absorb some of its power.

After saying goodbye to Chief Togbe Asor II, I made up my mind to drive on to Kpandu, one of the principal centres of the resistance movement. In the preceding days, abandoned training camps had been found in the bush round Kpandu, and several caches of weapons and explosives had been unearthed. This was March 5th – eve of Independence Day – and it was feared that despite the precautions taken to send military units into the area, rebellion might break out at any moment. There were few signs of life in the villages we passed through. Houses and shops were shut up, and there were no decorations. The driver, who was understandably nervous at the possibility of running into a battle, took to stopping at every village to inquire about the situation along the road immediately ahead. This meant a de rigueur call on the chief and his council and a certain amount of punctilious time-wasting.

Dzolokpuita stands out in the memory. Dzolokpuita was a pretty little Italianate-looking cluster of neat stone houses built on rust-red earth and shaded by flame trees in full blossom. Here the opposing factions had withdrawn to opposite ends of the village and were waiting, so the chief told us, with their cudgels and knives, ready for the coming of night. This chief was a rare pro-Government one – that is to say, he was pro integration with Ghana, and he was in fear of his life because his party was in the minority. He was the poorest chief I had so far met. He received me on the veranda of his hut, seated in a deckchair with a replica of his sacred stool at his side. A child's chamber-pot had been hurriedly pushed out of the way underneath the stool. The chief's linguist was literally dressed in sackcloth, although, when the council of wing-chiefs came scrambling in, I noted that some of them wore old French firemen's helmets – a suggestion that they had seen better times. The wing-chiefs were scared stiff – they expected to have their throats cut that night – and they fidgeted and peered nervously about while the interminable routine of formal questions and answers was being got through. It was clear to me that even in the shadow of bloody revolt the chief wasn't going to be balked of a prolonged exchange of the courtesies. After I had asked him how many children he had begotten, and he had gravely replied, 'They are numerous,' he was going on with the full recital of their names, together, so far as he could remember, with those of their mothers, until he was stopped by cries of protest from his thanes. An army truck with a soldier crouched purposefully behind a Bren gun rolled into the square, and a wing-chief went rushing out to demand its protection; but the driver hastily accelerated away again, leaving the wing-chief waving his helmet frustratedly after it. 'We shall all die, tonight,' the paramount chief said. He asked me to bring their desperate situation to the notice of Queen Elizabeth. After that a sackcloth-clad official poured a libation of locally distilled bootleg gin, and I was allowed to get away.

I went up to Kpandu, and back through this brilliant and menaced countryside. There were soldiers drilling in little groups of threes and fours in the open spaces of small towns, with the passion and dedication that West Africans bring to their military exercises. Where there were no soldiers, there were lurking groups of cudgel-armed men. The market in Kpandu was nearly deserted and dreadfully malodorous. Here they sold millions of tiny sun-dried fish, and smoke-cured cane rats, which filled the air with a fierce ammoniacal stench. You could also buy lovely ancient-looking beads copied from Phoenician models, spurious amber made in Japan, short-swords used in the north for protecting oneself from hyenas, pictures of Princess Margaret and Burt Lancaster, and a clearance line of portraits of Dorothy Lamour in her sarong. While I was mooching about, a small, spruce soldier arrived with a portable gramophone, wound it up and put on a tune called 'Ghana Land of Freedom' which, while serving as an unofficial national anthem, has the unusual advantage of being a high-life, and is danced to as such (the other side of the record features Lord Kitchener in 'Don't Touch Me Nylon'). While the record was played to the ostentatiously turned backs of the few nearby traders, the soldier stood to attention. A moment later, what was clearly a local man of substance came up. He was dressed in Accra style in toga and sandals, and after offering me his hand in the easy genial way of unspoilt Africa, he nodded at the back of the retreating soldier and we exchanged knowing smiles. 'I fear, sir, he is batting on a sticky wicket,' the new arrival said. I was inclined to agree with what was clearly an Achimoto University man. And although that night, to most people's astonishment, passed off peaceably, and no one slit the throats of the chief and council of Dzolokpuita, it was a verdict that I was worried might be applied to the nascent State of Ghana as a whole.

'Tubman bids us toil'

The mild optimism shown by Liberians at the end of this article wasn't justified. After Lewis's visit, the grotesque William Tubman continued as president for another fourteen years until his death in 1971.

In a New Yorker article (July 19, 1998) John Lee Anderson describes how Liberia's misery continued. During a horrifying civil war between 1989 and 1996, some two hundred thousand people were killed, and more than three-quarters of the population was displaced. Gangs of teenagers ran wild – raping, pillaging and slaughtering at will. Many gangsters arbitrarily executed civilians, decorating their checkpoints with human heads and entrails.

More recently, Liberia has known peace. In the 2017 election George Weah, a former professional footballer, considered one of the greatest African strikers of all time, was elected president. His inauguration marked Liberia's first fully democratic transition in seventy-four years.

First published in the *New Yorker*, January 3, 1958

ON THE WHOLE Liberia has had a poorish press. Back in the 1930s Graham Greene gave the impression, in his travel book *Journey Without Maps*, that he found it a sad and sinister place. John Gunther, writing in 1955, summed it up as 'odd, wacky, phenomenal, or even weird'. Lesser authorities in between have produced books with supercilious titles like *Top Hats and Tom Toms*, while some of us with tenacious memories can still recall the startled headlines in 1931 when a League of Nations committee published a report proving that slavers still hunted down their human prey in the Liberian hinterland. To me in the spring of 1957, Liberia still sounded potentially a traveller's collector's piece; so on my way home by slow stages from the Ghana independence celebrations, I decided to break my journey at Robertsfield airport and see something of the country.

It happened that I was seated in the plane next to the only other passenger getting off at Robertsfield, an American rubber man who had also read John Gunther's account: he spoke of what faced us with the kind of macabre relish sometimes found in old soldiers and world travellers. 'If anything, Gunther soft-pedalled the situation,' he assured me. 'And don't, by the way, run away with the idea that this place is a kind of American colony: they push us around like anyone else. Step out of line and you pretty soon find yourself in gaol.' I asked the rubber man if he knew Monrovia well, but he said no, they had a pretty comfortable set-up on the plantation, and he'd only been down there once or twice. And that reminded him, we should be fingerprinted at the airfield. They would take our passports away, and we should have to go to the police headquarters in Monrovia to get them back. As a final warning, he recommended me to keep out of arguments with Liberian officials, and to submit with good grace to the going-over that awaited us in the customs.

These predictions proved to be ill-founded. It was an hour before dawn when we touched down. The immigration officer, yawning, stamped our passports wordlessly and disappeared as if dematerialised. The customs man put his mark on our bags and waved us away. A moment later my American friend, claimed by a colleague with a waiting car, was borne off to the security of his comfortable set-up, and I was left alone in the dimly lit customs shed until a boy of about fourteen appeared and offered to show me where I could get what he called 'morning chop'. There was a wait of three hours before the daily DC-3 plane took off for Spriggs Payne airfield, Monrovia. So I went with him. We walked about a hundred yards before reaching a long building raised on piles, looking like an Indonesian long-house, which the boy said was the airport hotel. Here, he said, I should have to leave my luggage, which it was forbidden to take to the restaurant. I was suspicious of what seemed to me a possible manoeuvre to dispossess me but, before I could argue, a zombie came out of the hotel, took my bags from the boy, went in, and shut the door. The restaurant was a little farther on: a chink of light showing in the black shutters of the forest. The boy pointed it out and told me that he would come and fetch me when it was light. He shook hands with me and went off. At the time, this perfunctory service struck me as peculiar; but after I had been in Liberia a few days I realised that small boys preferred not to walk about by themselves in the dark.

The only other occupants of the restaurant made up a conspiratorial group, muttering in Spanish at a nearby table. I was served an American-style breakfast by a taciturn waiter. While I was busy with this, a young Liberian in a flowered shirt wearing a snub-nosed gangster's pistol in a shoulder holster came in and gave me a quick, power-saturated, policeman's stare. The muttering Spaniards looked up hopefully. I paid one dollar and twenty-five cents, and went outside again. Dawn was rising in total silence like grey smoke among the trees. A thick coverlet of mist lay along the low roofs of the airport buildings. There was a pharmaceutical smell coming out of the forest like the odour of dried-up gums, medicinal roots and benzoin. The air was flabby as if breathed in through a mask holding the moist warmth of one's last exhalation. I could see the boy squatting distantly by the door of the airport hotel, waiting for the daylight, to cross the no-man's-land between us.

It was a fifteen-minute flip in the DC-3 across forty miles of swamp to reach Spriggs Payne airfield, on the outskirts of Monrovia. From the air, the capital looked gay and dilapidated like a Caribbean banana port. It was crowded onto a peninsula outlined in yellow beach, with a hard white line of surf on the Atlantic side. The cheerful mossy green of the bush came unbroken right up to the neck of the peninsula. Big ships floated in the port, and as the plane came down you could see a few vultures drift past over the rooftops and the tangle of traffic in the central streets.

Visitors to Monrovia have complained that it is short of public transport, that the telephone system is uncertain, that there are regular breakdowns in the basic services, that most of the streets are unpaved, and that until recently the appropriation for brass bands exceeded that for public health. Mostly they have been stolidly impervious to the city's faded charm and its colour. The ex-slaves who were the original settlers here built Monrovia in the time-improved image they carried in their minds of the American South. They built with a touching and preposterous affection for Greek columns, porticoes, pilasters and decorative staircases; and a century of Liberian sun and rain has reduced their creations to splendidly theatrical shacks. The bright, slapped-on paint no longer serves to keep up pretences, although Van Gogh would have been in his element here among all the sun-tamed reds and blues and browns. There is a carnival cheerfulness about all the sagging, multi-coloured façades beneath which the citizens of Monrovia promenade with the senatorial dignity of a people whose ancestors have carried burdens on their heads in a hot country. It was Monrovia that taught me the beauty and the interest of corrugated iron as a building material, when suitably painted, with its rhythmical troughs of shade. And in Monrovia it is in lavish use.

By night, especially if there is moonlight to put back a little of the colour, the effect is strikingly romantic. The city becomes fragile, its buildings cracked and seamed with the pale internal light of Halloween candles. There are visions of interiors crammed with Victoriana, and walls hung with holy pictures and framed diplomas. As the Liberians, although shy, are polite and sociable, one is continually greeted by a soft guttural 'How-do-you-do?' spoken by an invisible watcher behind a shutter. A little music is spun out thinly into the night from aged gramophones playing in barbers' shops: 'Bubbles', 'Alice Where Art Thou?' Cadillacs, festooned with fairground illumination and bearing their dark-skinned grandes dames and white-tied cavaliers, swish over the snow-soft laterite dust of the streets. A congregation of silent worshippers collects outside the Mosque – a yellow building with its walls edged with broken lager bottles, possessing a minaret like a fat section of drain-piping. Here, where the Faithful have cleared a space, they spread their personal prayer-rugs on a small oasis of clean white sand amid the urban debris. If it is Sunday there will be the sound of ecstatic hymn-singing from the direction of many nonconformist Houses of Worship. Every night at about ten the town's *belles de nuit* begin to lurk in the neighbourhood of the cinemas, where the last performances are coming to an end. They are dressed in the style introduced by the missionaries of the last century: blouses with leg-of-mutton sleeves, voluminous frilled skirts, and – it is said – honest calico underclothing. They also carry parasols.

It is by day that you notice the squalor bred from the problem of the relative indestructibility of modern waste. At Byblos and at Sidon the domestic debris of a thousand years may be compressed into a yard of dust. In Monrovia there are towering middens of imperishable rubbish; of iron, rubber and plastic that are the legacy of barely two generations. Most Monrovian houses are raised on piles, and the space under each house serves for the concealment of old engine blocks, back axles, radiators and batteries. In the gardens you sometimes see one car-chassis piled upon another, their members entwined with flowering convolvulus and transfixed by the saplings of self-sown tropical fruit trees. Almost every side-street is littered with abandoned vehicles, many of them recent models which may at first have been immobilised by some small mechanical failure, but then subjected to nightly piracy for spare parts until only the bare bones have remained. The citizens of Monrovia have not yet learned to clear up their debris as they go along. Even the fragments of basic rock blasted out over a hundred years ago to level the ground for the original buildings are still left just as they fell.

Old prints of Monrovia suggest that basically the town has changed little in its appearance from the days when over a century ago the first settlers ventured to leave their tiny stronghold on Providence Island in the Mesurado River estuary, establishing themselves on the mainland. These pioneers were negro freedmen returned to Africa from the United States under a scheme promoted by a philanthropic body known as the American Colonisation Society. Their first decade was precarious. They suffered from disease, semi-starvation, and the attacks of slavers who probably felt that the success of this experiment in resettlement might establish a disastrous precedent for the business. The first Liberians were supported entirely by shipments of provisions from the USA and protected by the guns of American and British warships. Their numbers were strengthened by further batches of Africans released from slave-running ships; but from 1822 – the date of the first settlement – until 1840, they were ruled by white governors appointed by the American society, and the country had no official existence in international law. In 1846 the first black president, Joseph Jenkin Roberts, prepared a constitution, and Great Britain recognised Liberia as an independent republic, although the USA did not follow until 1862. The first century of Liberia's existence has been called by Liberians 'The Century of Survival'. Considerably more territory than Liberia's present extent of 43,000 square miles was originally claimed by the settlers, but this, before the days of exact surveying, and despite the frowns of the USA, was constantly nibbled at by the adjacent British and French colonies.

It is said that the Liberian pioneers included many skilled tradesmen who were in fact responsible for the building of Monrovia. This spadework having been accomplished, later generations seem to have been content to relax. A social order recalling that of the American plantations soon developed, with the freed slaves and their descendants playing the part of pseudo-aristocratic and leisure-loving masters, leaving all manual work to be done by such native Liberians as could be induced or compelled to do it. An elaborate social ritual was built up, from which Liberia has never fully recovered, and which sometimes seems to the foreign eye to achieve the opposite of the dignity at which it aims. All professions fell into social disrepute except those of the law, politics and diplomacy. Liberians developed into a race of astute politicians, but there were no native craftsmen, doctors, technicians, engineers – and there are few even today. In the meantime the hinterland, occupied by its twenty-odd tribes, remained roadless and neglected, and a concealed oppression of native Liberians by their African brothers returned from servitude gradually developed, until it was fully exposed by the League of Nations committee in 1931, with the ensuing worldwide scandal. The fact still remains that in spite of all reforms that have since been carried out, Liberia has been, and remains in practice, a species of colony in which about two million tribal Africans are governed by a minority of 150,000 English-speaking Americo-Liberians from which they are totally separated by barriers of race, religion, language and way of life.

At the present time there is a drive towards the integration of the tribal people into what is called 'the social life of the nation'. This unification policy is a favourite enterprise of President Tubman, Liberia's eighteenth president and probably the most able and energetic figure ever to appear on the Liberian political scene. President William Vacanarat Shadrach Tubman performs the considerable feat of leading a parliamentary democracy in which no official opposition is permitted to exist. The president was elected in 1943 to serve a term of eight years, re-elected in 1951 for a second term of four years, and in 1955 once again (although he was reluctant, the *Liberian Year Book* informs us) for a further four-year term. The national jubilation at the enormous majority obtained by the president in 1955 was marred by an attempt on his life. Since this occurrence the official opposition has ceased to exist, its members having withdrawn into exile, died suddenly, or been converted with equal suddenness to the policies of the True Whig Party of which His Excellency is the leader. It is said that President Tubman, in spite of the geniality and exuberance of his character, is resentful of criticism. When the official opposition crumbled and fell, such journalistic masterpieces of divergent public opinion as the *Friend* and the *Independent* also collapsed.

The yearbook tells us that they were suppressed as 'irresponsible'. The two remaining newspapers, the *Listener* and the *Liberian Age*, wholeheartedly support the president's point of view. With the object of emphasising the unanimity of the country's acclaim for the president, these papers would sometimes publish eulogistic tributes from ex-opponents newly released from prison. At odd times an appreciation in poetic form may be slipped into their pages. Here is an example from a recent *Liberian Age*, which in its complete form runs to eight happy verses:

Tubman bids us toil
(*Tune: Jesus Bids Us Shine*)
By John N. George
Public Relations Officer, Sinoe County

Tubman bids us toil at the Nation's Plan,
With the Lone-starred banner building every clan
As he ever trusts us we must work,
So in your small corner don't shirk and lurk!

Tubman bids us toil in the gleeful way,
Saving every moment of the precious day;
Whether big or little we must work,
So in your small corner don't shirk and lurk!

Etc.

After a century of stagnation, in which Liberia lagged far behind the adjacent areas under white colonial domination, the country has begun to move rapidly ahead under President Tubman's firm, paternal guidance. Liberia assumed strategic importance during the last war. It could at any time provide a base in traditionally friendly territory for American armed forces defending the South Atlantic, and, along with Brazil, it is the only source of vital natural rubber bordering the Atlantic Ocean. The president's canny exploitation of these factors has conjured up such evidence of prosperity as the new Free Port of Monrovia, five hundred miles of new roads, a three-million-dollar bridge over the St Paul River, a sprinkling of new hospitals in the hinterland, an air-conditioned hotel with a magnificently eccentric Spanish lift, taxis, telephones, piped water and modern sewage disposal for Monrovia, and a fairly elaborate yacht for the president himself. The president's 'open door' policy has attracted foreign capital to Liberia and an assortment of American, Swiss, German and Spanish firms who now share with the Firestone Rubber Company – the great monolithic pioneer – in the considerable natural wealth of the country.

The main problem confronting these concessionaries is Liberia's acute shortage of unskilled labour. The Liberian tribesman has always been accustomed to gain the mere necessities of life with a minimum of effort. At the most he will consent to clear and burn a little virgin bush, and then leave it to the womenfolk to plant the 'dry' rice and cassava forming the basic diet. Even the women's agricultural work is very light. No hoeing, weeding or watering is done. The family simply waits for the crop to come up, and supplements its diet by harvesting a few tropical fruits. The Liberian countryman will eat anything. There are no sizeable wild animals left in the country to hunt, but the chance windfall of a serpent or a giant snail, the seasonal manna of flying ants, and palm grubs – all are joyfully accepted for the cooking-pot. The result of this catholic appetite is a well-balanced diet and a good physique. The amount of leisure enjoyed by a Liberian villager – especially a man of substance with a full quota of three wives to wait on him hand and foot – is quite beyond the comprehension of modern civilised man. It is natural enough that such a villager is extremely reluctant to exchange this lotus-eating existence for that of a plantation labourer working up to twelve hours a day for a wage of thirty cents, and what are called 'fringe benefits', i.e. free housing, medical supervision, and so on. When, indeed, he is driven by force of circumstances into the plantations, he will sometimes pathetically attempt to emphasise the transitory and separate nature of his life as a labourer by adopting a temporary name which often recalls the brighter side of plantation life, such as Dinner Pail, T-shirt, Pay Day, or Christmas. In these circumstances, labour is simply obtained by a system of bonuses paid to local chiefs – whose word is more than law. There is nothing furtive or shamefaced about this procedure, and the amounts paid duly figure in company balance sheets submitted for stockholders' approval.

Firestone, which controls a labour force of 25,000 men to operate its million-acre rubber concession, and which sets the pace in these matters, pays $1.5 per man, per annum, and its 1955 balance sheet discloses a total of $90,000 expended in this way. 'In addition [I quote from *Case Study of Firestone Operation in Liberia*, published in the Nation Planning Association series] a regular scale of non-monetary gifts from Firestone to the paramount, clan, and occasionally town chiefs, has also evolved.' This regular scale of non-monetary gifts for the supply of labour goes under the dignified title of the 'Paramount Chiefs Assistance Plan', and it was developed, we are assured, with the full knowledge and consent of the Liberian Government, as well as being adopted by other foreign companies.

I learned, by the way, that it was considered highly unethical to outbid one's competitors in this extremely restricted labour market. Just as a successful tradesman may consider it a good thing to contribute occasionally to the local police benevolent fund, foreign companies operating in Liberia are also notably generous in their support of charitable, educational, cultural and religious institutions in Liberia.

The minimum wage of thirty cents a day, which is between one-third and one-fifth of wages paid for equivalent labour in the adjacent colonies of the British Sierra Leone and the French Ivory Coast, is explained in the publication already quoted, as a device for keeping inflation in check. More cogently it is argued that Liberian employers of labour could not afford substantial pay increases. Liberians have been quick, in fact, to convert themselves into plantation owners, and as soon as a new road is completed it is lined on both sides with the plantations of prominent Liberians who act as small subsidiaries of Firestone. These native plantation operators obtain free seedlings from Firestone, and as long as their labour costs remain cheap, and their land can be obtained under 'advantageous' terms from tribal communities who have never heard of title deeds, they seem to be onto a very good thing. Occasionally in the current scramble for land, someone oversteps the mark, and there is a rumpus in the Liberian press. While I was in Liberia, a tribe actually dared to take a foreign company to court for the illegal enclosure of its tribal land, and no one was more astounded than the tribesmen themselves when they won their case.

Considering the obvious subservience of the Liberian Press, it is extraordinary how much self-criticism can be found in its pages, combined with extreme sensibility to adverse comment from anyone outside the True Whig Party family – especially foreigners. All the private scandals of Liberian government: the corruption in the judiciary, the oppression of tribal people by district commissioners, the bribe-taking by persons in high places (with the exact amount of the bribe), are ruthlessly exposed to the foreign eye. Some of these revelations, in fact, such as the account published in the *Listener* of May 14th, 1957, of organised highway robberies on one of Liberia's two main roads – which had then been in progress for over two months, and had the backing, the paper thought, of 'top interior officials' – make almost incredible reading. Yet the same papers explode with indignation on the slightest foreign comment that might be taken as injurious to national pride. Such outbursts are sometimes lacking in a sense of proportion. Recently the *Listener* came out with banner headlines: 'Stamp Dealer Says Liberia Owns Savage Cannibal Tribes'. About one quarter of the space normally allotted the news was devoted to mulling over this slander, and there was a further orgy of wound-licking in a long editorial headed 'Please Treat Us Kindly Next Time'. It turned out that an obscure stamp dealer in Boston had had the enterprising notion of printing a little geographical information on the packets he sent out. Naturally this highly coloured stuff about cannibal tribes was intended to excite the interest of the children, whom one presumes would be his principal customers; but to the Liberian inflamed sensitivity it was a monstrous calumny that overshadowed any international crisis, such as the current one at the Suez Canal.

There are of course no 'cannibalistic tribes' anywhere in Africa but the fact that cannibalistic practices do exist in Liberia is abundantly clear to anyone who reads the very uninhibited Liberian Press. Cases of 'medicine' murders by 'Human Elephant Men', 'Snake People', 'Water People', an organisation with the macabre official title of the Negee Aquatic Cannibalistic Society, and various other criminal secret groups, are regularly reported in the newspapers. These often contain gruesome anatomical details, and are sometimes accompanied by a journalistic-reactionary demand for the reinstitution of trial by Sassywood – which in its pure form means that the suspect drinks deadly poison, brewed from the bark of the sassy tree, *Erythrophloeum guineense*, from which he is supposed to recover if innocent. Here is an example extracted from the editorial published in the *Liberian Age* of April 30th, 1956, which incidentally explains the motive behind ritual murders.

Trial by ordeal

In the last few weeks the *Liberian Age* reported that two men and a child have been murdered to make medicine. One was to invoke the blessings of the gods so that there will be a plentiful harvest in the rice season and the others were for reasons far more dubious.

The Government might do well in the circumstances to put a check to these unwholesome and superstitious practices by reinstating trial by ordeal, commonly known as trial by Sassywood.

Admittedly, Sassywood is a pagan cult and in a Christian State pagan cults should be frowned upon and eliminated. But the fact remains that in order to check these pagan practices we must employ the one method in which practitioners of paganism have an abiding faith, namely, the Sassywood trial.

In the Revised Statutes and in the Administrative Regulations trial by ordeal is forbidden except in minor matters and under licence of the Interior Department.

But the Constitution also provides that government should use every possible method to protect the life of the citizen and to punish the guilty of wilful murder. In such cases where life is endangered, the Government would be perfectly justified in using any legitimate method in bringing to account persons with pagan proclivities who are in the habit of so destroying life for foolish ends. The case for trial by ordeal even becomes stronger when the ordinary process of law becomes powerless in finding the guilty, due to the fact that persons who normally engage in such practices belong to some society or the other which gives them protection.

The 'medicine' referred to in this article is sometimes called 'borfina'. It is manufactured from the organs of a murdered person, and as well as being employed in the ancient magical ceremonies common to primeval humanity to promote rainfall and to influence the growth of crops, it is in brisk demand by those who dabble in witchcraft for their own ends. Borfina is in common use, not only in Liberia, but in most of West Africa, and it is reported that rich men will offer as much as one hundred dollars for a scent bottle full of the grisly stuff. In Liberia it is obtained by professional 'heart-men' who usually work at night and prefer women and children for their victims. It is a sinister fact that the 'heart-men' are more active at the period of the Christian festivals of Christmas and Easter, when they are believed to invade even the capital itself in search of their prey. At these times Liberian countrymen go armed and in pairs along the jungle paths, and the women working in the fields keep in touch by calling to each other at frequent intervals. It is practically unknown for a white man to be the victim of a medicine murder, as it is believed that medicine obtained from a white man is of little or no value.

Trial by ordeal, I soon discovered, although not practised by colonial regions of Africa, is an everyday occurrence in Liberia, and I had only been in the country about a week before I had the opportunity of seeing how it worked. Wanting to learn as much as I could of the interior – a little of which had become accessible in the last few years by the completion of new roads – I hired a taxi in Monrovia and drove across country to the frontier of French Guinea and back, a journey taking three days. I carried with me a letter of introduction to Mr Charles Williams, District Commissioner of the Bgarnba, where I hoped to stay the first night, and it was at Bgarnba that I encountered this survival of medieval justice.

Mr Williams was in court when I arrived. I sent in my letter, and in a few minutes the commissioner came out to welcome me. He was a tall handsome man, with a reserved, almost melancholic expression. He mentioned that he still had a large number of cases to try, and asked whether I would be interested in seeing the district court in operation. I was naturally more than interested. As we strolled back to the courthouse Mr Williams softly whistled a bar or two of 'Through the Night of Doubt and Sorrow'. I later discovered that he was a devout Episcopalian, with a great affection for *Hymns Ancient and Modern*.

The court was held in a large circular hut. About fifty members of the public were present, seated on rows of benches. The atmosphere was relaxed and informal. Most of the women had their babies with them, which they fed intermittently at the breast. The soldiers who brought in petty offenders from time to time hung about in wilting attitudes until they were dismissed. A pair of counsels, nattily dressed in sports clothes, kept up a crossfire of legal repartee. Seated behind his desk Mr Williams looked mildly judicial and perhaps a trifle sardonic. Once in a while he picked up his mallet and brought it down with a crash, sometimes to restore order, sometimes to deal with a tsetse fly that had alighted on his desk. Most of the complainants and defendants did not speak English, and as Mr Williams was a member of the Liberian ruling class and therefore spoke nothing but English, the services of an interpreter were often necessary. The interpreter translated from the tribal languages into a kind of Liberian pidgin, which I found impossible to understand. Even Mr Williams was often in difficulties, and called on the interpreter to repeat a sentence.

The examination of witnesses began with the routine question put by the commissioner, 'Do you hear English?' – followed, if the witness did in fact hear English by a second question, 'You Christian man?' Three out of four of those appearing before the court were not Christian men, and in these cases Mr Williams ordered the administering of an oath by 'carfoo' – a liquid concoction, or medicine, prepared by a witch-doctor, which although normally innocuous, is supposed to be fatal to the pagan perjurer. Although restrained in his manner at most times, Mr Williams seemed unable sometimes to control an outburst of genial contempt when he noticed a tendency on the witness's part to hang back at this moment of the oath-taking. 'Come, drink carfoo and lie, so that you may die tonight,' was a typical invitation roared at a minor chief who showed some reluctance when the witches' brew was put into his hands. When a Christian witness avoided touching the Bible with his lips, the commissioner leapt to his feet and pushed the book into his face. 'Come on, man. Kiss the holy book, unless you are determined to lie.'

Most of the civil cases arose out of what is known in Liberia as 'woman palaver'. Mr Williams explained to me that no man of standing would have fewer than three wives, each having been purchased for the standard bride-price of forty dollars, paid to the girl's father. A rich man bought wives as an investment. They worked his land for him without expecting to be paid, and they produced valuable children into the bargain. The local paramount chief, he mentioned, probably had a hundred wives, each one decently housed in her separate hut in his compound. Unfortunately the tendency was for a man's wives to increase in number as he himself advanced in years, and – well – you knew what it was – the women sometimes found themselves with a fair amount of time on their hands. This meant that they were inclined to get into hot water, and although most possessors of large harems took a pretty civilised view of wives forming subsidiary friendships, there were a few narrow-minded and litigious husbands who went to court, particularly to sue for the return of the forty dollars paid, when a wife ran off with some other man. In dealing with these 'woman palaver' cases, one of Mr Williams's chief difficulties was his evident distaste for coarse language. When a man complained that his wife refused to sleep with him, Mr Williams winced and put this blunt statement into the more elegant and evasive English of Monrovia. 'Your husband alleges that you refused to accord him the privilege of meeting with you,' was how he reworded this delicate circumstance, when cross-examining the wife.

The only European to appear in court that day was the Italian overseer in charge of a gang of labourers working on a bridge-construction project nearby. His offence was that in dismissing one of his men for malingering he had referred in a burst of anger – as Italians will – to the man's wife, using at the same time a four-letter word. This was a grave matter indeed in a country where a European can be heavily fined and deported for calling a man a 'nigger', and all work on the bridge stopped while the whole gang of workmen were brought to the court to testify. Mr Williams, after first ordering English-hearing women to leave the court, asked for the actual word complained of to be repeated. It was spoken in a stunned silence. The Italian spread his palms and smiled apologetically. One English word was like another to him. He genuinely didn't understand what the fuss was about. In the end Mr Williams read him a long lecture on vulgarity and let him off with a caution, and the Italian went away still mystified, shaking his head.

Shortly after this a woman was brought in by her husband, who charged her with infidelity. She had confessed to five lovers – or as Mr Williams put it, to granting intimate favours to five men other than her lawful husband – and in accordance with Liberian law the husband had been awarded damages of ten dollars against each man. The trouble was that he now claimed that the names of other lovers had been concealed. Witnesses and counter-witnesses were produced, there were charges of perjury, and it was clear that this had all the makings of a lengthy and endlessly complicated case, when the woman agreed to submit to trial by ordeal. With evident relief Mr Williams ordered this to take place next morning immediately after dawn.

I slept the night in the commissioner's house, and at the appointed hour next morning I went over to the local lock-up, outside which the trial was to be staged. I found the calabozo of Bgarnba to consist of a long, thatched hut. On its veranda several female prisoners, faces plastered with white cosmetic clay, were reclining in hammocks, under the apathetic guard of a soldier of about sixteen years of age. A witch-doctor – previously referred to by Mr Williams as 'a mystical man' – had arrived, and was lighting a small fire of twigs. He was a foxy-looking old fellow dressed in a fairground mountebank's purple robe. Mr Williams was not present.

As soon as the fire was well alight, the mystical man produced from the folds of his robe a metal object like a large flattened spoon, engraved with Arabic characters, and put this to heat in the heart of the fire. This was to be a version of the ordeal by the burning iron. In another variant of this type of ordeal, a heated sabre is brought into contact with one of the limbs. Other ordeals in common use involve the insertion of small pebbles under the eyelids, or the thrusting of needles into the flesh.

A few minutes later the wronged husband and the errant wife came on the scene. Both had dressed very carefully for the occasion – the man in a sort of yellow toga and the girl in a bright cotton frock printed with a pineapple design. They were accompanied by the clerk of the court, who wore a sports blazer with a crest on the breast pocket, and had a pencil stuck in his thick, woolly hair. A young soldier carrying a rifle trailed behind them. No one spoke or showed the slightest interest in the preparations. Liberians, other than the citizens of Monrovia, are trained by their long and rigorous years of initiation in the bush to maintain an attitude of formal unconcern in the face of all such crises. Later, I discovered that the woman had not been held in custody overnight and may have had the opportunity to visit the head of the local women's secret society, the Sande, then in session, who might have prepared her with some 'bush-medicine', or even induced a hypnotic state, for her forthcoming ordeal.

Chairs were fetched, and the couple took their seats facing the fire, which was now burning briskly. They sat only a few feet apart, stolidly oblivious of each other, like bored life-partners awaiting the serving of an uninspiring meal. The mystical man pulled out the iron, tested it with his spittle, and pushed it back into the fire. There was a short wait, and at a nod from the witch-doctor the girl put out her tongue. He bent over her and there was a faint sizzle. The witch-doctor went closer, peering at the girl's mouth like a conscientious dentist. He dabbed again with the iron. Nothing moved in the girl's face. Her husband looked glumly into space. The witch-doctor picked up a mug that stood ready, containing water, handed it to the girl, who filled her mouth, rinsed the water round, spat it out, and thrust out her tongue again for inspection. The witch-doctor, the clerk and the soldier then examined it closely for condemnatory traces of burning. 'Not guilty,' said the clerk in a flat voice. He took the pencil out of his hair, wrote something in a notebook, and the whole party, their boredom in no apparent way relieved, began to move off. Justice had been done.

The bush society which may well have taken a surreptitious hand in these proceedings is probably the feature of Liberian life which has most impressed – or appalled – foreigners who have visited the hinterland. African tribal life from the southern limits of the Sahara Desert to the borders of the Union of South Africa is dominated more or less by secret societies, but it is in Liberia, where European influence has been least felt, and the original fabric of tribal life therefore best preserved, that the secret societies are most strongly entrenched. There is a society for the men called the Poro, and one for the women, the Sande. These are in session alternately, each for several years. Every member of the tribe must enter the society and the prestige of the society is so great that, outside the control exercised by government officials, it is the de facto ruler of the area, with the grand-master of the society as a kind of undercover opposite number of the government-appointed district commissioner. When the women's society – the Sande – takes over from the Poro for its normal session of three years, an actual power passes to the women. All major decisions relating to tribal life are decided by them, and it is customary for men to dress in symbolical homage as women, and in this guise to apply for admission to the Sande – which is of course refused.

Exact information about African secret societies is extremely difficult to obtain, even by anthropologists, but it is clear that their real purpose is to perpetuate the tribe's highly complex way of life, by the communal education of its youth, which at the same time is physically and mentally prepared for the hard life of savannah and jungle. Both societies impose a spartan, even terrifying, discipline on their initiates. The boys must in theory – even if the practice has fallen into disuse – be transformed into warriors, must learn to defend themselves against savage animals, to take part in successful raiding parties, and to frustrate the attacks of tribal enemies. To achieve this result they are subjected to a more than military discipline; starved, flogged, made to sleep in the rain, to take part in gladiatorial combats, attacked and wounded superficially by human beings disguised as wild beasts, finally 'swallowed' by the totemic animal of the tribe, after which they are 'reborn' – in theory with no memory of their past lives – as fully initiated tribal members. The training of the girls is less arduous, but may be even more painful since it includes processes of beautifying by cicatrising, tattooing, and sometimes actually carving the flesh with knives, and finally that scourge of so many African women: clitoridectomy – performed with crude surgery, and without anaesthetics.

Many African races seem to have decided that only supernatural sanctions can induce human beings to submit to such a course of self-improvement: therefore teachers in the bush-schools

153

are masked and regarded by their pupils as spirits. These are the celebrated 'bush-devils' of Liberia, who vary in their importance according to their function, and who are presided over by a kind of super-devil who is a combination of headmaster, sergeant-major and ghost – as well sometimes as judge, and even executioner – and who projects a power so devastating that merely to catch sight of him as he walks in the moonlight can be death to an African. Not all this aroma of terror is consciously a disciplinary device. The devils, who are high-ranking members of the bush society, are believed by adepts to be controlled at certain times by powerful spirits, including the tribal ancestors – a belief which may well be shared by the devils themselves. Anthropologists in neighbouring French Guinea, where such aspects of tribal life are more easily observed than in Liberia, believe that masked dancers often pass into a kind of trance, on ceremonial occasions – or sometimes as soon as they put on their masks, which in themselves are supposed to possess a kind of separate life, and to require 'feeding' with blood.

Remarkably enough, the life of the bush-school is often popular with Africans. After initiation – which corresponds to graduation in the West – people frequently return to the bush on a voluntary basis to take further courses, and success in these 'postgraduate courses' is recognised as a stepping-stone to advancement in the hierarchy of the secret societies, and carries with it at the same time much social prestige.

African art is seen at its best in the production of cult objects and masks for the Poro or the Sande, and Liberia is one of the last strongholds of vigorous, untainted African art. Because the masks worn by the principal bush-devils possess a kind of sanctity, it is not easy for a foreigner even to inspect one, let alone to purchase one. The men who carve the sacred masks – who are usually high-ranking adepts of the Poro – say that they do so only when under the influence of an inspirational dream. While I was staying in one of the villages in the bush with an American anthropologist I shall call Warren, the local tribe's best carver dropped in to pay one of the formal calls which are a part of the complex social ritual of African village life. The carver came in smiling, shook hands, with the characteristic Liberian snap of thumb and finger, accepted a glass of cold beer, and picked up an illustrated book on African art that had just arrived from the United States. 'Why you no come before, man?' Warren asked him. 'I'm vexed with you because you no come.' The mask-carver said he hadn't been able to dream for weeks, and as his inspiration seemed to have dried up, he'd gone off to look for diamonds – a popular occupation at present in the area adjoining the Sierra Leone frontier. Warren was relieved. He was afraid that he had unwittingly offended the man in some way. The elaboration of Liberian tribal etiquette makes it quite bewildering to a white man, and although Africans will make intelligent allowances for the foreigner's ignorance of good manners, it is sometimes difficult to avoid giving offence.

The mask-carver turned over the pages of the book, giggling slightly, and Warren asked him what he found funny. It was the African's turn to tread warily now. He'd probably done a six-months course in the bush-school, learning, the hard way, how to avoid hurting people's feelings, and he clearly didn't want to tell Warren that he found this collection of masterpieces chosen from the whole African continent pretty poor stuff. The mask-carver, having worked in the plantations, had picked up a fair amount of English, so eventually Warren got him to express his objection. 'I no see the use for these things.' Non-Liberian African art, in fact, was as extravagant – as grotesque even – to him, as African art mostly tends to appear to the average untutored Westerner. He just couldn't see what purpose these distorted objects could serve. The idea of art for art's sake was completely foreign to him. He flipped over the pages of the book, making a well-bred effort to disguise his contempt. None of these objects could be used in his own tribal ceremonies, so they were useless – and ugly. He was like a diehard admirer of representational painting asked to comment on the work of, say, Braque. The point was that his own work, which both Warren and I readily accepted as great African art, was as exaggerated and distorted in its own way as were all the rest in this book: except of course that all these diversions from purely representational portraiture had some quasi-sacred meaning for him. Warren had managed to buy a single mask from this man. He had made it to be worn by a woman leader of society, who for some reason had not taken delivery. The mask was kept out of sight, covered with a cloth. It was dangerous because it was sacrilegious to have it in the house, and it was destined for an American museum unless the Liberian Government suddenly decided to clamp down on the export of works of art – which this certainly was.

The village of the mask-carver was the cleanest 'native' village I have ever seen in any part of the world, as well as being very much cleaner than the average village of southern Europe. Silver sand had been laid between the neatly woven huts, and there were receptacles into which litter – including even fallen leaves – had to be put. While I was there, a tremendous hullabaloo arose because a stranger from another village had relieved himself in a nearby plantation instead of taking the trouble to go to the proper latrine creek in the bush. This was an exceedingly grave offence by Liberian country standards. The man was hauled before the town chief, and as he had no money and therefore couldn't be fined on the spot, he was sentenced to ignominious expulsion from the village – a sentence which was carried out by a concourse of jeering children.

It was in this village too that I heard of the eerie sound of the head woman bush-devil coming out of the sacred bush for a rare public appearance. The devil's attendants acted as female lectors, as well as administering mild beatings to anyone who happened to cross their path. We could hear the cries of these attendants, first faint and then coming closer, as she came down the jungle path leading to the village. A neighbour popped in to tell us that she was on her way to supervise the clearing of a creek by the women's society. Then something happened and the bush-devil failed to appear. Perhaps she had been informed of the insalubrious presence of a stranger in the village. We heard the warning cries of her attendants grow fainter again, and then stop. The men pretended to be relieved.

It was while I was in Liberia that an economic use in the modern scheme of things was found for the bush-devil, and the sophisticates of Monrovia were as happy as if they had hit upon a method of extracting cash from some previously discarded industrial by-product.

Liberia possesses two predominant flourishing industries: rubber, and the mining of the extremely high-grade iron ore. Business heads on the look-out for further sources of national income recently thought of the tourist trade, which has been the economic salvation of far less viable countries than Liberia, and there was some talk even of developing tourism as a third industry. Accordingly plans were laid, and in March this year Monrovia received its first visit from a cruising liner, the *Bergensfjord*, a luxury Norwegian ship carrying 350 passengers, most of whom appeared from the passenger list to be presidents of US banks and insurance companies, along with their wives.

Unfortunately the *Bergensfjord* docked on a Sunday, which in Monrovia is surrendered to a zealous nonconformist inactivity, the silence only disturbed by the chanting of hymns and the nostalgic quaver of harmoniums in mission halls. The town was shut up – 'like a clam', as the *Listener* put it. Liberia's new industry was in danger of dying stillborn, when someone thought of the bush-devils, and a few fairly tame and unimportant ones were hastily sent for. Even when the tourists finally landed, the situation was in the balance. Although they had already been given handbills describing the traditional Liberian entertainment that awaited them, they found their path barred by a large and determined matron in a picture hat who was determined to protect them from such pagan spectacles as they had been promised. When asked where the devil-dancing was to take place, she smiled indulgently and said, 'In Liberia we do not dance on Sunday. We remember the Sabbath day, to keep it holy.' She would then recommend various places of interest which might be visited by taxi, such as the Capitol building, the lighthouse, the nearby Spriggs Payne airfield, and the Trinity Pro-cathedral.

Most of the passengers succeeded in escaping the clutches of this well-intentioned lady, and led by an organiser of the Bureau of

Folklore in a jeep, they were taken in a taxi-caravan to the vacant lot behind a garage, where the dancing was to take place. There were half a dozen assorted devils whose bodies were concealed by mantles of raffia, and their faces by inferior masks. Three little bare-breasted girls, who had just finished their initiation were, despite the presence of a mob of camera-brandishing tourists, still plainly timid of the devils. But the little girls did a rapid, sprightly dance, and the devils whirled and somersaulted diabolically in their manes and skirts of flying raffia. When the dancers stopped, the tourists clapped enthusiastically. They lined the girls up, took close-up portraits of them with miniature cameras, asked them their ages, shook hands, and gave them silver coins.

Next morning the Liberian Press wallowed in its usual self-criticism. They dug up the failure of Vice-President Nixon's visit, when the town's lights and telephones, and much more had failed. On this occasion the newspaper recounted stories of tourists being carried off on enormous purposeless drives by taxi-drivers who didn't understand English and who charged them extortionate fares, and of others stuck in the City Hotel's Spanish lift. 'We did it again,' wailed the *Listener*. '...Here was a chance to impress some of these big business tycoons and draw their capital here some day – but we did it again.'

In the paper's next edition, however, the situation wasn't looking quite so bad. The wife of a president of a Boston safe-deposit and trust company was reported to have said she loved the country and wanted to come back. Liberia's latest industry may have got off to a hesitant start, but at least it was on the move.

Human Flesh

It seems that a small amount of cannibalism may have continued in West Papua subsequent to Lewis's visit. In September 2006 the Smithsonian *magazine published an article by the journalist Paul Raffaele, in which he went deep into the territory of the Korowai tribe. Many of the tribe, famous for living in tree houses, claimed that occasionally they still eat human flesh, but never as a routine source of food.*

First published in the *New Yorker*, August 15, 1999

ALTHOUGH HUMAN FLESH may substitute in times of shortages for normal foodstuffs, cannibalism – as in the case of the Emperor Bokassa's deep freeze – is not necessarily the product of hunger. The matter was clarified for me back in the sixties by Mr Williams, an official in Liberia, when I called on him during a journey through his country. 'Many of our countrymen are rarely able to fill their stomachs to satisfaction and may thus be tempted to turn to food resources prohibited by law,' Mr Williams said. 'More common are crimes committed by persons of standing and education in search of what I will call tonic effect.' He was referring to a case reported in that day's *Listener* – one of several similar accounts I had collected – in which a corpse, which had been recovered in the early-morning hours from Monrovia's main street, was found to be short of certain body parts, including the fleshy covering of the forehead and the palms of the hands. 'This,' said Mr Williams, 'is the work not of hungry men but of collectors of medicines.' He listed several of these, including the Human

Elephant Men, Snake People, Water People, and the most publicised and exclusive tonic-effect collectors of them all, the Negee Aquatic Cannibalistic Society, vigilant watchers of the waters of Monrovia for incautious swimmers. 'In all such cases, very little is taken as food,' Mr Williams insisted. 'Maybe just a snack. Not more.'

For all Mr Williams's opinions, however, real hunger lies behind the cannibalistic acts that are frequently reported in Central Africa. The insoluble food problem in this region was illustrated on a journey through the interior of Togo, when my driver admitted to me that he felt faint with hunger. The rare markets we passed offered nothing but malodorous snails twenty times the normal size, which he said he couldn't face. Finally, he managed to buy a large rat, which he cooked by the roadside and consumed; he was back at the wheel in about fifteen minutes, visibly transfigured by the experience.

After Africa, Irian Jaya, in Indonesia, comes next in the hunger stakes. In 1991, I flew to Emdomen, the farthest airstrip visited by missionary planes, where I expected to gather first-hand information on the subject of cannibalism in the area of a much-publicised local massacre in 1974. At that time, the local tribes were very much under pressure by evangelists, who found it less trouble to install airstrips if they could start with levelled areas on which villages had been built. It seems that some villages were therefore demolished, and in a counterattack more than a dozen evangelist deacons and a prominent preacher were killed and eaten. Mr Williams's tonic effect of light snacks did not come into this operation – every scrap of the victims was ravenously consumed.

With what delicacy I could muster, I put my questions to a local tribesman through an interpreter.

'They tell me you live largely on fleshy insects, caterpillars, sago, and roots. Do you ever eat meat?'

'If the rainy season has been good, we may kill a pig. This is very rare.'

'And human meat. Do you eat that, too?'

'The police will not allow it.'

'But everyone knows you did in the past.'

The tiny, naked, smiling, supremely courteous man was eager to help in any way he could. 'Sometimes,' he said. 'After a battle.'

'Which did you prefer? This or the pig?'

'The pig does no work,' he said. 'Its meat is white, soft, and sweet. A man works. His flesh is yellow and hard. It is better than nothing.' He was studying me speculatively and with what might have been modest approval. 'If you had come here ten years ago,' he said, 'we would have eaten you.' I suspected that this was to be taken as a compliment.

Village of Cats

In this anthology, 'Village of Cats' is the only piece originally published as part of a full-length book – Lewis's superb Voices of the Old Sea. *The text published here is almost the same as that published a year earlier by* Granta *magazine (Granta 10. Travel Writing. Winter 1983).*

Both in the article and in the book, the village is identified as 'Farol', but its real name was Tossa de Mar. Lewis first discovered the village in 1951, and subsequently spent three summers there.

First published in *Voices of the Old Sea*
(Hamish Hamilton, 1984)

WHEN I WENT TO LIVE IN FAROL, the grandmother who owned the house gave me a cat. 'Don't feed it,' she said. 'Don't take any notice of it. It can sleep in the shed and it'll keep the rats away.' Farol was full of cats, for which reason it was often called Pueblo de los Gatos – Cat Village. There were several hundred of them living in whatever accommodation they could find in the village, and in caves in the hill behind it. They were an ugly breed, skinny with long legs and small, pointed heads. You saw little of them in the daytime, but after dark they were everywhere. The story was that Don Alberto, the local landowner, who was also a bit of a historian, claimed that they had always been there, and produced a fanciful theory based on some reference made to them by an early traveller that they had some connection with the sacred cats of Ancient Egypt. Mentioning this, the fishermen of Farol would screw their fingers into their temples and roll their eyes in derision as if to say, what will he come up with next? Their version was that the cats had been imported in the old days to clean up the mess left when they degutted fish before packing them up to be sent away. No one in this part of the world would ever kill a domestic animal, so their numbers soon got out of control. In addition to scavenging round the boats, they hunted lizards, frogs, anything they found edible, including fat-bodied moths attracted to the oleanders on summer evenings, which they snatched out of the air with their paws. Whenever a cat became too old or sick to have about the place, it would be put in a bag and taken to the cork forest and there abandoned. The people who owned this part of the forest lived in the village of Sort, about five kilometres away. They had no cats but were overrun by dogs, and as they, too, were squeamish about taking life, they brought down unwanted animals, borrowed a boat, and left them to die of hunger and thirst on an island a hundred yards or so offshore.

It soon became clear that the grandmother was a person of exceptional power and influence in the village. All the domestic aspects of life – and largely the financial ones, too – came under the control here of the women, 'dominated', to use the local word, by the grandmother, just as the males were dominated by the five senior fishermen owning the major shares in the big boats. In each case the domination was subtle and indirect, a matter rather of leadership accorded to experience and vision.

The grandmother had gathered a little respect in deference to her money, but most of it was based on sheer spiritual force. She was large, dignified and slow-moving, dressed perpetually in black, with the face of a Borgia pope, a majestic nose and a defiant chin, sprouting an occasional bristle. A muscular slackening of her right eyelid had left one eye half-closed, so that she appeared at all times to be on the verge of a wink. Her voice was husky and confiding, although in a moment of impatience she was likely to burst into an authoritarian bellow. Everything she said carried instant conviction, and the villagers said that she was inclined to make God's mind up for him. Whenever people left a loophole of doubt about future intentions by adding the pious formula, 'if God is willing', she would decide the matter there and then, with a shout of, 'Sí que quiere' – of *course* He's willing.

As a matter of routine the grandmother meddled in the family affairs of others. She provided instruction on the mechanics of family planning, investigated the household budgets of newly married couples to decide when they could hope – if ever – to afford a child, and put forward a suitable name as soon as it was born. All the names suggested for male children were taken from a book she possessed on the generals of antiquity, and the village was full of inoffensive little boys called Julio César, Carlos Magna (Charlemagne), Mambró (Marlborough), and Napoleón.

Above all, she was an expert on herbal remedies and the villagers saved on the doctor's fees by prescriptions provided after a scrutiny of their faeces and urine. Quoting a saying attributed to Lope de Vega, 'Mear claro y cagar duro' (clear piss and hard shit), she claimed that these were at the base of health and prosperity. She also offered a sporadic supply of the urine of a woman who had recently given birth, locally regarded as effective in the treatment of conjunctivitis and certain skin ailments – although in a village where the birthrate must have been one of the lowest in the world, it was rare for a donor to be available.

My room in the grandmother's house was odd-shaped and full of sharp edges, with a ceiling slanting up in four triangles to a centre point, and dormer windows throwing segments of light and shade across walls and floor. In Farol they were nervous about using colour, so it was all stark white. Living in this room was rather like living inside a crystal, in which the grandmother, when she came on the scene, appeared as a black, geometrical shape.

On my first visit to the house, I was taken into the garden to admire another feature of the accommodation: three strands of barbed-wire twisted together round the top of the wall, cut from a roll the grandmother had bought as an extravagance. Beyond the wall a rampart of sunflowers, besieged by goldfinches, hung their heads. Through their stalks I could see a rank of purple and yellow fishing boats, leaning on a beach of glossy, translucent pebbles glittering among coarse limestone chips. I asked the price of the room, and the grandmother's eyes became misted with introspection. She passed her tongue very slowly in a clockwise direction around her teeth, and said, 'Five pesetas a day. Here,' she continued, 'you will enjoy great tranquillity.'

This proved true; to find the place had been an immense stroke of luck. I had been attracted to Farol by its reputation of being the least accessible coastal village in north-east Spain, and I had spent my first week being driven out of the *fonda* largely by the smell of cat. The inn was run by two shy, silent brothers who I never saw except at mealtimes, when one or other of them would bring the food, drop the plate on the table, head averted, and scuttle away. The food was always hardboiled eggs and tinned sardines – a luxury in this place where they caught fresh sardines sometimes by the ton. The brothers kept sixteen cats in their cellar, and had taken four more away and left them in the cork forest only the week before I arrived.

My room in the grandmother's house had been occupied until a few days before my arrival by the grandmother's eldest daughter, her son-in-law and their two small children, who – as I was later told – had been hugely relieved after some years of living in the shadow of the grandmother's personality, to be able finally to make their escape.

There were fifty or more such houses in Farol built in an irregular and misaligned fashion into a narrow zigzag of streets, and a few more squeezed where space could be found among the semicircle of massive rocks almost enclosing the village. Standing aloof were several mansions originally belonging to rich cork merchants, who came here for their holidays at the end of the last century; all the mansions were in varying states of decrepitude, and decorated with stone coats of arms to which their owners had not been entitled.

Farol catered for basic needs with a small, decayed church, a ship's chandlers, a butcher's shop, and a general store selling a wide range of goods, from moustache wax to hard black chocolate that had to be broken up with a hammer and was kept in a sack. One single book was for sale: Alonso de Barros's *Eight Thousand Familiar Sayings and Moral Proverbs*, published in 1598, of which almost every house possessed a copy and by which people regulated their lives. The bar offered thin acidulous wine for half a peseta a glass, and was notable for its display of the mummified corpse of a dugong, known locally as 'the mermaid'. This grotesquely patched and repaired object, with its mournful glass eyes, sewn-on leather breasts and flap covering the sexual parts, was believed to vary its expression, whether pensive, sceptical or malicious, according to the weather, and it was noticeable that strangers who took refuge in the bar from the horrors of the *fonda* – generally agreed to be probably the worst in Spain – seated themselves so as not to be depressed by the sight of this macabre trophy.

The village enjoyed its own brand of democracy, and an absence of status-seeking, which resulted from a manageable shared-out poverty. All the same a few influential people emerged in addition to the grandmother.

The formal head of the community, the Alcalde, an outsider who had been inflicted on the village, had almost been forgiven for serving in the Nationalist Forces in the Civil War. He had convinced the villagers that he had been a Nationalist not by choice, but by the geographical accident of having been born in territory taken over by the Nationalists. Shopkeepers in Farol acted as bankers, supplying goods on credit throughout the winter in anticipation of sardines and tunny to be caught in summer, and were therefore entitled to some grudging respect. Inevitably, the butcher wielded power through his control of the rare meat supplies – more importantly, of the blood hot from the veins of slaughtered animals, given to sickly children. My next-door neighbour attracted attention to himself in a community that hardly understood the usages of property, through his marriage to a rich peasant girl, who had brought him some fields and trees he had never seen. Five senior fishermen expected to be listened to when any matter relating to the village weal came under discussion. The survivor of the great storm of 7 January 1922 carried in his small boat almost to Italy before being picked up, was never allowed to sit in a bar alone, owing to the belief that his luck was communicable by physical contact. Don Ignacio, the priest, in so far as he could be considered a villager, was well thought of, because he had lived quite openly with a mistress, and had learned to mind his own business.

The other person of consequence would have been seen by most outsiders as a prostitute, although a villager might have pretended, or even felt, surprise at such a suggestion. Sa Cordovesa, possessor of a delicate beauty and charm, had arrived as a child refugee from Andalusia, and now conducted multiple affairs with discretion, even dignity, behind the cover of making cheap dresses. By common consent the community wore blinkers in this matter, a posture of self-defence adopted to cope as painlessly as possible with a situation in which most men could expect to reach the age of thirty before they could afford to marry. Taking refuge in self-deception, Farol invested Sa Cordovesa with a kind of subjective virtue. She had allies – such as the grandmother – by the dozen, and was made welcome in any house. It was not long before I discovered that there had been a succession of Sa Cordovesas in the past. Farol had solved a social problem in its own unobtrusive way.

This, then, was Farol, cut off more by secret human design than by the accidents of nature, ever since the narrow, winding and precipitous road leading to it had, within living memory, been dynamited to keep outsiders away. By reason of its continuing isolation it remained a repository of past customs and attitudes of mind. Life had always been hard – an existence pared to the bone – and local opinion was that it was getting harder, purely because mysterious changes in the sea were directing the fish elsewhere. In most years catches were a little sparser than the year before, but there were optimists who believed that the decline was not necessarily irreversible, and they awaited in hope the end of the cycle of lean years.

The fishermen were totally absorbed by the sea, almost oblivious of the activities of those who lived by the land, wholly ignorant of the fact that only a few miles away a catastrophe was in the making. Three miles back from the shore the cork-oak forest began – hundreds of thousands of majestic trees, spreading their quilt of foliage into the foothills, and up and over the slopes into the low peaks of the sierra. The great wealth of cork belonged to the days before the invention of the metal bottle-top, but even now with slumped sales and low prices the oaks provided a livelihood for hundreds of tree-owning peasant cultivators of Sort, village of dogs, as well as many other forest hamlets.

In the year before my arrival, people in Sort began to notice that something was happening to the trees, that the early spring foliage had changed colour and was withering. Word of their neighbours' alarm reached Farol, but the fishermen shrugged their shoulders and went on preparing their lines or mending their nets. It was impossible for them to understand that their destiny could be in any way linked with that of peasants with whom they had little contact and from whom they were separated by huge differences of temperament and tradition. For the fishermen of Farol the peasants of Sort might have been the inhabitants of another planet, and they found it difficult to interest themselves in their fate, whatever misfortune might have befallen them.

My next-door neighbour, Juan, was the only man in Farol who should at least have had some slight interest in the fate of the oaks, for his wife, Francesca, had brought fifty oaks as her dowry to this dowry-less village, and another seventy had passed to her on her father's death. She was a lively, high-stepping, intelligent woman who wore a silk dress on all occasions, and had strutted about in high-heeled shoes until the Grandmother had warned her in a tactful fashion that all articles made from leather were taboo in the village. Her gaunt but imposing young husband with his seer's face, who always seemed on the verge of prophetic utterance, had expressed doubt about the morality of property acquired in the way his had been. Juan salved his conscience by neglecting to visit the trees and the few barren acres that had gone with them, although he agreed to accompany his wife on mushroom-hunting trips in the vicinity, from which they returned with basketfuls of Caesar's mushroom, *amanita caesaria*, celebrated for its delectable flavour, and a favourite of early rulers of the Roman Empire.

Francesca confirmed that the trees were ailing, about half those on her property being affected. She was worried about the possible loss of revenue from cork, but even more so by the fact that only about half the normal crop of mushrooms had come up the last autumn. Like the rabbits, the mushrooms needed cover and shade. Sort was full of men who had spent their life with trees, and knew all that was to be known about cork oaks, but nobody had ever seen anything like this before. Juan and the rest of the fishermen withheld their sympathy. It was firmly believed that every peasant had a boxful of thousand-peseta notes buried under his floor. 'They'll never go short of anything. Let them live on their fat,' was the general verdict.

The first signs of hard times in Sort was that their dogs were clearly getting even less food than usual and were therefore becoming more venturesome in their forays into Farol territory, where they managed to catch and devour not only an occasional cat, which no one grudged them, but a chicken here and there, which was a grave and unpardonable offence.

Whereas the cats of Farol needed no more than the presence and companionship of man, the dogs of Sort were not wholly independent in the matter of feeding themselves. Their function was to hunt game in the forest, and they were rewarded with the skins, the heads and the feet of the rabbits they caught. Apart from that, they had to make do with the sparse offal to be picked up around the village, and rare cannibal feasts when one of their own kind perished through accident or disease.

Unlike the people of Sort, who were individualists, those of Farol, accustomed to the communal enterprises of the sea, lost no opportunity to work as a team. In both villages women helped to make ends meet by keeping chickens. These, in Sort, would be shut up in cages at night, suspended from trees to keep them out of the reach of the dogs, or the rare fox that ventured into the village once in a while. In Farol, although this kind of protection was less essential, a communal coop had been built for the use of the aged and infirm, and a week after my arrival a pack of famished dogs from Sort managed to break into this and carry off many of the hens.

This was a calamity for which there was no redress. Sort denied responsibility. A peasant from the dog village who had driven a cartload of vegetables over to Farol to barter them for fish, was tackled in the Alcalde's bar about what was to be done. His reply was, 'How do you know they were our dogs? You can't tell one dog from another.'

The fishermen, who were given to informal meetings, held one on the spot; after which they told the man he could take his vegetables back. At a second meeting reprisals were decided upon. The view was that if the Sort people were not prepared to cut down on their dog population, the fishermen would have to take their own measures to reduce their numbers. But how? It was impossible to conceive of anybody taking an axe or a club and killing a dog outright and the idea of using rat poison went against the grain. The final solution was to procure several dried sea-sponges, and fry these in olive oil to provide a flavour irresistible to dogs. When, a few days later, the animals had recovered from the surfeit of chickens, the sponges were put out for them on the periphery of the village. It was a time-honoured method, and as ever, successful. The dogs gorged themselves on the dried sponges, which swelled up as they absorbed the gastric juices, until in the end the dogs' stomachs ruptured. A dog that had come on the scene too late to partake of the fatal meal, was trapped and then, as a traditional gesture of defiance and contempt, castrated and sent home with a black ribbon tied round its neck. The black ribbon symbolised cowardice.

After that, the Sort people kept their dogs under control by fastening them to heavy logs which they had to drag about wherever they went. From this time on, the relationship between Sort and Farol – never more than a watchful neutrality – fell into decline, and the shadow of the vendetta fell across the villages. Despite the annual visit of a clairvoyant who cast horoscopes, consulted the

Tarot cards, and thus directed their affairs, the people of Farol had no way of knowing that the great shoals of fish of the past would not return, and the predicament they faced was at least as great as that threatening the villagers of Sort. It was a time for enmities to be put aside, and alliances cemented wherever they could be found, but Farol and Sort turned their backs on one another, and went their separate ways towards an obscure fate.

A Letter from Ibiza

For most of the 1950s, Lewis arranged his life to spend the summers in Spain and winters writing in England. The rapid arrival of tourism eventually encouraged him to migrate to the still-unspoilt Ibiza in 1955.

He spent his first summer in Ibiza with family, soon discovering Santa Eulalia, where they then spent three summers. Lewis was enraptured by the island. He told Ito, his son by his first wife, that although usually he was 'perpetually a mildly sad person', in Ibiza he was 'euphoric'.

At 'Farol' Lewis had already become obsessed by fishermen and their culture. Ito told Lewis's biographer that 'My father was infatuated with Spanish fishermen, totally infatuated... [he] spent hours and hours and hours talking to them. They used to have these little sayings, and some of the sayings are quite spectacular. He would repeat them again and again to us. "Don't you think this is wonderful?" he would say. Eventually a fisherman used to take us out, long distances into really faraway places, and this guy used to be with us all day while we fished, and my father would be in conversation with him for hours. This guy used to be totally drained by the time we got back... the information my father used to get – he was still like an Intelligence Officer. He had notebooks all over the place which he could never utilise because there was so much material.'

First published in the *New Yorker*, March 2, 1956

I SPENT FIVE CONSECUTIVE SUMMERS in Spain, migrating farther south every year before the tourist invasion from the northern countries, which by 1954 had provoked the building of thirty-two hotels in my favourite Costa Brava village, with its native population of about one thousand.

In 1955 I crossed the hundred miles of sea separating Ibiza, the smallest and southernmost of the Balearic Islands, from the mainland, and took a house for the season in the coastal village of Santa Eulalia, about fifteen miles from the island's capital, also called Ibiza. By a stroke of luck of the kind that turns up occasionally in the lotteries of life, this was the house I had always been vainly looking for: a stark and splendidly isolated villa, on the verge of ruin, with an encroaching sea among the rocks under its windows. I paid instantly, and without question, the extortionate price of 3,000 pesetas (about £23) demanded for a season's tenancy – I never dared admit to my Spanish friends to paying more than half the sum – and settled down to my annual courtship of the brilliant and infallible Spanish summer.

The Casa Ses Estaques (House of the Mooring-posts) happened also to be the port of Santa Eulalia – or at least, its garden was. Its original owner had been allowed to build in this superb position among the pines on a headland commanding all the breezes, only by providing in the rear of the premises, as a quid pro quo, several small well-built shacks in which the fishermen stored their tackle. This house turned its back on the basic amenities. The water supply came from an underground *deposito*, normally replenished from rainwater collected on the roof, but now dry; the alternative to the clogged and ruined installation in the lavatory was a broken marble throne among the rocks overhanging the sea.

In spite of this it offered many advantages from my point of view, not the least of these being the unique vantage-point for the study of the ways of Ibizan fishermen. These were a sober and softly-spoken breed – quite unlike the boisterous hearties of the Catalan coast. The fishermen expected the stranger to make the first move when it came to opening diplomatic relations, and only occasionally indulged in an accumulated craving for violence and noise by ritually exploding one of a store of oil drums they had recovered from the sea.

The House of the Mooring-posts had been built in the Thirties to foster, it was said, the adventures of a gallant bachelor from the mainland, and it was full of the grandiose vestiges of a thwarted ambition. There were ten rooms, all fitted with basins and taps through which water had never run. Wires, undoubtedly intended to connect the lamps in elaborate chandeliers, curled miserably from the centre of every ceiling, and the only illumination provided was by four oil lamps of the kind carried by the Foolish Virgins in children's illustrated Bibles. Of the original furniture, stated by the fishermen to have been sumptuous, only a colossal mirror remained, which must have been placed in position before the roof went on, since it would have been impossible for it to pass through the door. For the rest, there were ten simple beds, all broken in the middle, a table which could only have served for a dwarfs' tea party, or for reclining orientals, because it was impossible for normal human beings to get their legs under it, and a country auctioneer's collection of wickerwork chairs which, whenever they were sat in, sprayed the beautiful, polished-stone floor with the fine white powder of their decay. The only pictures left on the walls were seven framed engravings of early steamships, and three damp-stained lithographs of the predicaments of Don Juan. When the windows were first opened – they opened inwards – a number of nestlings which had been hatched in the space between the glass and the exterior shutters flew in and perched on the pictures. The garden was thickly coated with pine needles, and in certain lights it glistened as if gem-strewn with the fragments of the gin bottles that the fishermen claimed had been hurled from the flat roof at imaginary enemies by the previous tenant, a Turkish princess. The fishermen recounted that she had never thought much of the place, and that friction had arisen between her and them as a result of their practice of drying their nets on the front doorsteps as well as stringing fish up to be sun-cured between convenient pine-trunks in the garden.

The Turkish princess's tenancy, which had preceded mine, with an interval of six months, had provided an episode certainly destined for incorporation in the permanent folklore of the island. About nine months before I arrived she had gone off on a jaunt to Madrid, leaving her beautiful seventeen-year-old daughter in the charge of a trusted maid. The daughter had promptly fallen in love with a young fisherman, and in keeping with the traditions of the house, which had been architecturally designed with this kind of adventure in mind, she had succeeded in receiving him in her room at night, without the maid's knowledge. Returning from Madrid to learn the worst, the mother had placed her daughter in a convent in Majorca, and given up the house. But in the course of time the girl suddenly turned up again in Santa Eulalia and went to live with the young fisherman's parents. The civil guard were called up to intervene, but in Spain a romance is never abandoned as hopeless on the mere grounds of an extreme disparity in the social position of the parties involved, and eventually, notwithstanding her mother's wrath, the marriage took place. The couple are now in the process of living happily ever after – their first child has already arrived – on a fisherman's average income of twenty-five pesetas (or about four shillings) a day.

One of the pleasures of Ses Estaques was the contemplation of archaic modes of fishing, which were always graceful and unhurried, and not very productive. Soon after dawn every day a boat would be visible from the terrace of the house, gliding very slowly over the inert water, with an old man rowing, who stood up facing the bow. This was one of the six Pedros of the port, known as 'he of the octopuses'. At intervals he would lay down his oars, pick up a pole with a barbed iron tip, jab down into the water, and snatch out an octopus. He appeared never to miss. Pedro a gaunt, marine version of Don Quixote, had dedicated his whole life to the pursuit of octopuses, which he sold to the other fishermen to be cut up for bait. He had developed this somewhat narrow specialisation to a degree where every man who fished with a hook depended upon him. He could see an octopus lurking where another would have seen nothing but rocks or seaweed through the wash and flicker of surface reflections. Pedro, whose daily activities were circumscribed by the light-shunning habits of his prey, also gave a spookish flavour to the early hours of the night – particularly when there was no moon – by moving afreet-like about the black-silhouetted rocks with a torch with which he examined the pools and shallows.

Another picturesque adjunct to the scene was Jaume, an artist in the use of the *raï*. The *raï* is a circular, lead-weighted net, in use in most parts of the world, which in the Mediterranean is thrown from the shore over shoals of fish feeding in the shallows. Usually these are *saupas*, a handsome silver fish with longitudinal golden stripes, considered very inferior in flavour, but highly exciting to stalk and catch. Jaume's routine was to patrol the shore when, in periods of flat calm, certain flat-topped rocks were just covered by the high tide. Schools of saupas would visit these to graze like cattle on the weed which had recently been exposed to the air and, as there were only a few inches of water, would thrash about in a gluttonous orgy, their tails often sticking up right out of the water, completely oblivious of Jaume's pantherish approach. Jaume had been doing this for thirty years, and, just like Pedro, he never missed. At the moment of truth his body would pivot like a discus-thrower's, the net launched on the air spreading in a perfect circle, then falling in a ring of small silver explosions, with Jaume's arm still raised in an almost declamatory gesture in the second before he sprang forward to secure his catch. Sometimes he caught as many as thirty or forty beautiful fish at one throw, but they were worth very little in the market. Jaume also fished with a kind of double-headed trident with a twelve-foot haft called a *fitora*, usually at night, spearing fish by torchlight as they lay dozing in the shallows after rough weather. This kind of fishing was also unprofitable, depending, as it did, too much on time and chance; this type of fishermen were usually bachelors without mouths at home to feed. They had an aristocratic preference for the sport as opposed to profit.

The great aesthetic moment of any day was when, all too rarely, Pedro and Jaume appeared together in the theatrical seascape laid out under our windows, Pedro passing like an entranced gondolier while in the foreground Jaume stalked, postured, and invoked Poseidon with a matador's flourishes of his net.

These were the dedicated artists in our community. Besides them there were others who fished with hook and net, and thereby wrested a slightly more abundant living from the sea within the limits imposed by their antiquated methods and tackle, as well as their superstitions, their hidebound intolerance of all innovation, and their lack of a sound commercial outlook. Even the hooks these men used were exact replicas of those employed by the Romans, to be seen in the local museum, and when these were in short supply nothing would ever persuade them to use others of foreign origin having a slightly different shape.

Only three of the Ses Estaques men, working as a team, made anything like a living by Western standards. They fished all night, putting down deep nets at a conflux of currents off a distant cape, and at about nine in the morning their boat would swing into sight round the headland, its lateen sails slicing at the sky. All the citizens of Santa Eulalia with a fancy for fish that day would be gathered in our garden awaiting the boat's arrival, which would be heralded by three long, mournful blasts on a conch shell. Each day this little syndicate landed between six and twenty kilograms of fish, about half of which would be of the best quality – mostly red mullet. Within a few minutes the catch would be sold, the red mullet at the fixed price of sixteen pesetas a kilogram, while the rest, gurnets, bream, mackerel and dorados, fetched about ten pesetas. In summer there was never enough fish to go round, but there was no question of raising the price to take advantage of this situation. Ibiza may well be unique in the world in that here the laws of supply and demand are without application. Whatever the catch, the price is the same. The system by which in Barcelona or Majorca, for example, prices are advanced to as much as sixty pesetas a kilogram when hauls are scanty is considered highly immoral, although this strange island morality of Ibiza can hardly survive much longer in the face of the temptations offered by the defenceless and cash-laden foreigner.

From this it will be understood that no fishermen of Ses Estaques have ever made money to free himself – even if he wanted to – from the caprices of wind and tide. There is no question of his ever rising to the bourgeois level of a steady income from some small enterprise, nights of undisturbed sleep, and a comfortable obesity with the encroaching years. If he leaves the sea at all, he is driven from it by failure, not tempted from it by success. This is regarded as the worst of catastrophes. The life of a fisherman is a constant adventure. He realises and admits this, and it is this element of the lottery that attaches him to his calling. In the long run he is always poor, but a tremendous catch may make him rich for a day, which gives him the taste of opulence unsoured by satiety. The existence of a peasant, with its calculation and lacklustre security, and that of the generous, improvident fisherman, are separated by an unsoundable gulf. For an ex-fisherman to be condemned to plant, irrigate and reap, bound to the wheel of the seasons, his returns computable in advance to the peseta, is considered the most horrible of all fates.

The village of Santa Eulalia lay across the bay from Ses Estaques. It was built round a low hill which glistened with Moorish-looking houses and was topped by a blind-walled church, half fortress and half mosque. The landscape was of the purest Mediterranean kind – pines and junipers and fig trees growing out of red earth. Looking down from the hilltop, the plain spread between the sea and the hills was daubed and patched with henna, iron rust and stale blood. The fields were curried more darkly where newly irrigated; the threshing-floors were paler with their encircling beehives of straw; where the farm-carts passed, the roads were smoking with orange dust. From this height the peasants' houses were white or reddish cubes, and the cover of each well was a gleaming egg-shaped cupola, like the tomb of an unimportant saint in Islamic lands. The course of Ibiza's only river was marked across this plain by a curling snake of pink-flowered oleanders. Oleanders, too, frothed at most of the well-heads. A firm red line had been drawn enclosing the land at the sea's edge. Here the narrow movements of the Mediterranean tides seemed to submit the earth to a fresh oxidation each day, and after each of the brief, frenzied storms of midsummer, a bloody lake would spread slowly into the blue of the sea, all along the coast.

The sounds of this sun-lacquered plain were those of the slow, dry clicking of water-wheels turned by blindfold horses, the distant clatter of women striking at the tree branches with long canes to dislodge the ripe locust beans and almonds, the plaintive cry, '*Teu teu*' – like that of the redshank – with which the farmers' wives enticed their chickens; and everywhere, all round, the switched-on-and-off electric purr of the cicadas. The whole of Santa Eulalia was scented by great fig trees standing separated in the red fields, each spreading a tent of perfume that came not only from the ripe fruit, but from the dead leaves that mouldered at their roots.

Down in the village, life moved with the placid rhythm of a digestive process. The earliest shoppers appeared in the street soon after dawn, although most shops didn't close before midnight. By about 9 a.m. the first catch of fish was landed, and the fisherman who sold it arrived on the scene blowing a conch shell, a solemn, sweet and nostalgic sound, provoking a kind of hysteria among the village cats, who had grown to realise its significance. After that, nothing much happened in the lives of the non-productive members of the population until 1.30 p.m., when the day's climax was reached with the arrival of the Ibiza bus amid scenes of public emotion as travellers who had been absent for twenty-four hours or more were reunited with their families. Between three and five in the afternoon, most village people took a siesta. Shutters were closed, filling all the houses with a cool gloom redolent of cooking pots and dead embers. The venerable taxis, Unics, De Dions, Panhards, crowded into the few pools of shade by the plaza. The only signs of life in the streets were a few agile bantam cocks which appeared at this time, to gobble up the ants, and some elderly men of property who, preferring not to risk spoiling their night's sleep, gathered pyjama-clad on the terrace of the Royalti bar to play a card game called 'cao'. At seven o'clock in the evening the water cart came to fill up its ex-wine-barrel at the horse-trough in the square. With a sprinkler then fitted, it would pass up and down the only street that mattered, spraying the roseate dust. The horse's name was Astra – by which name most goats are also called in Santa Eulalia – and the driver, who was very proud and fond of it, used to urge it on with gentle, coaxing cries in what was just recognisable as Arabic that had become deformed by the passage of the centuries. This was the same water-cart which would replenish my drinking tank at Ses Estaques, this time with what was guaranteed to be river water, usually containing one or more drowned frogs.

At the weekends things brightened up. Saturday evening saw an invasion from the countryside of farm-labourers and their heavily chaperoned girls. The farm-labourers worked all the hours of daylight for eighteen pesetas – or about three shillings – a day. On Saturday nights they paraded the principal street of Santa Eulalia, which does not possess a single neon sign, until it was time to go and dance at Ses Parres. Drinks at Ses Parres cost only three pesetas and the purchase of a round entitled the patron to watch the floor show and also to dance all night. About a third of the girls still sported the local costume, which is voluminous in an early-Victorian way, a matter of many petticoats and an abundance of concealed lace, worn with a shawl like Whistler's mother, pendant earrings, and long-beribboned pigtails. Many still wear the *paesa* costume because it is insisted upon by their husbands or future husbands. A friend, a prosperous small farmer, told me that of a family of eight girls, only his wife retained the paesa dress, the pigtail and the tight side curls. He had insisted on this and made it a stipulation of the marriage, as he thought it improper that another man should see his wife's legs. Women dressed in paesa style are allowed to wear 'modern' *ciudadana* clothes and rearrange their hair style if they leave the island – usually on a visit to a medical specialist in Palma.

Sunday mornings in Santa Eulalia always produced a curious spectacle. As the growth of the village away from its defensive position on the hill had left the church rather at a distance from the centre, people had taken to going to mass in the chapel of a tiny convent tucked away among the grocers' shops and the bars in the main street. The sixty women, or thereabouts, who attended seven o'clock mass filled this building to overflowing, so that the men – who in any case were separated from them by custom – were obliged to form a devotional group on the other side of the road. Here, divided from the rest of the congregation by the flow of morning traffic, they followed the service as best they could. There were usually about twenty of them, and, as in Catalonia, I noticed that no fishermen were present. The fishermen of Ibiza are, and have probably always been, almost savagely anti-Catholic. This antagonism does not arise merely from recent conflicts over attempts to compel fishermen to attend mass or to join in religious processions, but appears to be rooted in some ancient resistance never completely overcome, to Christianity itself. It is unlucky to see a priest, or to mention the name of God unless coupled with an obscenity, and fatal, indeed, to the day's luck with the line or nets to overhear Christian prayer. One of my fisherman friends, Vicente, told me that his daughter, whom he had been obliged to send to the nuns to be taught her three r's, took advantage of this fear of his, to blackmail him into taking her fishing. If he refused, all she had to do was to threaten him with the Lord's Prayer. The Lord's Prayer for him was a malefic incantation of terrible power which would bring the dolphins to ruin his nets.

'And then of course,' Vicente said, 'you'll have heard of the Inquisition. They used it to try to get the better of us. All this happened somewhat before my time, fifty or sixty years ago. It was our wives they were after. Every priest's house had a hole dug in it with iron hooks on the sides and a trap door. If they took a fancy to your wife they ordered you to take her to their house for some reason or other, and you can be sure that it wasn't many minutes after you got there, before the priest had your wife, and you were down the hole.'

Ses Parres bar, dancing and cabaret, functioned on both Saturday and Sunday nights. The floor show was innocent entertainment, intended to provide something typical for foreign visitors, and usually consisted of a group of local artists performing Ibizan dances. However unexciting this might have been for the peasants in the audience, it at least did nothing to dissatisfy them or endanger their cultural integrity by potentially corruptive spectacles from the outside world. In these dances of Ibiza – so unlike the bouncing *jotas* and *sardanas* of the neighbouring regions of Spain – anything that is not Moorish is pre-Moorish, or perhaps even Carthaginian, in origin. The woman twists, turns, advances, recedes, eyes cast down with resolute unconcern, body uncompromisingly stiff, feet twinkling invisibly beneath the sweeping skirt. The effect is that of an oriental doll moved by an exceptionally smooth clockwork mechanism. Her partner is more active. He postures at a distance, arms raised, hands clacking castanets, and swoops deferentially to the rhythm of flute and drum. Sometimes the rhythmic beat may be accentuated by striking a suspended sword. Occasionally the entertainers at Ses Parres are persuaded to sing those strange songs – the *caramelles* – each line of which ends in a sobbing, throaty ululation. The caramelles are properly sung before the altar on high feast days, and nobody knows anything about them, except that there is nothing to be heard like them anywhere in the world, and that their antiquity is so great that they no longer sound like music even to the most imaginative ear.

Ibiza's un-European flavour is, simply enough, the product of the island's geographical position, of which its history has been almost the automatic consequence. It is on the nearest sea route between Spain and the two conquering North African civilisations of the past – those of Carthage and of the Moors. It was taken and colonised by Carthage only 170 years after the foundation of the mother city herself in 654 BC. For the Moors it was the indispensable halfway port of call – in the days when a fair proportion of galleys never reached their destination – between Algiers and Valencia, the richest city of Moorish Spain. These were the two civilising influences in the island's early history and the thousand years in between were full of the pillagings of Dark Age marauders: Vandals, Byzantines, Franks, Vikings and Normans. In 1114 Ibiza was considered by Pope Pascual II a sufficiently painful thorn in the Christian side to justify the organisation of a minor crusade in which five hundred ships were necessary to carry the loot-hungry adventurers normally employed on such expeditions. But after Ibiza's final recapture from the Moors in 1235 its strategic importance was at an end. It was no more than a remote and inaccessible island, with no natural wealth to attract Spanish settlers, and soon deteriorated into a hideout for corsairs, while being pillaged indiscriminately by Christians and Arabs. Within a few years of the recapture, the population had declined to five hundred families.

Much of the island's distinctive style, and those special and subtle flavourings which differentiate it from the other Balearic Islands, and also from the adjacent mainland, are likely to have been formed during the two breathing spaces in antiquity of peace and plenty. The Carthaginians taught the natives almost all they knew about agriculture, including such basic Mediterranean techniques as how to grow olives. They also instructed them in the making of garum, the most famous of Carthaginian dishes, which consisted of the entrails of the tunny fish beaten up with eggs, cooked in brine and left for several months to soak in wine and oil – a modern version of which, *estofat del buche del pescado* (tunny-fish stomach stew) is still prepared. They struck enormous quantities of coins bearing the effigy of their god Eshmun, shown as a bearded, dancing dwarf, and built cave temples for the worship of Tanit, the Carthaginian Venus, who in spite of her appearance, which in her statuettes is as sensible as a Dutch barmaid's, had a sinister reputation for demanding young children as sacrifices in time of national stress. The Carthaginians were extremely systematic in the disposal of their dead, which they buried in vast necropolises, which were as standardised in all their details as a modern block of flats. Although most of these must have been ransacked in the past, a few still remain intact, and one or two, with their inevitable yield of ivory charms, figurines and lachrymatories, are opened every year.

During and after the Carthaginian period, the island manufactured and exported great quantities of amphorae. The Ibizan product was esteemed throughout Europe for certain magical properties attributed to the clay from which it was made, including the talismanic power of driving away snakes. Many galleys laden with them foundered in storms when outward bound along the island's excessively rocky coast, and a minor modern industry has arisen as a result of the large number of amphorae which have been salvaged intact in the fishermen's nets. These amphorae fetch between 500 and 1,000 pesetas apiece in Ibiza, according to their size, shape, and the secondary interest of the marine encrustations with which they are covered. A local industry consists in 'improving' genuine amphorae with interesting arrangements of shells, which are cemented in position. It once took me several hours to remove those that had been stuck on a wonderful 2,600-year-old pot. The fraudsters also use another strategy, submerging modern amphorae in the sea until enough molluscs have attacked them so that they resemble genuine antiques.

The Arab contribution to the Ibizan scene is obvious and dominant. It persists in the names of all the most essential things of life – which tend to be prefixed with the Arabic definite article 'Al'; in the cunning systems of irrigation with which the Ibizan farmer sends water coursing in geometrical patterns all over his fields; in the semi-seclusion of the women; and above all in the architecture. An Ibizan farmhouse, which is as Moorish-looking as its counterpart in the Atlas mountains, is in its simplest form a hollow cube, illuminated only by its door. With the family's growth in size and prosperity, more cubes and rhomboid shapes are added, apparently haphazardly, although the final grouping of stark geometrical forms is always harmonious, and perfectly suited to its natural setting.

In recent years poor communications and austere standards of comfort on the island have fostered Ibiza's individuality. An air service was inaugurated in 1958, but when I was there the most direct route from Spain was by a grossly overcrowded ship sailing once weekly in winter and twice weekly in summer from Barcelona. It required long foresight and a fair amount of luck to obtain a passage on this; sailing times were sometimes changed without notice, and in my experience letters to the Compania Transmediterranea, who are the owners, were rarely answered. One's best hope of getting to Ibiza in the summer season was to arrive in Barcelona on the day previous to sailing, and to be ready to queue at the company's office soon after dawn on the following morning. The sea crossing still takes all night, and conditions probably parallel those of a pilgrim ship plying between Somaliland and the port of Jeddah. Decks are packed with the recumbent but restless forms of passengers doing their best to doze off under the harsh glare of lights, installed with the intention of reducing contacts between the sexes to their most impersonal level. This concern for strict morality gives the ships of the Compania Transmediterranea, as they pass in the night, an appearance of gaiety that is deceptive.

Island transport is by buses which are of a design not entirely free from the influence of the horse-drawn vehicle, by taxis which until recently were impelled by what looked like kitchen stoves fixed to their backs, and by spruce-looking farm-carts without much springing. The choicest spots in the island are only to be reached on foot, or with the aid of a bicycle, which has to be carried across flowery ravines.

Once, when I was temporarily interested in spear-fishing, I asked a Spanish friend on the mainland where to go with a reasonable chance of seeing that splendid Mediterranean fish, the mero, which has practically disappeared from the coastal waters of France, Spain and Italy. He said, 'That's easy enough. All you do is to look out for a place without things like running water and electric light ... a dump with rotten hotels, where no one in his right mind wants to go.' He thought for a moment. 'Ibiza,' he said. 'That's it. That's the place you're looking for.'

The description was most exaggerated and unjust. You can find a bleak, clean room in a *fonda* anywhere in the island, and if it happens to be in Ibiza town itself, or in San Antonio or Santa Eulalia, there may be a piped water supply, and almost certainly a small, naked electric bulb that will gleam fitfully through most of the hours of darkness. What can you expect for thirty pesetas a day, including two adequate – often classical – Mediterranean meals? Ibiza is very cheap. (I know of people who still pay rents, fixed in the early years of the last century, of one peseta a month, for their houses.) Resourceful explorers have found that by taking a room only, at five pesetas a day, and buying their food in the market, they can live for a third of this sum. The standard price for drinks in backstreet bars – whether beer or brandy – is two pesetas, as compared to five pesetas in Barcelona. The strong wines of Valencia and of Tarragona are sold at six pesetas a litre. The proper drink, though, of Ibiza, is *suisse* – pronounced as if the final 'e' were accented. This is absinth mixed with lemon juice, and costs one peseta a glass. At the *colmado* of San Carlos – a village once famous for excluding as 'foreigners' all persons not born in the village – you can see the customers on Sunday line up, a glass of suisse in hand, to receive an injection of vitamin B in the left arm, administered by the proprietress, Anita. The injection costs five pesetas, and is supposed to ensure success in all undertakings, especially those of the heart, during the ensuing week. These economic realities make Ibiza the paradise of those modern remittance men, the freelance writer who sees two or three of his pieces in print a year, and the painter who sells a canvas once in a blue moon.

Every year the Spanish police decide that they must cut down on the floating population of escapists, who regard the island as a slightly more accessible Tahiti, and a purge takes place. Deportation is usually carried out on grounds of moral insufficiency. A fair amount of laxness in the private life is tolerated in Spain so long as an outward serenity of deportment is maintained. A departure from this, whether it be a matter of habitual drunkenness in public places, or brawling, or obvious sexual nonconformity, becomes officially '*un escándalo publico*', and the perpetrator thereof receives a visit from the Commisario de Policia who, if the malefactor is a woman, will kiss her hand, before begging her to depart on the next boat. Annually, Ibiza's bohemian plant is pruned back to the roots, and with each new season it produces a fresh crop. Most of these Gaugins are both harmless and picturesque. In 1955 the beard came fashionable again and was adopted by all nationalities except the Spanish. It was no longer the furious growth inherited from naval service but a sensitive and downy halo worn on, or under, the chin in true *fin de siècle* style. The female of the species looked as if she might have woven her own clothes.

A fair number of these refugees from the left-bank cellars of the northern cities drifted up the coast to Santa Eulalia. Our prize specimen, of whom we were very proud, was an English actor who had embraced a strict yoga discipline, and who regularly reached phases of reintegration in our open-air café, El Kiosko. On one such occasion he sank to his knees, eyes lifted heavenwards, in the path of a bus just about to depart for Ibiza, and remained in this position for about five minutes, while the bus awaited his pleasure with the engine ticking over and a pair of civil guards sat at a nearby table eyeing him with a kind of grim connoisseurship. We also had with us for a short time, until he was removed to a madhouse, a genial American who in his less lucid moments believed himself to be Ernest Hemingway, while any evening after five it was unusual not to be accosted in one of the two popular bars by a Russian nobleman anxious to explain his solution of the problem of perpetual motion. Native – or perhaps I should say Spanish – eccentrics were comparatively rare, but they included a massive Catalan who strode through the streets perpetually cracking a stock whip, and an undistinguished bullfighter who had found a summer asylum in the house of a local lady of quality, accompanying her on long walks while holding an iron bar in his extended right arm to develop the muscles employed in skewering his bulls.

These were some of the transients who brightened our lives. We also had a small colony of permanent foreign residents who sometimes acted strangely by Spanish standards. The only American resident, for example, a charming lady who loved animals, built a tower just across the water from Ses Estaques to shelter a pony she had found with a broken leg. The tower, in faultless local style, harmonised with the several others that had survived in this majestic panorama from the Middle Ages, but the quixotry of the action was complicated by the fact that it had inadvertently been built on someone else's land. But the Spanish were as tolerant of all such foibles as if they had been Buddhists of the Hinayanist canon. No extravagances ever produced so much as a raised eyebrow. If they'd had the chance to travel, they'd probably have cut a comic spectacle in the foreign country too. That was how the man in the street saw it. The police sat and pondered whether or not yoga trances in the main thoroughfare constituted a public scandal, shrugged their shoulders and decided to refer the matter to higher authority.

The interest I developed in Ibiza's eccentrics, both of the present time and of the past, actually provoked me into making a pilgrimage, to the highly inaccessible village of San Vicente, where Raoul Villain – most notable of them all – spent his last years.

In July 1914, Villain succeeded in concealing himself in the French Chambre des Députés, and there shot dead the Socialist leader Jean Jaurès, who opposed France's entry into the First World War. This action was committed by Villain in the sincere belief that he was a reincarnation of Joan of Arc, charged with the mission of protecting France from the shame of a craven retreat. He spent a few years in a lunatic asylum, after which he was quietly released and smuggled out of the country by his relatives, who sent him to San Vicente on the north-east corner of the island. Here he lived quietly enough until the outbreak of the Spanish Civil War in 1936, when he was killed by the anarchists.

San Vicente is about eight miles up the coast from Santa Eulalia, and as it was said to be in surroundings of extraordinary beauty, I decided one day to make a trip there. When Villain's influential kinsmen had picked out San Vicente as being, so far as Europe went, the end of the world, they were undoubtedly well-informed. The hardiest of our *taxistas* excused himself from taking me in his 1923 Chevrolet, so I hired a bicycle – as usual, devoid of brakes –

which I hauled most of the way through a landscape of infernal grandeur. Peasant women robed like witches passed with a slow gliding motion over fields that were stained as if with sacrificial blood. Ancient isolated fig trees hummed and moaned mysteriously with invisible doves sheltering in their deep pools of foliage. The stumps of watchtowers stood up everywhere half strangled with blue convolvulus, and the air was sickly with the odour of locust beans. This was a scene that had not changed since Gimnesia, the Island of the Naked, was written about in *Periplus* – except that in the matter of clothing the people had gone from one extreme to the other.

I lost my way several times and was redirected by signs and gestures by the aged women who were permitted to appear at the doors of their houses, from one of whom I received a bowl of goat's milk. San Vicente proved to be a sandy cove, deep-set among mouldering cliffs, with a derelict house, a farm, a fonda, and a shop. The beach, which was deserted, had been carefully arranged with antique wooden windlasses and a frame like a miniature gallows, from which huge, semi-transparent fillets of fish hung drying against a violet sea. The quality and distribution of these objects in this hard, clear light, had clothed them in a kind of vitreous surrealism. This may have been the chief Carthaginian port in Ibiza. About a mile inland lies the cave temple of the goddess Tanit, called Es Cuyerám, which is only partially excavated, and from which in the course of amateurish investigations great archaeological treasures have been recovered, and most of them smuggled out of the country.

The derelict house had been built by Villain, but never finished. The walls were painted with faded fleur-de-lis, and there were black holes where doors and windows had been. The first inhabitant of San Vicente I encountered had known the exile well, and luckily for me he spoke Castilian Spanish. Villain, he said, had been much liked by the village people, among whom he had developed a kind of gentle patriarchal authority. He had been a bit funny in the top storey, my informant said, screwing his forefinger against his temple in a familiar Spanish gesture – but then, clever people like that often were. The villagers, it seemed, had shown no desire to argue when Villain propounded his doctrine of reincarnation, and had listened with interest and respect while he described episodes from his previous existence, and told them what it felt like to be burned at the stake. When the anarchists landed there soon after the outbreak of the Civil War in 1936, they all ran away except Villain, who, in his grand role, and carrying the standard of Joan of Arc, went down to the beach to meet – and perhaps repel – the invaders. My informant's belief – which is commonly held – was that he was on the anarchists' list for liquidation. This strikes me as highly unlikely. The truth of the matter probably is that they had as little sense of humour as they had regard for human life. At all events, Villain was shot twice and left for dead, lying on the beach. Two days later, when the villagers decided to return, he was still there, and still alive – but only just, and before a doctor could be brought, he was dead.

When you have seen enough of Ibiza's foreign birds of passage, all you have to do is to move out of one of the three centres already mentioned, where they concentrate between migrations. The interior of the island, which is 26 miles long, with an average width of about 9 miles, and has a population of 36,000, is not only unspoiled but mysterious: so much so that Don Antonio Ribas, the leading authority on everything appertaining to folklore in Ibiza, was unable either to confirm or to deny a rumour that a mountain hamlet exists in which women are still veiled in Moorish style.

The Ibizan peasant is the product of changeless economic factors – a fertile soil, an unvarying climate, and an inexhaustible water supply from underground sources. These benefits have produced a trance-like routine of existence, a way of life that in the absence of some social cataclysm might remain in a state of cosy ossification until doomsday. The peasant lives, on the whole, monotonously, with calculation and without surprise. He suffers from inbreeding, which produces a great deal of baldness in the women, an addiction to absinth (which in Ibiza is the real thing), and an abnormally high incidence of syphilis. Like the industrial proletarian, the

Ibizan peasant carefully separates work from play, and his many fiestas and ceremonies are the sauce for the long, savourless days of hard labour. Much of the remote past is conserved in the husk of convention, and archaic usages govern his conduct in all the crucial issues of existence.

Of the peasant's many customs the most singular are those associated with courting, called in the Ibizan dialect *festeig*. This has no parallel elsewhere in Spain – or probably in the world – and is at least an intelligent advance on the matchmaking system employed in most oriental and some semi-oriental lands. A marriageable girl's state is officially proclaimed by the act of her attending mass standing between her father and mother. Eligible young men may then present themselves formally at the girl's home and apply to her father for permission to take part in the festeig, which is staged in public. The father usually appoints an hour or more on three evenings of the week – Tuesdays, Thursdays and Saturdays – for this. Courting time begins at eight o'clock, to give the girl time to prepare herself after her day's work in the fields, and in the case of a girl who has a large number of suitors it may be continued until midnight, or even one in the morning. The highest number of suitors reported by my informant was fifteen. Three chairs are placed in the centre of the principal room, one each for the girl and her father, and a third for the suitor, and this is occupied in turn and for exactly the same number of minutes by each of the young men who have entered the amorous contest. While he puts his case the others look on critically. As soon as his time is up, he is expected to get up and leave of his own accord, and if he fails in this the suitor whose turn comes next is entitled to throw small stones at him as a reminder. If this warning is ignored it is taken as a deliberate insult, and in theory at least the injured party leaves the house and waits outside for his rival, with knife drawn. The festeig – now only to be found in remote parts of the island – has in the past been responsible for numerous killings.

While docile in all other things the Ibizan is traditionally pugnacious where the heart is affected. A girl who finds herself unable to accept any of the candidates presenting themselves at her festeig, or who takes too long to make up her mind, may be publicly stoned. Peasant society – though not the Guardia Civil – approves of an admirer showing his enthusiasm for a girl by firing his pistol at a point in the ground a few inches from her toes as she leaves mass. If rejected, he sometimes, and with public toleration, gives vent to his natural frustration by firing at the ground behind the girl. In either case she loses face if she displays anything but the completest indifference. This amorous gunplay has given the police some trouble in the past. Even now a civil guard rarely passes a young peasant who is not at work when he should be, without satisfying himself that he is not concealing a weapon. The commonness of feuds in bygone days arising from breach of courting and other customs is attested by the fact that even now, no Ibizan *paes* greets another after dark: originally this was to avoid the possibility of betraying his identity to an enemy.

Such customs as these – the miming and buffoonery at the annual pig-killing, and the elaborate feasting and dancing which accompany the communal ploughing and the harvesting of various crops and, above all, marriages – are on the point of disappearance. They can no more survive improved education and 'standards of living', technical progress, and the example of how the rest of the world lives as demonstrated by the cinema, than similar customs could survive these things elsewhere.

One other extraordinary custom survives, and in spite of the energetic disapproval of the Guardia Civil. This is the *encerrada* – which also continues to exist in off-the-beaten-track regions in Andalusia, and is well described by Gerald Brenan in his book *South from Granada*.

The Spaniards appear always to have felt an antipathy towards the remarriage of widows or widowers. There is evidence to suggest that in the Bronze Age the surviving partner was promptly killed off, since husband and wife appear to have been buried at the same time, squeezed into the same funeral jar. The encerrada is the public form taken by this disapproval, which varies very much and according to the circumstances of the case, between the extremes of noisy but harmless peasant horseplay and something very close to a lynching party. The Ibizans, who are scrupulous about the forms of mourning, consider it particularly scandalous to remarry within the year, the more so if either of the contracting parties has children. The encerrada in its mildest form consists in a party of neighbours collecting to keep the newly-weds awake all night on the first night of the marriage by a raucous serenade played on guitars and accompanied by the blowing of conch shells and the beating of tin cans. When a breach of custom has been unusually shocking, the encerrada may be prolonged for four or five nights and draw hundreds of participants from other parts of the island. An atmosphere of hysteria prevails and obscene verses are improvised and screamed under the windows. At this point the civil guard usually arrives, and the violence and bloodshed start. In 1950 at the village of Es Cana the police arrested the participants in an encerrada, all of whom spent fifteen days in gaol; but the encerrada still goes on. Police permission is actually given for an encerrada, so long as no obscene verses are sung. When permission is refused, the encerrada is still sometimes organised by the women only, in the knowledge that they will receive milder treatment from the civil guard, when they appear upon the scene, than would the menfolk if they too had been involved. The object of the encerrada, when it is seriously undertaken, is clearly to force the offenders to leave the neighbourhood, and in this it is usually successful.

On the feast of the August Virgin, which occurs on the 15th of that month, to celebrate the fiesta I gave a little lunch at Ses Estaques. The lunch was for the family of the woman who looked after the house along with her two daughters and a son. They had been born in the town of Ibiza, although their mother was a peasant from San José who would still intone, after a great deal of persuasion, the old warbling, elusive African melodies. In one generation the young people had moved forward a thousand years or so, even though they were still not quite modern Europeans. On this occasion they spent an hour or two happily searching for sea-snails, and prising limpets off the rocks to enrich the splendid ritual *paella* their mother would cook for the midday meal. Afterwards the boy went off to watch a football match, while the girls relaxed under the pines, half absorbed in novelettes of the kind in which servant girls marry the sons of rich men owning racehorses and yachts. At the same time I observed the girls giving half an eye to the antics of some French women who were disporting themselves (illegally) in bikinis on the rocks nearby.

This sight appeared to provoke a certain restlessness in the young women from Ibiza. It was as if they were the not wholly reluctant onlookers at the performance of some religious rite to which they felt they owed some concession of reverence, and after a while, and following a whispered conversation, both girls removed their frocks, and sat there somewhat defiantly in their petticoats.

During our little party, I saw the rather astonishing sight of one of our fishermen, another Pedro, taking his wife for a pleasure trip in his boat. This, by local standards, was definitely taboo. Wives stayed at home, and women in boats were almost as unlucky as priests. But Pedro had been taking out mixed parties from the local hotel, so his wife had probably told him that if it was all right for him to take out foreign females, then he could take her too. At this time, the height of the season, Santa Eulalia was crowded with fair strangers, many of them unattached. Their admiration for the hard-muscled, sun-bronzed fishermen, who took them out for boat rides at twenty-five pesetas a time, was sometimes manifested indiscreetly, and a few were said to have gone so far as to make advances when the time and place was right. One of the young fellows thus favoured had only recently confided his doubts to me. Was it not a fact that foreign ladies usually suffered from syphilis?

And so this golden day passed with its contrasts and its confrontations. The bikini-clad French ladies came and went, happy daughters of that full turn of the wheel where sophistication joins hands with innocence, oblivious of the Bishop of Ibiza's pastoral fulminations on the subject of decency in dress. Pepita and Catalina got badly sunburned in spite of the markedly olive undertone of their Mediterranean skin, and were thus chastened for their first cautious step forward into the full enlightenment of our times. The tide moved up a few inches, licking at the ruins of our private sea wall. This would be the last season that the house of Ses Estaques would embellish this shore with its patrician decay, because the land along the seafront had now gone up to forty pesetas a square metre, so the house was to be pulled down and replaced with the stark white cube of an hotel.

Just outside the fine ruin of the archway entrance to what was left of the garden, a family of peasants were gathered round their cart. They lived in a fortified farmhouse in the mountains in the centre of the island, which sheltered several families and was in reality a hamlet in its own right. At this time they were relaxing after a late meal of goat's flesh and beans. One of the men had invoked the fiesta spirit by blackening his face and dressing up like a woman, and the other, sitting apart, was playing a wistful improvisation on his flute. The sister had left them. After studying the French women and the Ibizan girls, she had pinned back her skirt from the waist so that it fell behind in a series of dressy folds, to show an orange silk petticoat, while she gleefully dabbled her toes in the edge of the tide.

The men spoke Castilian, and one of them told me that it had taken them half the morning to get down to Santa Eulalia which, because of the difficulty of the journey, they only visited once a year – on this day. But next year, he said, things would be a lot better. The roads were going to be made up, and the piles of flints were already there, awaiting the steamroller. With a good surface on the roads they could cover the distance in half the time, which meant they would be able to come more often.

Assassination in Ibiza

This episode happened in 1957 during the family's last summer at Santa Eulalia.

First published in *The Changing Sky* (Jonathan Cape, 1959)

A NY FOREIGNER WHO INSTALLS HIMSELF for the summer in Ibiza is certain sooner or later to be approached by an extraordinary dog. This will be an Ibicine hound on the lookout for temporary adoption. At first sight it may seem ludicrous to any of the island's visitors that they could ever be induced to cherish such an animal. The Ibicine hound is admittedly of ancient lineage. Many local savants believe that the Phoenicians introduced the breed when they had an important settlement in Ibiza, and there is even some romantic nonsense talked about it being related to the sacred dog of the ancient Egyptians. But aesthetically it is hard to accept. There is something haphazard and unplanned about its general outlines, suggesting the result of a union between a greyhound and the most depraved-looking Indian pariah. In colour it is brown and white. Its long, pointed face ends in a pale tan muzzle, and it possesses large, pink up-pricked ears, pink toes and amber eyes.

Until one gets to know the dog its expression, which is really mild and speculative, appears to be charged with a shifty imbecility. But above all the dog's condition is usually appalling. It will almost certainly be dreadfully emaciated, as a result of the local belief that to feed a hunting dog is to reduce its keenness. Most Ibicine hounds are kept tied up during the daytime, or at best chained to a heavy log which they drag painfully behind them. Only at night are they released, to hunt for rabbits. Despite this absence of immediate charm, the extraordinary fact is that the few dogs that make their escape and turn to beachcombing soon find someone to look after them. The secret may lie in the Ibicine hound's quiet tenacity of purpose, and the natural tact with which it finally wears away the repugnance engendered by its hideous presence. A ceremonial offering of food is all that is necessary to attach one of these wanderers to one's person and one's house. Thereafter the summer visitor never again feels himself a complete stranger. He has been formally adopted by a dog which will guard him and his possessions in the most unobtrusive way, and which will keep its distance and know its place. In fact a natural aristocrat of a dog.

This year, as usual, I passed the summer by the shores of an Ibizan bay round which five small stark Moorish-looking cottages had been put up by local enterprise for holiday occupation. By the time I arrived four of the five cottages had already been taken by miscellaneous foreign families, and each family had already acquired its dog. Within twenty-four hours I had mine too, an errant bitch known locally as Hilda, after the star in a recently shown film who had unknowingly given her name to about one-third of the female animals of Ibiza. Hilda was a normal Ibicine hound, silent and self-effacing in her disposition. Her only drawback consisted in her insistence on trotting in front of my car about twenty yards ahead – a custom carried over from the old days of the vendetta when the watchdog had to be on the lookout for enemies lying in ambush. When we made excursions together this reduced my speed to eight miles per hour. Otherwise I was reasonably well satisfied.

The local farmer too had got himself a new dog – a six-months-old puppy – but this had turned out to be far from satisfactory. It had what was known as 'el vicio' – that is to say it had never become resigned to its hunger – and this had caused it to devour several hens, as well as the farmer's cat, when it had been released at night. It now spent its days tied up miserably in the thin shade of a locust-bean tree fifty yards from my door. It was roped to a bough over its head in such a way that it could just manage to lie down but not walk about. For at least half the day it was in the full glare of the sun. The farmer had left a bowl for water which was dry when I inspected it, but it was clear that the dog was hardly being fed. Its neck was raw from straining at the rope when anyone came near it. I watched it unhappily for a whole day, then when night came I took a knife, and crept out and cut it free. I was a little nervous about this interference in local affairs, and I was afraid that the dog might give me away by barking, or might even attack me as I groped towards it in the darkness. These fears turned out to be groundless. It probably took me three minutes to saw through the enormously thick, home-twisted rope. While I did so the dog licked every exposed part of my anatomy that it could reach, and as soon as it was free it streaked off into the night.

Alas, I woke next morning again to the sound of its forlorn yapping. It seemed that at dawn it had surrendered itself to its master, and was now tied up even more dreadfully than before, in a tumbril-like cart standing just by the farmhouse door. At about 6.30 a.m. a woman called Pepa, who went round the cottages doing the odd chores, came and knocked on my door. Pepa was a fisherman's daughter, now middle-aged, who worked eighteen hours a day to bring up and to pay for medical treatment for the spastic child that had been left on her doorstep in the town twelve years before. She had a request to make on behalf of the farmer. Was there any chance of my making a trip shortly to Portinaix? Because if so the farmer would be glad to know if I would oblige him by taking the dog with me and abandoning it there. Portinaix was a lonely beach at the other end of the island to which I made occasional fishing trips. She added innocently that someone – certainly a foreigner – had cut the dog loose in the night, and it had slain more chickens. In that case, I suggested, what objection could there be to putting the animal out of its misery? Why not, for example, shoot it, rather than abandon it to starve?

The reply taught me how much, after four summers of life in Ibiza, I still had to learn about the island mentality.

'The peasants don't like to kill these dogs.' There was a hint of contempt in her voice when she spoke of the peasants.

'Not even the vicious ones?'

'No, they're too superstitious. They're afraid to kill a dog. If a peasant wants to get rid of a dog he takes it across to the other side of the island and lets it go.'

'And we get the dogs from San Miguel and Portinaix?'

'That's right. That's where Hilda came from. Mind you, if a farmer happens to be on good terms with a fisherman, he usually asks him to take the dog out to one of the islands and let it go. In that way he can be sure the dog will be all right. There are plenty of rabbits on the islands.'

The fate of this dog was now beginning to assume for me a most uncomfortable importance. I fed it several times that day, but there seemed no way of defeating its chronic and ferocious hunger. It was the poorest and most repellent specimen of its breed I had seen, with the face of a monster from the bestiary, and possessed of a kind of mad vitality. In its noisy, hysterical demonstrations, too, it was most untypical of the true Ibicine. I suddenly found that I was feeling the beginnings of an attachment for this appalling dog, and the knowledge frightened me a little. A few more days' acquaintance and I knew that I should find myself asking the farmer to give it to me. And then, what was to happen when I went back to England? I was relieved of this fear by the appearance of the only other English member of the colony: a middle-aged woman. She was on the verge of tears. 'That poor, poor animal,' she wailed. 'I haven't been able to sleep for nights, for worrying about it. And of course, it's simply ruined my holiday. All I want to do now is to get away from this dreadful island, and never set foot in it again.' What was important about this visit from my point of view was that she had come to ask me to see the farmer and find out whether he would sell her the dog.

The farmer of course hadn't the slightest objection to parting with the beast. Nor did he want any money. His only stipulation was that it must be kept tied up. So the dog was removed forthwith from its tumbril and tied to a fig tree at the back of the Englishwoman's cottage. The Englishwoman put penicillin ointment on its sores and bound the loop of the rope, where it touched the dog's neck, with a soft cloth. While she was tending it the dog struggled to lick her hands. It gulped down the quart of milk she gave it, and the moment it was left to itself it started its mournful barking again.

Next day Pepa brought incredible news. 'You won't believe this – the Englishwoman's going to have the dog killed!' It was the first time I had seen her shaken out of her stolid acceptance of the behaviour of foreigners in general. 'She says that the dog's hers now, so she has the right to have it killed. Please don't ask me to understand the mentality of people like that.' Pepa had shrugged off the berserk drunks, the betrousered women, the artists with their beards and sandals, the occasional nudist on the beach, and the actor who practised yoga exercises in the village square. But this was too much. To ask to be given the dog, only to have it destroyed!

But the intended mercy-killing turned out to be a harder project than the Englishwoman had expected. None of the local males could be persuaded to undertake the execution, and the village veterinary surgeon – under the pretext that he had run out of chloroform – succeeded in excusing himself too. Undismayed by this setback she took the seven o'clock bus next morning to Ibiza town, nine miles away, where she finally discovered a vet who had emancipated himself from the local superstition. He promised to come out next day on his motorcycle, saying that he would arrive at about one o'clock.

Next morning a shadow had fallen upon our little colony. The atmosphere was charged with mass emotion – a kind of mob-hysteria in reverse – that made us shrink from meeting one another. The dog leaped about in the shade of the fig tree as the foreigners, French, Germans and Catalans, slunk up with their last offerings of food.

Pepa, who was to cook lunch for me that day, fussed about the kitchen doing nothing in particular, and then at midday appeared with a strained face to say that she was off home.

'*Me pongo nerviosa.*' ('I'm feeling upset.')

'Didn't you tell me that you killed the pigs at the matanza?' I asked her, referring to the great autumnal slaughter when every village in the island is full of the shrieking of pigs.

'That's different. Anyway, I don't kill my own pigs.'

Just as she was leaving she was treated to the spectacle of the dog being given its last meal by the Englishwoman. This contained a fair amount of meat, which Spanish fisherfolk can only afford to give to sick children. Pepa commented on the seeming illogicality of this, in a voice which carried at least two hundred yards.

After that began the waiting. All the foreigners had closed the shutters of their windows facing the direction where the dog sat under the fig tree digesting its enormous meal, its ugly face twisted into a smirk of crazy beatitude. The members of the farmer's family, who at this season when the harvest was in, spent most of the day doing odd jobs about the farmyard, had disappeared from the scene. Even the mule-carts seemed to have stopped coming down the road. I went into a room overlooking the sea, for no clear reason locking the door, and tried to read, but listening all the time for the executioner's arrival. I could hear the dry clicking of the distant waterwheels sounding as though the landscape on which I had turned my back were full of ancient timepieces ticking off the seconds until one o'clock. Now I knew a little of the state of mind of prisoners confined to their cells awaiting the obscene moment when, somewhere under the same roof, the trap door of the gallows will be sprung. I had caught the Spanish horror of this cool and premeditated killing, and as a foreigner, I felt myself included in their disapproval.

The blindfolded mules turning the waterwheels ticked off the seconds. One o'clock passed, then one-thirty, and I was beginning to permit myself to hope that the vet from Ibiza too might have suffered from cold feet at the last moment. But at two o'clock death approached, with the feeble puttering sound of a two-stroke motorcycle bumping slowly up the terrible road. (Later I heard that the rider had stopped at the kiosk in the village to brace his nerves with a couple of absinths.) I went through to the bathroom in the front of the house and looked out through the shuttered window. The vet had leaned his motorcycle against my wall just below, and he was unpacking the kit strapped to his carrier. The Englishwoman came out, and they talked in low voices. 'I shall require someone to control the animal while I administer the injection,' the vet said. 'Rest assured, there will be no struggle – no sensation of pain.' The woman said that she would hold the dog. 'It is just as well that it appears to possess an affectionate nature,' the vet said softly, filling his syringe, '– although perhaps excessively excitable. I say this because it is not a good thing to be bitten by an animal of this kind, which is liable to carry various infections.' The woman reassured him in her halting Spanish. '*Es muy bueno. Tiene mucho cariño.*' ('It has much affection.')

They went off together, walking very slowly towards the fig tree. From my angle of vision through the slats of the shutter, I could see only the lower part of their bodies for a moment as they moved away, and then I saw no more of them, but I could hear the snuffling, whining excitement of the dog and the tug of the rope as it jumped towards them, fell on its pads and jumped again. And then I heard the woman's quietly comforting voice, in English. 'Good little doggie. Good little doggie. Now keep still there's a good boy. There's a good little chap. Good little doggie.' After that, as Lorca puts it, a stinking silence settled down.

The farmer buried the unsatisfactory Ibicine hound, being paid for this service the sum of ten pesetas – first, however, removing the rope, which was in good condition, and which he took away with him. In keeping with the discreet traditions of a people whose ancestors have suffered, on the whole silently, under many tyrannical regimes and alien people, I believe that he never commented again on this distasteful business. Pepa, who returned to duty that evening, also avoided mentioning the subject for some time. Several days later though, after a glass or two of wine, she was induced at the village stores-cum-tavern known as the *colmado* to discuss the foibles of foreigners – a subject on which she was considered by the villagers to possess expert knowledge. This time she had a new charge to add to her previous main objection about their lack of taste and common decency in matters of dress. 'They are frequently egotists,' she said. 'This applies in particular to the women, who are also spoiled. Take for example the case of the one who recently assassinated the dog. Do you really ask me to believe that she did it out of love or consideration for the animal? What nonsense! She was suffering from bad nerves through having too much money and too little to do. The dog barked at night, she was distressed by the sight of it, and she could not sleep. Therefore the dog had to die. And, by the way, a woman possessing real warmth of heart will not think so much of dogs but more perhaps of certain children who go hungry. But because the children do not come to cry at her door this woman has no bad nerves for them.'

The last sentiment was applauded by several of the regulars, and then, perhaps remembering the shocking spectacle of a condemned dog eating good meat that it would never have time to digest, Pepa was struck by an idea. Perhaps, after all, it is because the foreigners never see the misery of the children. Perhaps we should tell our children to go and weep where nervous foreigners can see them.

Bullfighting

Lewis visited this bullfight in 1957 during a trip he made to Spain with his motor-racing friend Arthur Baron.

First published in *The Changing Sky* (Jonathan Cape, 1959)

WHEN I FIRST LIVED IN SPAIN, I went occasionally to a bullfight. It used up an afternoon in one of the big vociferous cities when I had nothing better to do with my time, and although I saw the leading bullfighters of the day go through their smooth, carefully measured-out performances, I never witnessed any sight that nailed itself in my memory. The bulls came, shrewdly chosen for weight, horn-breadth and ferocity (not too much or too little), and they died in the correct manner at their appointed time; and the bullfighters, borne on the shoulders of their supporters to their waiting Cadillacs, went off with the stars of the nascent Spanish film industry. The bullfights used up some of the sad afternoons for me, but I never became a regular. I missed all the fine points, and in still shamefully enjoying seeing the man with the sword tossed – although not injured – I demonstrated a lack of natural passion for the art of tauromachy.

After that I moved to Catalonia, where the natives are seriously addicted to football but don't care for gladiatorial spectacles; so until the spring of 1957, when I found myself at a loose end on a Saturday in the southern town of Jerez de la Frontera, a period of many years had passed since I had sat on the sharp-edged *tendido* of an amphitheatre, witnessing with incomplete understanding this ancient Mediterranean drama of men and bulls.

I went to Jerez to arrange a visit to Las Marismas, the great area of desert and marsh at the mouth of the Guadalquivir, where the last of the wild camels, presumed to have been brought in from the Canary Islands, but first recorded in 1868 by a naturalist called Saunders, have only recently been captured and subdued to the plough. At Jerez it happened that the man who owned most of Las Marismas was away for two days on his country estate, so while waiting for his return I went on to Sanlúcar de Barrameda for the annual *feria* of the Divine Shepherdess. Sanlúcar is twelve miles south-west of Jerez, at the mouth of the Guadalquivir. It was the Las Vegas of Spain in the Middle Ages before the completion of the Christian reconquest, famous in particular for its homosexuals – a tradition which lingered until the Civil War, when the puritans on both sides used machine-guns to suppress entertainment by male dancers, who went in for long hair, women's clothing, and false breasts. Across the river from Sanlúcar there is nothing but desert and marshes almost all the way to the Portuguese frontier. The half-wild fighting bulls roam in the wasteland beyond the last house, and its men are fishermen and bull-herders as well as producers of splendid sherry. It is one of the many mysteries of the wine trade that an identical vine, growing in identical soil at Jerez de la Frontera, should produce a fino sherry, while at Sanlúcar it produces the austere and pungent manzanilla.

The road to Sanlúcar went through the whitish plain of the frontier land between the ancient Moorish and Christian kingdoms. A few low hillocks were capped bloodily with poppies; adobe huts sparkled in the distance; some bulls were moving quietly in the grassy places on short, stiff legs; and storks planed majestically overhead in the clean spring sky. The peasants, festive cigars clenched in their teeth, were coming out of their fields for the fiesta; hard, fleshless men in black serge and corduroy who bestrode the rumps of their donkeys with the melancholy arrogance of riders in the Triumph of Death. The countryside smelt of the sweet rankness of cattle, and the villages of dust, saddlery and jasmine. Sometimes, as the car passed a scarlet thicket of cactus and geraniums, a nightingale scattered a few daylight notes through the window.

Sanlúcar was a fine Andalusian town laid out in a disciplined Moorish style, white and rectangular, with high grilled windows and the cool refuge of a patio for every house. The third evening of the *feria*, which is spread out over four days, was flaring in its streets. There had been a horse show, and prizes for the best Andalusian costumes; and now family parties had settled round tables outside their house doors to drink sherry, and dance a little in a spontaneous and desultory fashion for their own entertainment and that of their neighbours. It was the time of the evening when handsome and impertinent gipsies had appeared on the streets with performing dogs, and a street photographer with the fine, haunted face of an El Greco saint had already taken to the use of flash bulbs. In the main square, where a great, noisy drinking party was in progress, trees shed their blossom so fast that it was falling in the sherry glasses. Sometimes a glass of sherry was thrown out on the decorated pavement, and a gypsy's dog rushed to lick at the sweetness, while sometimes a little white blossom remained on the lip of the drinkers.

Down by the waterside, where eight hundred years ago the first English ships arrived to buy wine from the abstemious Moors, a catch of fish had been landed and spread out with orderly pride on the sand. A hunchback chosen for his mathematical ability was Dutch-auctioning the fish at a tremendous speed, intoning the sequence of numbers so quickly that it sounded like gibberish. Girls frilled at the shoulders and flounced of skirt strolled clicking their castanets absently through the crustacean fug, while distantly the dancers clapped and stamped in all the waterfront taverns.

In Andalusia a spirited impracticability is much admired, and Sanlúcar had squandered on its fiesta in true patrician style. Tens of thousands of coloured electric bulbs blinked, glared, fused, and were replaced, over its streets. Every mountebank in this corner of the ancient kingdom of El Andalus had gathered to sell plastic rubbish, penicillin-treated wrist-straps, 'novelties from Pennsylvania and Kilimanjaro', hormone face-creams and vitamin pills. Only music and the dance were tenacious redoubts in the creeping uniformity of the modern world. The ancient orient still survived in the pentatonic shrilling of panpipes bought by hundreds of children, and although the professional dancers engaged to entertain the rich families in their private booths went in for sweaters and close-cut hair, in stylish reproof of the frills and curls of their patrons, they gyrated with snaking arms to Moorish pipe music and deep-thudding drums. The gestures of the dancers too, that trained coquettish indifference, that smile, directed not at the audience but at an inward vision, were inheritances from the palace cantatrices of Seville and Granada, not yet discarded with contempt.

The bullfight of Sanlúcar which was held at five in the afternoon of Sunday, the next day, was a *novillada*; a typical small-town affair of local boys and local bulls (which happened in this case to be formidable enough), fought in a proper ring and watched by a critical, expert, and indulgent public unable to afford stars but determined to have the real thing. Besides the formal *corrida de toros* – the bullfight seen by most foreigners – Spain offers many spectacles involving the running, the baiting, and even the ritual sacrifice, of bulls. At one end of the scale are the *corridas*, which are a matter of big towns, big names and big money, and at the other end are the Celto-Iberian Bronze Age ceremonies of remote villages of Castille and Aragon, sometimes involving horrific details which are properly left for description to scientific journals. In between come the *capeas* and the *novilladas*. The *capeas* are village bullfights, where the bull is rarely killed, for the village cannot afford its loss, but is played with capes by any lad who wishes to cut a public figure in a ring formed by a circle of farm-carts. The amateurs with the capes fight not for money but for the bubble reputation, sometimes receiving the bull's charge seated in a chair or in another of a dozen facetious and suicidal postures, and so many aspiring bullfighters meet their deaths in this way that often the newspapers do not bother to report such incidents. The small towns that possess a real bullring hold *novilladas* in which apprentice bullfighters, who are badly paid by bullfighting standards, fight bulls that have not reached full maturity. In theory these should be inferior spectacles to the *corrida*, but often enough this is not so, owing to the dangerous determination of the young bullfighter to distinguish himself, and the fact that the bulls, although perhaps a year younger, are often larger and fiercer than those employed in the regular *corrida*, where they prefer the bulls not to be too large or fierce. There is less money for everyone in a *novillada* and therefore less temptation for behind-scenes manipulations; but on a good afternoon you can see inspired fighting, and plenty of that kind of madness sent by the gods, and most of those who meet their end in the bullring do so at this particular type of fight.

Sanlúcar's *novillada* held the promise of unusual interest. In the hometown of the breeders of the great bulls of Andalusia – which dwarf those of northern Spain and of Mexico – it would have been audacious to present any but outstanding bulls; and these, fresh from the spring pastures, would be at the top of their condition. Moreover, the first of the five superlative bulls chosen by well-informed local opinion was to be fought by a *rejoneador* – a horseman armed with a lance instead of the matador's sword, and mounted on a specially trained horse of the finest quality, and not a broken-winded picador's hack supplied by a horse-contractor. The *rejoneador* in action is itself a rather rare and interesting spectacle, surviving from the days of the old pre-commercial bullfight, and in this case there was an additional interest in the fact that the horseman was a local boy, who it was supposed would be out to cover himself with glory on his home territory. Finally, one of the two *novilleros* who would fight the remaining four bulls on foot was already considered an undiscovered star, equal to any of the much-advertised and top-grade matadors, and certain to become one himself very shortly – if he didn't push his luck too far, getting himself killed during the present apprentice stage.

I spent the morning correctly, as all visitors to Sanlúcar are supposed to, tasting sherry in the different *bodegas*, and after a siesta, was driven stylishly in a victoria to the bullring, timing my arrival for half an hour before the fight began. The bullring was a small, homely structure of pink-washed brick, in the heat at the far end of the town. There was little refuge from the sun, which kept the storks, nesting on the thatched huts all round, rising stiffly to let their eggs cool off. Water-sellers with finely shaped jars were waiting at the entrances. When I arrived, a pleasant confusion was being caused by the three picadors, who were riding their horses at a creaking, shambling gallop into the crowd waiting outside, and practising bull-avoidance tactics on convenient groups of citizens. A woman protesting at being charged admission for a beautifully dressed little girl of five cried out with such passion that ripples of emotion and fury, dissociated from their origin, were stirring the fringes of the crowd a hundred yards away.

Over to one side of the plaza, small boys were running about under some pine trees, clapping their hands and uttering inhuman cries, in an attempt to dislodge the doves sheltering in the foliage above, driving them over the guns of a number of Sunday sportsmen whom we could see crouching like bandits in ambush wherever they could find cover. Occasionally one of the old sporting pieces was discharged with an enormous blast, and the girls screamed prettily, and the picadors, struggling to calm their horses, swore those terrible Spanish oaths denounced ineffectively in wall posters all over the country.

In due course the promising *novillero* arrived, in a veteran Hispano-Suiza with a tremendous ground clearance and lace on the seats. He was in full regalia, accompanied by his manager and by three aged women dressed stiffly in black. One of the old women was clutching what looked like a missal. The manager was a fat, gloomy and nervous-looking man, who wore a grey Sevillian hat. He and the driver lifted down the worn leather trunk containing the tackle for the fight, the swords and the capes. The manager opened the trunk and began to forage in its contents while the others stood by – the *novillero* smilingly indifferent and the women with practised resignation. Something was missing from the trunk. 'I told you to count them before you put them in,' the manager said fussily. 'I don't see why you couldn't have checked them from the list. It would have been just as easy.' He closed the trunk clicking his tongue, and the old woman with the religious book said, 'I made sure of the cotton. I brought it myself.' After that they went away to their special entrance.

The promising *novillero*, who had the rather fixed serenity of expression of a blind man, and who smiled into the sky, didn't look in the least like a bullfighter (bullfighters on the whole are dark, and a trifle saturnine in a gypsy fashion); he looked perhaps more like a cheerful and promising hairdresser's assistant. Inside the bullring the crowd had separated into its component castes. The townsmen in stiff, dark, bourgeois fashions, with their regal wives, had massed in the best shade seats. The cattlemen, drawn together, each on his hard foot of bench, were a solemn assize of judges in grey sombreros, ready to deliver judgment on what was to come. A hilarious clique of fisherfolk in gaudy shirts and dresses kept their own slightly tipsy company. Above them all, in the gallery, a posse of civil guards under their black-winged hats, brooded down on the scene, rifles held between knees. Only the girls in their splendid Andalusian costumes were missing. It turned out that they had gone off to watch the bicycle race, which was the competing attraction of the day, and something of a novelty in the bullfighting country.

Fifteen minutes after scheduled starting time, encouraged by the trumpetings of municipal band music and the exasperated slow-handclapping of the spectators, the *novillada* got under way. The *rejoneador*, Cayetano Bustillo, aged nineteen, handsome, pink-cheeked and open-faced, dressed in the Sevillian manner in short waistcoat, leather chaps and a flat-brimmed hat, made his entry on a superb horse, executing a graceful and difficult step known to the haute école as the Spanish Trot. Bustillo's mount was all fire, arched neck and flying mane, an almost mythological creature, and it would have needed only a background of fallen Grecian columns and sea, in place of the dull blood-red barrier fence and the sun curving on the wet sand, to turn this scene into a picture by Di Chirico come to life. Bustillo made a circuit of the ring, went out and returned on his working horse, a black Arab, spirited, more nervous than the first, with several small pink crescents left by old horn gashes on flanks and chest. The bugle was blown, and bull number one came through the open gates of the *toril*, shattering the tensed silence of the crowd. A kind of great contented grunt went up as they saw its size and speed. The bull came out in a quick, smooth, leg-twinkling run, at first not going dead straight but weaving a little as it looked from side to side for an enemy. Bustillo was waiting, his horse turned away, nervously across the ring and a little to the one side; his three peons – who were to work to his orders with their capes – had been placed equidistant at the edge of the ring by the barrier-fence, watching the animal's movements and trying to learn quickly from what they saw. The bull appeared not to see the horse, and selecting one of the peons it went for him, tail out, shoulder muscles humped and head held up until the last moment when it lowered it to hook with its horns. The peon thus chosen, Torerito de Triana by name, stood his ground instead of taking refuge behind the protective barrier, the *burladero* (which screens the entrance to the passageway), received the bull with what looked to a layman like an exceptionally smooth and well-measured pass with his cape, and turned it so sharply that the animal lost its balance and almost fell. He then proceeded to execute three more stylish and deliberate movements with the cape. The hard-faced experts all round me exchanged looks, and there was some doubtful applause from the better seating positions. The critic of *La Voz del Sur* in his somewhat sarcastic account of the fight, which appeared in next day's issue of the paper, said: 'Four imposing passes by Torerito – who was of course quite out of order in making them, as he had appeared solely in the capacity of a peon. But then, what can you expect? The poor chap can never forget the day when he was a *novillero* himself.' Torerito and the other two peons were middle-aged men with worried eyes, blue chins and fat bellies straining grotesquely in their tight ornamental breeches. You saw many of their kind sitting in the cafés of Jerez drinking coffee and shelling prawns with a quick skilful fumble of the fingers of one hand. These men had failed as bullfighters, remaining at the *novillero* stage throughout their long undistinguished careers. Now when their sun had set it was their task to attract and place the bull with their capes, to draw it away from a fallen bullfighter or picador, to place a pair of *banderillas* in the bull's neck, but not to indulge in performances competitive with that of the star of the moment.

Bustillo, who took the sideshow good-humouredly enough, now called '*Huh huh!*' to attract the bull, which at last seeming to notice the horse, left the elusive Torerito and went after it with a sudden, scrambling rush. This charge Bustillo avoided by kicking his horse into an all-out gallop that took him in a flying arc across the lengthening and curving line of the bull's attack, and then when, from where I sat, it seemed certain that the bull had caught the horse, although its horn-thrust had in fact missed by inches, Bustillo leaned out of his saddle and planted a pair of *banderillas* in its neck.

Bustillo repeated this performance several times, using more *banderillas*, and then the *rejón*, which is the javelin with which the *rejoneador* tries – usually without success – to kill the bull. The rising tension and the suspense every time this happened was almost unbearable.

Bustillo, racing away at a tangent from the bull's line of attack, his gallop slowed to the eye by the curvature of the ring, would seem to be forcing his horse into a last desperate spurt, and you saw the bull go scrambling after it over the sand as smoothly as a cat, the enormous squat bulk of head and shoulders thrust forward by the insignificant hind-quarters, short-paced legs moving twice as fast as those of the horse. After that the two racing masses would appear to fuse, the bull's head reaching up and the white crescent horns showing for an instant like a branding mark in the fluid silhouette of the horse slipping by. At this second everyone got up, moved as if by a single muscular spasm, and you found yourself on your feet with all the rest, keyed up for an intolerable sight – at the very moment when the two shapes fell apart and the tension snapped like the breaking of an electrical circuit. Everyone let go his breath and sat down. Judging from his report, the *La Voz* reporter remained immune to nervous strain. 'As for Cayetano Bustillo,' he wrote, 'let me say at once that as a horseman he appealed to me, but as a *rejoneador* – no. He was content to plant his weapons where best he could in an animal that soon showed signs of tiring. And what a slovenly trick he has of throwing down the hafts of the *rejóns* wherever he happens to be in the ring! Has no one ever told him that the proper thing to do is to give them to the sword-handler?'

In the end Bustillo's bull, tired though it may have been, had to be killed by a *novillero* substituting for Bustillo on foot. In the course of his action he gave what the critic described as several exhibitions of 'motorless flight', being caught and butted a short distance by the bull without suffering much apparent discomfort. The bull, which was too much for this *novillero*, died probably from fatigue and loss of blood resulting from several shallow sword-thrusts, of the kind delivered by a bullfighter nervous of over-large horns. Bustillo was accorded the mild triumph of a tour on foot round the ring, and several hats were thrown down to him, which he collected and tossed back to their owners, showing great accuracy of aim. The bugle then blew again, the doors of the *toril* were thrown open, and in came the second bull.

Bull number two was prodigious. It was the largest bull I had ever seen in the ring and it brought with it a kind of hypnotic quality of cold ferocity that produced a sound like a gasp of dismay from the crowd. The three peons who were waiting for it worked in the troupe of the *novillero* Cardeño, a man in his thirties, whose face whenever I saw him was imprinted with an expression of deepest anxiety. The peon's function in this preliminary phase of the fight is to test, by the simplest possible passes, the bull's reactions to the lure of the red cape. Torerito, whose flamboyant behaviour with the first bull had caused unfavourable comment, was present again, and it was perhaps lucky for him that the bull decided on one of his colleagues, thus relieving him of the temptation to indulge in any more of the stylish bullring pranks of his youth. The peon chosen by bull number two, who was also a middle-aged man of some corpulence, was prudent enough to hold his cape well away from his body. The bull ripped it from his hands, turned in its own length, and went after the man who had started to run as soon as the bull passed him, and with a remarkable turn of speed for a man of his years and weight, reached the barrier fence and vaulted it perhaps a quarter of a second before the bull's horns rapped on the wood. Each peon in turn tried the bull but taking great care to keep very close to the *burladero*, behind which the man skipped as soon as the bull had passed. Five minutes were spent in this way, and the bugle sounded for the entry of the picadors.

'The Luck of Spears', as this business with the picadors is picturesquely called in Spanish, is one of the three main phases in every bullfight that is conducted in the Spanish style in any part of the world; the other two concerning the work of the *banderilleros*, and of the man with the sword whether *novillero* or full matador. It is the part of the fight which upsets most foreigners as well as many Spaniards in the past, although in the last twenty-five years the horse has been fairly effectively protected by padding. No longer seen is the spectacle, so repellent to D. H. Lawrence and, in a defensive way so amusing to Hemingway, of a horse completely eviscerated trotting obediently from the ring.

The purpose of the picador on his aged steed, and of the *banderilleros* who followed him, is to tire and damage the bull's neck muscles in such a way that, without his fighting impetus being reduced, he will hold his head low and thus eventually permit the swordsman, lunging forward over the lowered horns, to drive home to the bull's heart. These picadors are placed at more or less equal intervals round the ring, and each of them, if things go as they should, sustains one or more charges which he does his best to hold off by leaning with all his strength on the *pica*, jabbed into the hump of muscle at the base of the bull's neck. A metal guard a few inches from the *pica*'s point prevents this from penetrating far and inflicting a serious injury.

In this particular case, bull number two, supplied by the Marqués de Albaserrada, when lured by the capes to the first horse, showed no inclination to attack. When finally it did, it turned off suddenly at the last moment, ripping with one horn the horse's protective padding, in passing, and completely avoiding the *pica*'s down-thrust. This sent up a shout of astonishment which became a continuous roar when the bull performed the same manoeuvre a second and a third time. A short discussion on strategy followed between Cardeño and his men, after which the bull, enticed once again to the horse and, hemmed in by the four men with capes, charged for a fourth time, this time, however, changing its previous tactics and swerving in again when it had avoided the *pica*, to take the horse in the rear. Horse and rider went over, carried along for a few yards by the impact and then going down stiffly together like a toppled equestrian statue. Cardeño, rushing into the mêlée to draw off the bull with his cape, was tossed into the air with a windmill flailing of arms and legs. He picked himself up and straightened immediately, face emptied of pain. Great decorum is maintained in the ring in moments of high drama. The bullfighters accept their wounds in silence, but the crowd screams for them. As Aeschylus witnessing a boxing match remarked to his companion, 'You see the value of training. The spectators cry out, but the man who took the blow is silent.' It was at this point the man from *La Voz* seems to have realised that he had something on his hands justifying a report twice as long as he wrote the next day about the regular bullfight that opened the season at Jerez. 'This bull turned out to be an absolute Barabbas,' he wrote, 'one of the most dangerous I have ever seen. *It gave the impression of having been fought before.*'

This sinister possibility also appeared to have suggested itself to the public, and to the unfortunate men who had to fight the bull. The first picador was carried off to the infirmary with concussion – a limp and broken figure on a board; while the others, refusing to play their part, clattered out of the ring – an almost unheard-of action – receiving, to my surprise, the public's full support. Most of the two or three thousand spectators were on their feet waving their handkerchiefs in the direction of the president's box and demanding the bull's withdrawal. The bull itself, monstrous, watchful, and terribly intact, had placed itself in front of the *burladero*, behind which Cardeño and his three peons had crowded wearing the kind of expression that one might expect to see on the faces of men mounting the scaffold. Occasionally one of the peons would dart out and flap a forlorn cape, and the bull would chase him back, groping after him round the corner of the *burladero*, with its horn, without violence, like a man scooping unhopefully with a blunt finger after a whelk withdrawn into the depths of its shell.

The crowd was on its feet all the time producing a great inarticulate roaring of mass protest, and the bullfight had come to a standstill. A bull cannot properly be fought by a man armed only with a sword until it has been *pic*-ed and has pranced about a great deal, tiring itself in its efforts to free itself from the *banderillas* clinging to the hide of its neck. The sun-cured old herdsman at my side wanted to tell all his neighbours, some of whom were mere townspeople, just how bad this bull was. 'I knew the first moment I set eyes on him in the corral. I said someone's been having a game with that brute, and they've no right to put him in the ring with Christians... Don't you fight him sonny,' he yelled to Cardeño. 'You're within your rights in refusing to go out there and have that devil carve you up.' That was the attitude of the crowd as a whole, and it rather surprised me. They were sympathetic to the bullfighters' predicament. They did not want the fight to go on on these terms; and when the four men edged out from behind the *burladero* and the bull charged them and they threw their capes on its face and ran for their lives, the girls screamed, and the men cursed them angrily for the risks they were taking. The crowd hated this bull. Bullfight regulars, as well as most writers on the subject, are addicts of the pathetic fallacy. Bulls that are straightforward, predictable, and therefore easy to fight, are 'noble', 'frank', 'simple', 'brave'. They are described as 'co-operating loyally' in the neat fifteen-minute routine which is at once the purpose, climax and culmination of their existences. They often receive an ovation – as did bull number one on this particular afternoon – from an appreciative audience as the trio of horses drag them, legs in air, from the ring. Hemingway, a good example of this kind of thinker, tells us in *Death in the*

Afternoon that an exceptionally good bull keeps its mouth shut even when it is full of blood – for reasons of self-respect, we are left to suppose. No one in a Spanish audience has any affection for the one bull in a thousand that possesses that extra grain of intelligence. The ideal bull is a character like the British Grenadier, or even the Chinese warrior of the last century, who is alleged to have carried a lamp when attacking at night, to give the enemy a sporting chance.

In the next day's newspaper report this bull was amazingly classified as 'tame', although it was the most aggressive animal I had ever seen. When any human being appeared in the line of its vision, it was on him like a famished tiger, but tameness apparently was the professional name for the un-bull-like quality of calculation which caused this bull not only to reject the cape in favour of the man but to attempt to cut off a man's flight by changing the direction of its charge. The sinister and misplaced intelligence provoked many furious reactions. I was seated in the *barrera* – the first row of seats behind the passageway. Just below me a Press photographer was working with a Leica fitted with a long-focus lens, and this man, carried away by his passion, leaned over the barrier fence and struck the bull on the snout with his valuable camera. A spectator, producing a pistol, clambered down into the passageway, where he was arrested and carried off by a plain-clothes policeman and bullring servants. The authorities' quandary was acute, because the regulations as laid down prevented them from dismissing a bull on any other grounds than its physical inability to fight in a proper manner, or the matador's failure to kill it within fifteen minutes of the time when he takes his sword and goes to face it. But physically this bull was in tremendous shape, and although half an hour had passed, the third episode of the fight, sometimes referred to in Spanish as 'The Luck of Death', had not yet begun.

The outcome of this alarming farce was inevitably an anti-climax, but it taught me something I had never understood before: that bullfighters – at least some of them – can be brave in a quite extraordinary way. Black *banderillas* had been sent for. They are *banderillas* of the ordinary kind, wrapped in black paper, and their use imposes a kind of rare public degradation on the bull, like the stripping of an officer's badges of rank and decorations before his dishonourable discharge for cowardice in the face of the enemy. The peons, scampering from behind cover, managed to place two of the six *banderillas*, one man hurling them like enormous untidy darts into the bull's shoulders while another distracted its attention with his cape. After that, Cardeño, shrugging off the pleadings of the crowd, took the sword and muleta – the red square of cloth stretched over wooden supports that replaces the cape when the last phase of the drama begins – and walked towards the bull followed by his three obviously terrified peons. Although Cardeño had been standing in the shade for the last ten minutes, his forehead and cheeks were shining with sweat and his mouth was open like a runner after a hard race. No one in this crowd wanted to see Cardeño killed. They wanted this unnatural monster of a bull disposed of by any means, fair or foul, but the rules of the bullring provided no solution for this kind of emergency. There was no recognised way out but for Cardeño to take the sword and muleta and try to stay alive for fifteen minutes, after which time the regulations permit the president to order the steers to be driven in the ring to take out a bull which cannot be killed.

Cardeño showed his bravery by actually fighting the bull. Perhaps he could not afford to damage his reputation by leaving this bull unkilled, however excusable the circumstances might have made such a result. With the unnerving shrieks of the crowd at his back he went out, sighted along the sword, lunged, and somehow escaped the thrusting horns. It was not good bullfighting. This was clear even to an outsider. Good bullfighting, as a spectacle, is a succession of sculptural groupings of man and beast, composed, held, and reformed, with the appearance almost of leisure, and contains nothing of the graceless and ungainly skirmishing that was all that circumstances permitted Cardeño to offer. Once the sword struck on the frontal bone of the bull's skull, and another time Cardeño blunted its point on the boss of the horns. Several times it stuck an inch or two in the muscles of the bull's neck, and the bull shrugged it out, sending it flying high into the air. The thing lasted probably half an hour, and, contrary to the rules, the steers were not sent for – either because the president was determined to save Cardeño's face, even at the risk of his life, or because there were no steers ready, as there should have been. In the end the too-intelligent bull keeled over, weakened by the innumerable pinpricks that it had probably hardly felt. It received the *coup de grâce* and was dragged away, to a general groan of execration. Cardeño, who seemed suddenly to have aged, was given a triumphant tour of the ring by an audience very pleased to see him alive.

After that the *novillada* of Sanlúcar went much like any other bullfight. The stylish young *novillero* who had arrived in the Hispano-Suiza killed his bulls, which were big, brave and stupid, in an exemplary fashion. This performance looked as good as one put on by any of the great stars of Madrid or Barcelona, and it was clear that the old Hispano would soon be changed for a Cadillac. The bulls did their best for the man, allowing themselves to be deluded by cape flourishes and slow and deliberate passes of supreme elegance, and the *novillero* tempted fortune only once, receiving the bull over-audaciously on his knees and being vigorously tossed as it swung round on him for the second half of the pass. Miraculously all he suffered was an embarrassing two-foot rent in his trousers, and was obliged to retire, screened from the public by capes, to the passageway, for this to be sewn up, probably with the very cotton the old lady in black had remembered to bring. The crowd didn't hold this against the bull, and it was accorded a rousing cheer, when five minutes later it was removed from the ring.

With this the fight ended, to the satisfaction of all but the critic of *La Voz*. The two *novilleros* were carried back to their hotel on the shoulders of their supporters, followed by a running crowd of several hundred enthusiasts. Just before the hotel was reached they unfortunately collided with another crowd running in the other direction who were honouring the winner of the bicycle race; but the bicycle racing being an alien importation with a small following in this undisturbed corner of Spain, the bullring crowd soon pushed the others into the side streets, smothered their opposition, and fought on to reach their objective.

When I passed the bullring half an hour later the old Hispano-Suiza was still there. The enthusiasts had pushed it about twenty yards and it had broken their spirit; a man with a peaked cap and withered arm stood by it waiting to collect a peseta from whoever came to drive it away. Otherwise the place was deserted, and the circling storks had come down low in the colourless evening sky.

Among the Bulls

This visit to Spain took place in 1987. A gap of twenty-eight years separated it from Lewis's previous article on bullfighting.

The bullfighter Tomás Campuzano, born in 1957, was at the peak of his career at the time of Lewis's visit. He was one of five brothers, all bullfighters. His first fight was in 1974, but his peak period was between 1984 and 1990. In 1987 he participated in seventy-five bullfights before his retirement from most bullfighting in 1999, after which he concentrated on mentoring young matadors.

First published in *Departures*, March/April, 1988

'WHEN THE HORN WENT IN I felt absolutely no pain,' Tomás Campuzano said. 'I suspected this animal of defective vision from the first but failed to take proper precautions. It was like being hit by an express train. I was airborne, somersaulted and landed face down, shocked and acutely surprised. I rolled over, saw one of the boys take the bull away with the cape, while my blood was fountaining out. Still no pain. They shot me up with morphine in the sick-bay and then took me to Zaragoza hospital, where I spent a month.' Tomás showed me the tremendous scar left by this close encounter with death, scrawled like an undecipherable signature up the inside of the thigh from knee to stomach. He joked continually. 'If you are going to suffer a *cornada*, then Zaragoza is a good place. They have the best horn-wound surgeons in the country.'

Among the bullfighters of Spain, Tomás Campuzano is accepted as the readiest to tackle 'difficult' bulls, the euphemism for those with exceptionally large horns or suspected by the experts of potential unpredictability in action. There are few who have received more horn-thrusts (five to date) from the terrible Andalusian bulls with which, as a fully-fledged matador, he is so often called upon to match himself. He takes part in up to fifty fights in a season. Last year was outstandingly successful. A bullfighter who has given an impressive display with a bull may be awarded as trophies one ear, both ears – or, in exceptional cases, even the tail of the vanquished animal. In the 1986 season, despite a wound that nearly dislocated his sword arm, Campuzano collected a grand total of eighty-six ears and eight tails for a series of uniformly brilliant performances.

He started informal training at the age of seven at whatever bull-farm could be persuaded to allow him to practise his cape-passes with a calf, and appeared as a professional in the ring at the legal minimum age of 17. Now aged 30, and earning about £7,500 per fight, he has reached the height of his career, a modest, friendly man who smiles a great deal, and has remained unspoiled by success.

Tomás was born in Gerena, about 10 miles from Seville. It is the archetypal Andalusian hilltop village, put together from stark, white, geometrical shapes, raised above a prairie of pale wheatfields, patched here and there with great brassy spreads of sunflowers. In Gerena the narrow streets are calm and immaculate. Dignity of appearance and personal style is much cultivated. Men walk slowly, held erect, and few women are to be seen. It is a spare, silent place, a refuge of the Spain of the past. Almost the whole of the hill's summit is occupied by the low-lying, blind-walled palace of José Luis García de Samanieco, the Marqués of Albaserrada, who owns all that is visible from his rooftop of the almost Siberian landscape of this region of Andalusia. He is also one of the great *ganaderías* of fighting bulls.

Tomás, whose father was once a shepherd on the estate, has moved down with his family to take over one of the large new houses at the bottom of the village. It is a place to which he returns continually between fights, and where he is a living legend, a poor boy who has shot to the top of what in rural Andalusia still remains the most glamorous, and the most honourable of professions.

The new Campuzano house is an extended and softened version of the early peasant dwellings that present austere profiles to the village from the top of the hill. A big sitting-room holds modern furniture of the best quality, gathered under a vast chandelier; but with the retreat from simplicity there has been a loss of strength. Otherwise custom prevails. When I visited, the voices of women and children could be heard faintly beyond the ornate doors, but only men with a certain solidity were present: Tomás's father, still moving as if in control of sheep; an exceedingly genial brother who manages Tomás's affairs; a couple of old sun-cured uncles leaning upon their sticks. The mother flustered in with coffee on a tray, flashed a nervous half-smile before withdrawing. Tomás's wife – clearly, from her photograph, a beauty of the highest order – did not appear.

This, in some way almost oriental, gathering was dominated by the huge mounted heads of two of Tomás's most difficult and memorable bulls, whose challenging eyes it seemed hard to avoid. Tomás said that they were masterpieces of the taxidermist's art, and that the facial expression – different in every bull as in every man – had been most successfully preserved. He invited me to join him on the landing half-way up the staircase, at a point where the most fearsome-looking of these animals, Abanico by name, could be viewed from precisely the angle at which Tomás had been exposed to its stare six years before in the ring at Málaga. 'I'm off to Madrid on Monday,' Tomás said, 'and whenever I go on a trip I stand here and look into this brute's eyes, and tell myself, at least they can't throw anything at me worse than this one.'

At this point the subject of fear came up. It seemed a doubtful one to raise with a man generally accepted as among the most courageous of all bullfighters, but he cut across my attempts at tact. 'Was fear something you could come to terms with?' I asked. 'No,' he said, 'never.' The fact was that it got worse and worse, strengthening with each increase of a man's responsibilities. From the day he got married the fear increased, and now that his wife was expecting a child, it was closer again and more insistent. In summer, he said, when he could be fighting twice a week, a bullfighter's family was constantly overshadowed by fear. While the fight was on, no telephone calls to the house could be made by friends – to keep the line clear for any emergency – and only close relatives were invited into the home, to maintain what amounted to a silent vigil. They were exceedingly devout; crucifixes and rosaries hung everywhere about the home. Another minor bullfighter who had drifted in said, 'We take our troubles to the Virgin of Macarena. She's a Sevillian – almost a member of the family you might say. Imagine two fights in two days. Naturally you're praying half the time.'

The estate house of the Albaserrada bull-breeding farm is two miles up a country road from Gerena; a clean-cut example of purest Andalusian architecture. Decoration is forbidden, the atmospheric quality of these surroundings depending upon white, crystalline façades and the blue mossy shade of cactus and eucalyptus. Adjacent is a small, high-walled ring in which the cows from which the bulls are bred are subjected to a series of tests, known collectively as the *tentadero*. *Tentaderos* take place at frequent intervals during the summer months and have come to be treated as a social event, inevitably watched at Albaserrada by the Marqués and a few of his intimates. It is explained that the number of those present on such occasions is kept to a minimum to avoid distracting the animals under test.

When I arrived the testing was already under way. I looked down from the rim of the small arena at a rider on a padded horse, steel-tipped pole in hand, waiting on the far side of the ring for the entrance of the next cow under test. The Marqués had just expounded the bull-breeder's theory that taurine courage is transmitted through the female of the species, and that the male only adds strength. For this reason, only two-year-old cows are subjected to serious testing, and they are certainly no less fierce than the bulls.

The wall of the ring was painted a most profound and refulgent yellow, with the overhead sunshine rippling and showering down its uneven surface. The wall colour was intensified by that of the sand, and there was a yellow reflection in the faces of the onlookers. After a while the unearthly quality of the light seemed even to affect the mood, endowing this scene with a feeling of separateness from the surrounding world. A religious hush had fallen; the spectators were motionless and silent. An element of ritual was discernible here, a flashback perhaps to Celto-Iberian days and sacrificial bulls.

A small black cow came tearing out into the ring, slid to a standstill and swung its head from side to side in search of an adversary. It was big-horned, narrow of rump, all bone and muscle; faster in the take-off than a bull, quicker on the turn and with sharper horns. 'Ugly customer,' a herdsman whispered approvingly in my ear. The horseman thwacking the padding of his horse with the pole, called to the cow and it charged, crossing the ring at extreme speed, head down, horns thrust forward in the last few yards, thumped into the quilting over the horse's flanks and threw it against the wall.

Time and again, it skewered up ineffectively with its horns while the horseman, prodding and shoving down with the shallow, testing *pic*, scored the hide over its shoulders. Failing to get through the padding it trotted off, then turned back for a second charge. The watching herdsman noted points in their books under four headings: courage, speed, reflexes, staying power; and communicated what might have been approval or disdain with inscrutable signs.

It was this performance with the horse and the cow's indifference or otherwise to the prickings of the *pic* that sealed its fate; but when the serious business was at an end, fun for all followed with the cape. Tomás Campuzano had arrived to help the local boys add polish to their technique, conducting a series of passes with a mathematical exactitude that seemed sometimes to border on indifference. The onlookers smiled dreamily. Those that followed the master seemed agitated by comparison, and a young Venezuelan bullfighter who had come along appeared a little out of his depth with a beast of this kind, or perhaps the cow was learning quickly from its mistakes.

Surely, I asked myself, the keen-eyed selectors could ask for nothing better than this animal with its limitless vigour and thirst for aggression? But the experts detected weaknesses overlooked by the outsider, therefore rejection followed the completion of its trial. And so in the course of the morning six aspirant cows came and went. It was a spectacle providing its own special brand of addiction, preferred by many enthusiasts to the commercial bullfight itself. Spain's leading painter of bullfighter posters, present on this occasion, later admitted that he never missed a *tentadero* if he could help it. Both he and Don José Luis, although a little stiffened by middle age, gave brief but confident displays with the cape and came off intact, although the Marqués's boxer dog (always addressed in English) broke into the ritual calm with yelps of hysterical anguish at the sight of its master exposing himself to such danger. Of six cows, four rejects were subjected to the ignominy of having a few inches lopped from the end of their tails after the test. In this way they were marked for the slaughterhouse. The two accepted, to be kept for breeding, had the dangerous ends of their horns removed. Both operations – the second performed bloodily with a saw – were carried out forthwith and in view of the onlookers.

From the ring we moved back to the estate house for a snack served in the yard. This took traditional form: thick, solid potato omelettes cut into cubes to be eaten with the fingers, slivers of hard farm cheese, white wine of the last year's vintage (still a little murky) from the estate vineyard. Spurred on by Don José Luis's assurance that it contained only five degrees of alcohol, guests downed the winelike water. The informality of such occasions is much appreciated in Andalusia – and referred to approvingly as *simple*. To this slightly feudal environment Tomás Campuzano had been admitted as an admired friend. Part of the reward of a famous bullfighter is an escape into the nirvana of classlessness.

The bulls inhabit an untidy savannah of old olives, thorn and coarse grass entered a few hundred yards from the estate house. There are upwards of six hundred of them kept in two separate herds, the four-year-old *novillos* and the five-year-old bulls in the full vigour of life. Throughout the summer months their numbers dwindle steadily as the bulls are sent off, six at a time, to fight in the big city rings where the management can afford to pay for the best. For a corrida of six four-year-olds, the Marqués expects to be paid three million pesetas; for the five-year-olds the price is four million. He loses money on the bulls, he complains, but keeps afloat on the slight profit the estate makes from sunflower oil, wheat and olives.

All guests are taken as a matter of course to inspect the herds. They ride in a trailer drawn by a tractor from which fodder is distributed in times of dearth, and which is therefore acceptable to the bulls. The trailer has high steel sides and is heavy enough not to be turned over by a charge. The tractor's engine is always kept running because the bulls have learned to associate its sound with food. Still, the excursion is not quite in the same bracket as a trip through a safari park because a fighting bull is more aggressive than anything encountered in the wild and, if annoyed, is liable instantly and unforeseeably to charge the offending object, whether animate or otherwise. A Spanish treatise on the subject of bulls speaks of the bull's docility on the ranch. 'It is more than likely', it says, 'that the vast majority of fighting bulls would allow themselves to be stroked. To attempt this one must put away the almost insuperable fear that their presence and proximity inspires.' None of those present on this occasion seemed inclined to put the author's theory to the test.

Chugging behind the tractor into the bull pastures was accepted as a minor adventure. The bulls stood, heads lowered, a few yards away, to watch our approach with steadfast, myopic eyes. Their relative invulnerability has relieved them of the necessity to develop acute vision, but their hearing is exceptionally acute. Thus we probably appeared no more than a vague, invasive shape, but lulled by the soft clatter of the diesel and its promise of mash, they made no objection. It rains hardly at all in summer, so the bulls spend the day in ceaseless foraging for pasture, moving always very slowly and with great, ponderous dignity. The animals in each herd settle quickly to mutual toleration, undoubtedly realising that an inbred policy of no-surrender means the death of one of the disputants of any quarrel that is allowed to arise. They learn quickly. When Don José Luis's boxer decided to try his luck with a five-year-old, the animal soon realised that the dog was too fast to be caught by the horns, so, adopting an invitingly passive stance, he lured the boxer within easy reach and removed several of its teeth with a kick.

Under the protection of the tractor and its soothing noises no scene could have been more arcadian, and nothing more appropriate to this Andalusian setting than the bulls, viewed either in majestic silhouette against the green-grey wash of olives, or as they wandered ruminatively, deep in the strong tide of sunflowers that had burst through the fences of their enclosure.

'Whatever the financial loss, the bulls are my life,' the Marqués said, having taken us at the end of our visit to the palace for an inspection of his most treasured possessions. Once again I found myself confronted with mounted heads. These were of two Albaserrada bulls 'pardoned' following extreme bravery shown in the ring; one in Madrid in 1919; the other in Seville in 1965 – both unprecedented events. Despite some reluctance on the part of the traditionally minded authorities of the Maestranza (as the ring at Seville is known), they were obliged by the insistent demands of the crowd to break their rule. And so the Marqués's bull *Laborioso* (hard-worker) was returned, appropriately fêted and garlanded to the herd. It had taken seven thrusts of the *pic*, and had overturned seven horses, three of which had to be replaced. Its wounds were healed through massive injections of penicillin, and it lived on until 1976.

Bullfighting, practised in one form or another since Celto-Iberian times, not only in towns but in innumerable villages throughout Spain, began to fall into decline in the post-war period. Spanish attitudes were much changed by the tourist influx. New rings were opened in many northern areas where bullfighting was previously unknown, but the uninstructed demand of an overwhelmingly foreign audience was for pure spectacle. Bullfighting taken straight was seen as tedious, so buffoonery was often added with the provision of clowns and dwarves in bullfighting gear who threw custard pies in each other's faces; or the procedure might be livened up by performing dogs. Since the foreigners hardly knew one bull from another it was an opportunity for disreputable breeders to supply substandard animals at cut prices, encouraging instances of low quality, underpaid bullfighters who refused to tackle bulls without artificially shortened horns. This was the period when ambitious but inexperienced youngsters (known as *capitalistas*) were paid small sums to invade the ring and join the fight, sometimes with tragic results. Bullfighting began to suffer from the competition of football, while promising village boys aspired to become popstars rather than matadors.

Rock-bottom may have been reached in 1981 when the concluding corrida of the Seville season had to be put forward a day because it coincided with a home match by Seville FC, the promoters realising that otherwise the ring would have been empty. Of this melancholy occasion a leading newspaper critic wrote: 'Thus the present decadent season draws to its end. It has offered little but boredom for the public, and bad business for the promoters, with half the seats unsold.' The bulls, said the critic, had been uniformly atrocious: small, lame, numbed-looking, and inclined to totter about like calves on shaky legs. The sad and insipid bullfighters spread boredom like a disease. 'When they trundled on the sixth bull, I said to my colleague, "Perhaps I'll take a nap. Wake me up if anything happens." He didn't because he, too, fell asleep.'

From this disastrous year, there was a steady recovery. The commercial backers had come to understand that it was a matter of drastic reform or, for them, the end of the road. They paid more for their bulls and for their bullfighters; they got rid of the clowns, suppressed the circus antics of such as El Cordobés and his imitators, thus attracting a new generation of matadors. As a result, 1985 was adjudged 'brilliant' and 1986 'excellent'. As part of this renaissance, a bullfighting school opened near Seville in January 1987 with 16 young pupils ranging from 9 to 16 years of age. The event was sufficiently important for it to be attended by the representative of the Ministry of Culture responsible for what is officially entitled 'the taurine art', who spoke enthusiastically of the performance of the children in their encounter with the bull-calves of appropriate size.

Alcalá de Guadaïra, site of the bullfighting school for promising boys, is a small, white pyramid of houses dominated by a vast Moorish fortress defending the old approach to Seville. Otherwise it is notable for mining the high-grade sand supplied to bullrings that can pay its expensive price, and for the richly chromatic earth that provides the yellow paint for Seville's baroque buildings, and for the interior of the Marqués de Albaserrada's miniature amphitheatre. This place suggests the persistence of an ancient half-submerged bull-cult, for once again the mounted heads are everywhere, and every tavern and bar is glutted with bullfighting posters, photographs and prints. Within hours of arriving in Alcalá I received an invitation from an olive-growers' association: 'We're having a bit of a fiesta up at the inn this evening. Just a few friends. No more than a glass of wine and a sandwich. We'll probably kill a bull.'

The school is in the honeycombed building next to the ring itself; a single dim room cluttered with scholastic objects, exercise books, desks, a blackboard, plus piles of harness, plastic matador's swords, and a miscellany of horns. The largest of these is fixed to the front of a formidable contraption like a handcart on bicycle wheels used for chasing would-be *banderilleros*. Special respect, and possibly some magic virtue, attaches to these particular horns, as they were removed from a bull killed in a fight with another bull.

Students receive instruction for two hours a day on three days a week, about a third of this time being devoted to ring tactics and the rest to practice in the ring. Funds have been allocated to the school by the Ministry, but, through 'delays in legalisation', these have not yet arrived. Since it costs the equivalent of £75 to hire a second-class fighting cow for two hours (a first-class cow costs twice this sum), there is little practice with animals, and a great deal of make-believe in which masters, horns in hand, pursue their pupils all over the empty ring.

Joselito Ballesteros, aged 9 but looking hardly older than 7, gave an impressive demonstration with the cape, shaking it in a taunting fashion, and making defiant bullfighter noises at his father, one of the teachers. The father, shoulder hunched, head down and horns thrust forward, scraped with one foot in the sand in the manner of a bull who is about to attack. In due course he was despatched as Joselito lunged forward with his imaginary sword, an *estocada* loudly applauded by the bystanders. In another part of the ring the mature student of sixteen skipped aside to avoid the charge of an instructor manipulating the simulated bull on wheels, raising himself on tip-toe to plunge the pair of *banderillas* into a padded leather surface where the neck muscles would have been.

'In this profession, as in others,' the father of Joselito said 'everything depends on an early start. Joselito started training at the age of five. At the moment he can hardly see over an animal's back, but he could be giving private performances by the time he's fourteen. He can't be accepted as a professional for four more years after that.' Ballesteros described the principles inculcated by the school. 'The art of the ring is wrapped up with moral attitudes,' he said. 'We keep a close check on their behaviour in the day school as well as in the home. We are engaged in the development of artists and believe that art is inseparable from life.'

'How many children like this can expect to become great bullfighters?' I asked.

'Five percent.'

'And how many will die in the ring?'

The question startled him, and his face crumpled.

'When they're properly trained, as these boys will be, there's nothing to worry about. It's the old-timers trying for a comeback, and the kids that will do anything to get a start. The bulls cut them down, but they don't make the papers. How many of them go that way? There's no knowing.'

He seemed depressed at the turn our talk had taken. Perhaps it was something he wanted to put out of his mind. The boys were dancing round us with their capes, striking attitudes of defiance, sizing up phantom bulls, coming close to the imaginary horns. Now the master's attention was taken up, a little horseplay had been brought in.

'Above all we teach our boys to master fear,' Ballesteros said. 'That's the most important thing of all.' Almost plaintively he added: 'You see, the horns are very sharp. It's bad for them if they get scared.'

The more I saw of these Spaniards of the deep south the more it became clear to me that it was a misapprehension to believe that their feeling for bulls was anything less than an almost obsessional admiration and respect. Near Alcalá a handful of olive-growers had clubbed together to pay an enormous price for a bull to be killed at their annual fiesta. Whatever their excuse, this could not have been anything but a sacrifice to ensure a good harvest, and understandably, only the most splendid of animals could be offered to the gods. Noble is the adjective never out of Spanish mouths when they speak of the bulls, to whom they frequently attribute such human qualities as candour and sincerity. This is F. Martinez Torres, himself a bullfighter, on the subject of courage: 'The bull is the only animal in creation that is not daunted by any wounds he receives. He does not possess the treacherous or bloodthirsty instinct of other animals that crouch unseen and spring on their prey from behind. He attacks nobly from the front. Face to face, there is no animal that can beat him.'

This is only part of the story, for the bull is capable of enduring friendship, and never forgets a face – or a voice. 'There have been some bulls,' Torres tells us, 'which during the fighting, on being called by the herdsman they knew, have broken off the fight and trotted meekly over to the place where their former custodian stood, allowing him to stroke them from inside the fence, or at times in the arena itself. When he has finished doing this, they have returned to fight with the same fierceness as before.'

The hard-bitten professionals of the Spanish press, bull-lovers to a man, are saturated with the pathetic fallacy. Here is a passage from Antonio Lorca's account in *El Correo* of a *novillada* for young bulls in Seville at the time of my visit. Lorca cannot stomach vulgarity and believes that the bulls feel the same. 'Such a bull as this demanded at least a token authority to direct its noble charge. A real fighter would have provided inspiration for the bull but, faced with a tasteless fidgeter, it showed indifference, even impatience.'

Manuel Rodriguez of *ABC*, the oldest national newspaper in Spain, also covering this event, noted that the fourth (inevitably 'noble') bull gave the matador Fernando Lozano two warnings of the danger he placed himself in through misuse of the cape. Rodriguez too, abhorred vulgarity: 'Elsewhere they might have thrown cushions. Here in Seville we correct such lapses with an icy silence.'

Back to Lorca. 'Four bulls received huge applause from the crowd as they were dragged from the ring. As for the fighters, it was a mediocre harvest of a single ear. Again I ask myself the question, is a bullfighter born, or made? From what we saw yesterday I can only conclude that he is born. Nevertheless, it is his duty to himself and to us to continue to grow. Shame it was to see great bulls thrown away in this fashion.'

Seville Revisited

In 1981 The Sunday Times commissioned Lewis to write an article on Seville. He was now in his early seventies, and his biographer, Julian Evans, suggests that he was struggling to come to terms with his age. On this occasion he failed to appreciate Seville, finding it depressing and seedy in all the wrong ways. No article was published. In 1984 The Sunday Times sent him again, and his attitude totally changed.

According to Evans, Spain was Lewis's 'recurring and default destination', and Seville his favourite Spanish city.

First published in *The Sunday Times*, June 2, 1985

THERE ARE FEW HOTEL BEDS to be had in Seville when it acclaims the spring in its inimitable fashion. I found a place to stay out of town, dumped my baggage, and travelled in by taxi. At the San Telmo bridge over the Guadalquivir we ran into snarled-up traffic dominated by a woman policeman with a dark, ecstatic face. She ran, leaped and cavorted, unmeshing and directing the crawl of cars with competence and with artistry, causing the drivers to open their windows to shout their applause. 'That's a gypsy,' the taxi driver said. 'She used to dance at the Arenal. If we pull over and wait till the traffic clears, she'll do a seguidilla for us.' But the policewoman snapped her fingers at us and threw back her head, and we joined the queue on the bridge. There was another jam on the further side and the driver had a suggestion. We turned left into the wide and relatively quiet Avenida Colon that follows the river, then stopped and he pointed to a narrow street entrance across the road. 'Why don't you do the rest on foot?' he asked. 'The centre's only a couple of hundred yards up there.'

I walked up the Calle Dos de Mayo, as directed, a calm and almost countrified street of white walls, and windows draped in sumptuous folds of baroque plaster, picked out in sober yellow. Orange blossom bespattered the cobbles, and there was a champagne sparkle of May in the air. A Victoria of the most fragile elegance – a 'milord' of the kind introduced by Edward VII – passed with a clip-clop of hooves and a soft rumble of wheels. Blind white cubist shapes were piled round a Moorish battlement at the end of the street, and above and beyond, the Giralda Tower, once the greatest of all the minarets of Islam, possessed the sky. A wall panel in ceramic tiles showed the Calle Dos de Mayo as it was nearly a century ago, and there has been little change. The panel was put up by a soap-maker and it is one of the many magnificent tiled advertisements decorating the city's walls, for such things as the first Kodaks, gramophones with horns, cough mixtures, mustard plasters, and forgotten motor cars. All the products promoted in this charming fashion have one thing in common: they no longer exist.

At the end of this short cut to the centre, I was confronted with the grey, fortress shape of the Cathedral. 'Let us build a church so big that we shall be held to be insane,' a member of the Chapter urged as soon as the great mosque had been levelled and the building of the Cathedral began. The Emperor Charles V, most human of the Spanish monarchs, who gardened and kept parrots in his modest quarters in the Alcazar nearby, would not have approved; but he was too late upon the scene. 'You have built here what you or anyone might have built anywhere,' he said, 'but you have destroyed what was unique in the World.' Fortunately he was in time to save the great Mosque in Cordoba.

The Cathedral of Seville is vast, high and very dark, with visitors wandering a little apprehensively in the gothic twilight, like travellers lost in a foreign railway station. Light invests the gloom from a side-chapel in which a Madonna with the tear-streaked face of an unhappy fourteen-year-old girl broods over piled-up church treasure. A thousand candles flicker while the great organ crashes and booms.

I had hoped to see the tomb of Pedro the Cruel, but it turned out that this, which is in the crypt, could be visited only on one day in the year. Of the doings of this monarch, identifiable to his trembling subjects, as he stalked the streets at night, by his creaking knee joints, a single episode illustrates the utter foreignness of the medieval mind. When rejected by a celebrated beauty, Señora Urraca Osorio, the King had her burned to death. Having studied his record, this does not surprise us. What does – what remains beyond the compass of our mentality – is that the one thing that seems to have been of importance on this occasion was that Señora's modesty should not have been placed in jeopardy, and that to prevent this possibility her maid leaped into the flames so as to screen her mistress from any such exposure.

Pedro is supremely the 'bad' king of Spanish history. St Ferdinand, conqueror of Seville from the Moors in the 13th century, whose 'incorruptible' body lies in a silver casket in the Cathedral's Capilla-Real, is perceived as the King who meant well. The saint, a rigorous pietist who died eventually through excessive fasting, was the scourge of heretics, setting his people an example of righteous severity in one instance by lighting in person the pyre on which an assortment of dissenters of one kind or another were to be incinerated. 'His Majesty wore a rough gown tied by a rope, and carried a large cross. He ordered all those who had come to the place to kneel and pray, and imposed upon them a penance. Before taking the torch from the hand of the executioner he kissed the cheek of each of those who were about to suffer.'

The Cathedral expresses conquest and domination in architectural terms of sheer mass, and it comes as a surprise to learn that it is not the largest building in Seville, being second to the nearby tobacco factory, now the University, only exceeded in size by the Escorial. It was here that Don José had his first encounter with Carmen. Five thousand girls were employed to make cigars, and the Victorian British visitor found sexual interest in the sight of a girl rolling a cigar on her thigh, which she could be persuaded to do for a payment of ten centavos, equivalent of a penny. Murray, of the celebrated guidebook, viewed the scene with both unction and disapproval, although he clearly found it hard to tear himself away. He found the girls handsome but smelly, and 'reputed to be more impertinent than chaste'.

My visit to Seville in the spring of 1984 had followed one in the autumn of 1981, and I was delighted and relieved to discover the town transformed by change and renewal. In 1981 it had seemed dirty, depressed and anarchistic, a prey still to moral confusion and lack of guidance following the disappearance of the dictatorship. Sevillians had shown themselves at a loss with managing civil liberty, and some excesses had stirred up a sullen reaction. The walls were covered with resentful graffiti. 'Democracy is a lie, democracy kills,' they said. Franco's face had been stencilled everywhere, accompanied sometimes by the supplication, 'Come back to us. We can't carry on without you. All is forgiven.'

Municipal workers – like so many Spaniards at that time – had been on strike for months, and the streets were piled high with rubbish, visited by innumerable rats. Pornography had arrived – a stunning experience for a straitlaced people – with ubiquitous porn cinemas, horror video cassettes for sale, and window displays of the Kama Sutra in booksellers previously specialising in devotional manuals or the lives of the saints. For nearly forty years prior to Franco's death in 1975, Spanish lovers had been forbidden to kiss in public, and in small-town cinemas a priest stationed himself at the projectionist's side ready with a square of cardboard to be held over the lens where a romantic episode might be held to weaken public morality. With the return of democracy there was time to be made up. In 1981 public courtship had become a ritual, and in Seville couples fell into each other's embrace in any square that provided suitable benches, and lay locked together while the hours passed and the rats scuttled through the rubbish under their feet.

Package-deal trips to London were advertised to deal with unwanted pregnancies, at the all-in charge including two nights in the British capital, of £250. The restraining influence of the Catholic Church seemed to have collapsed along with that of the State. The city filled up with mystic carpet-baggers eager to fill the vacuum, with mediums, sun-worshippers, 'Cosmo biologists', whirling dervishes, American fundamentalists howling for Armageddon, and sects dating from the pre-Christian era, including one, as the newspapers reported, that sacrificed chickens, drawing omens from a study of their entrails.

All the symptoms of a society in advanced decay were present, yet suddenly Sevillians had managed to work the poison out of their systems, and the phase was at an end. Now all appeared sweetness and light. The streets had been swept and garnished, flowers were in bloom in all the parks, Pied Pipers had lured all the rats away, and the walls had been cleansed of nostalgia for Franco, and hate-filled graffiti. Some Sevillians were of the opinion that the Virgin known as La Hiniesta had come to their aid in response to the conferment to her image – despite the steadfast opposition of Communist councillors – of the City Medal of Honour. Now she was junior only to the Macarena Virgin, who not only held the medal but had been promoted in 1937 by General Queipo de Llano to Captain General of the Nationalist Armed forces.

I looked up an old Sevillian friend. 'What finally happened about Cristina?' (Three years ago he had been in despair after his only daughter, aged twenty, had gone off to live with a married man aged forty-four.) He seemed surprised that the matter should have been brought up. 'Oh, that's a thing of the past. She's settled down now with a nice chap who works in the Banco de Espana. Daughter aged two and expecting their second.'

The talk turned to the condition of Seville, and some mention was made of the four bank robberies on the previous day.

'It's nothing,' he said. 'Something they picked up from *Butch Cassidy and the Sundance Kid*. A boy walks into a bank, pulls out a gun, tries to speak Spanish with an American accent, and says "esto es un robo" just like they do in the film. The customers line up with their hands up against the wall, the cashier pushes a few thousand pesetas through his window, and the kid picks it up and gets out. It's a phase. Not a thing to take seriously. Nobody gets hurt.'

This relaxed view was in part shared by Charles Formby, the British Consul in Seville. 'In this town they snatch handbags and break into cars,' he said. 'We have specialists called *semaforazos* who break the windows of cars held up at traffic lights and grab what they can. But this is not a truly dangerous city. No one gets violently robbed. You're safer walking the streets of Seville at night than you are in London.'

Charles Formby supplied the statistic that, at thirty-eight percent, Seville's unemployment is the highest in Spain, and that twelve percent of the unemployed are university graduates. Although Seville is the centre of a depressed agricultural area, he didn't agree that petty criminality was a product of this economic situation, but believed it was a matter of obtaining money to buy drugs.

Don Rafael Manzano, Director of the Alcazar of Seville, agreed with him, adding his conviction that Spanish society was now the most permissive in Europe. Since the departure of Franco, the police were no longer feared, persons found to be in possession of small amounts of drugs weren't charged, and prison sentences were light except for serious crime. He added that on May 10th – the day of the four bank robberies – a young man only released from prison a few hours previously had been arrested in the act of throwing packets of cannabis and heroin over the prison wall.

One of Seville's problems seems to lie in its nearness to the point of entry of drugs from North Africa. 'I happened to be down at Algeciras, the other day,' said a reporter on *El Correo de Andalucia*, 'and watching all these pregnant women getting off the ferry from Tangier, I wondered how many were really great with child, and how many with bundles of hashish.'

So the drug problem remained, but otherwise Sevillians seemed to be ridding themselves of the social sickness mild or grave, largely transmitted through the cinema. The likely lads of Spain were no more immune than the youth of any other country from the cultural revolution inspired by such films as *The Wild Ones*, and the first motorbikers had appeared ten years ago. But they were a dwindling cult, and the Sevillian Hell's Angels Chapter was down to a fraction of its former membership. I found a group from the Chapter tinkering with their Yamahas in a square on the edge of town. The majority were gypsies with sensitive Asiatic faces and melancholy eyes, and one was quite frank about their problems. 'It never really took off here. In my view to be a real *Angel del infierno* you have to wear the right leather gear, and how can you expect anybody to do that when the temperature never drops below ninety in the shade?'

In 1981 there had been a few poorish imitations of punks and skinheads about, but now they were out of fashion, and were no more to be seen. On the other hand, youngsters in plenty thronged the well-lit squares and avenues far into the night. It was something new in Spain, a phenomenon the beginnings of which I had observed on the earlier visit, when bands of well-dressed and well-behaved children ranging in age between twelve and sixteen, managed to turn the nocturnal streets into a playground.

Back to Don Rafael in his cell of an office, tucked away behind the scenes in the magnificence of the Alcazar. 'We are suffering,' he said, 'from the side effects of the levelling process. Until recently, only rich people who didn't have any work to do could stay up enjoying themselves all night. Now everyone tries to. At the time of Holy Week it's understandable. The processions are going on all night, and it's something you have to see. No sooner is Holy Week over and we're into the Spring Fair.' The upper class rent chalets in the fairground and give parties that finish at six in the morning. Nowadays, however bad the unemployment crisis, people were determined to defend their democratic rights, one of them being to stay up as late as the rich. If the parents stay up all night, said Don Rafael, so do the children, their argument being that democracy knows no age limits. As a result they fall asleep over their books at school. 'We as a nation,' he said, 'have lost the ability to say no.'

All the old Spanish practices were staging a comeback, including bullfighting. In the autumn of 1981 so gloomy was the outlook for the national spectacle that the concluding *corrida* of that year had to be put forward a day because Sevilla FC would be playing at home on the Sunday originally planned. Of this overshadowed occasion, the *Correo de Andalucia's* sportswriter said, 'So the present decadent season draws to its end. It has offered little but boredom for the public, and bad business for the promoters, with about half the seats unsold. In this last *novillada* we saw underweight bulls, the fourth with a marked tendency to slip away, and the last two virtually calves, tottering about on shaky legs. There was great protest from the crowd that the fifth was lame. It wasn't. It was numb from being shut up in a pen for so many weeks.'

The aficionados blamed it on the quality of the bullfighters themselves. The spirit had ceased to breathe upon them, leaving them cold and cautious, and their performance a tawdry commercial transaction in which a minimum is returned for money received. Fighters of old such as Joselito drew more crowds to Seville than a visit by the King. Half his personal fortune went into the purchase of four emeralds for the Macarena Virgin, protectress of bullfighters (and, unofficially, smugglers), and when, at the age of twenty-five, he died in the ring in exemplary fashion, the virgin's image was dressed by her attendants in widow's weeds in which she remained for a month.

Could there ever be a return to those days? Even that seemed possible. The bullfights accompanying the Spring Fair this year suggested a long-hoped-for renaissance, and at a time when the enthusiasm for football seemed to be on the wane, all seats in the Seville rings were sold out. At last, we were assured, the bullfighters had put their house in order, and meek bulls with shaved horns were a thing of the past.

It was too soon to hope for another Joselito, but there were plenty of newcomers of promise, including Manuel Ruiz Manili, a rising star of the old order, already nicknamed *El Jabato* (the young wild boar), who put on a near-suicidal display. 'This man,' said one critic, 'really hurls himself into the fray.... If only he lasts!' He made another sportswriter shiver. 'I could *feel* those horns scrape his cuticles. Number two was as wicked a beast as I've ever seen. "Watch me cut him down," Manili had said. "They'll either carry me out on their shoulders, or I'm going to hospital." They carried him out. Delirium.'

Manili, accorded the title of *El Triunfador* (the triumphant) at the Spring Fair,was a gypsy, and he could have been the brother of the stylish policewoman I had seen directing the traffic at the Puente San Telmo. In 1984, as ever, the gypsies remain a little mysterious, uncharted human territory on the fringes of Spanish society. There was not a gypsy bank manager in the whole of Spain, but the best bullfighters, the best musicians and the best dancers were, and always had been, gypsies. For many foreigners, and quite a few natives, too, the mental image called to mind that typified the attraction of Spain was that of the arrogant dancing gypsy.

The capital of gypsy Spain had been the Sevillian suburb of Triana, birthplace of the Emperor Trajan, but the developers laid siege to it, and it fell. In Triana of old, fountainhead of the popular music of Spain, the gypsies lived in extended families in tiny, immaculate communal houses around a courtyard cluttered with flowers. These were called corrals, and now of the hundreds of such nuclei of which Triana had been composed, one only remained. It was occupied by twenty-three perplexed families lost among the upsurge of modern buildings, their main problem seeming to be how to pay for the water for the four hundred pot plants which turned their courtyard into a tiny Amazonian jungle.

For all the changes, stout-hearted pockets of resistance remained. In Triana, the secluded patios where dances – dating in all likelihood from before the days of Trajan – had been taught and practised, had all been swept away, but the dance went on. Anita Domingo's Academy of Spanish Art, where little girls go to stamp their heels and click their castanets, while Anita thumps out a *Sevillana* on the stiff piano, has a longer waiting list than ever. Half the pupils are gypsies, neat and small-boned, with flashing black eyes and the classic profiles stamped on ancient coins. Ancient customs refused to die. Newspapers reported the case of a gypsy working in a Triana department store, who was involved in a dispute over the granting of three days' leave of absence to allow her ritual abduction, the essential preliminary to a gypsy marriage.

Holy Week was at an end and, graced by the Royal Family's participation, it was spoken of as the greatest success since the days before the Civil War. The Spring Fair had been – to use a fashionable adjective – a marathon one, but here there was an undertow of caution in the deluge of praise. It was noted that only fifty years back three days had been considered ample for civic festivities, originally based on the conviction that they were essential to the production of spring rains. Now it was commented upon that not only had the fair been increased to seven official days, but that further prolongation was threatened by tacking two more days onto the front 'to test the illuminations'.

With these excitements at an end, we had slipped into the calm aftermath of the 'Easter of Flowers' when the city turns back with reluctance to the routines of normal existence. There was nothing at this time, I was assured, likely to be of interest to the visitor, except relatively unimportant processions when local Madonnas were carried on tours of their zones of influence, as if to be allowed to see for themselves that all was well. Arrangements were informal. Sometimes the little cortege might stop at the door of a particularly devout household to allow its head to be summoned into the presence, bowing a little anxiously to report on the doings of the family. Such outings were collectively known as 'Her Majesty in Public'.

In the Easter of Flowers a light diet and early to bed were the orders of the day. 'Good fun while it lasted,' was the general view of the fiesta. 'Now let's catch up on some sleep.' Released from the treadmill of pleasure, people escaped out from the bustle of the centre to take a quiet stroll by the river or settle for a nightcap and a chat in one of the cafés on the Avenida de las Constitucion, among night-scented flowering trees and within sight of the great and ancient places of their history.

Sitting with them, I noticed the pull exercised by these grandiose surroundings; how, as if moved by the tug of a magnet, they would shift their chairs for a better view of the floodlit palaces, the spires, the domes and the impeccable profile of the Giralda, the contemplation of which by night is claimed by some Sevillians to foster a lucid tranquillity conducive to untroubled sleep.

The last of the milord carriages passed homeward-bound, ghostly against the sash of mist that marked the course of the Guadalquivir. An exceedingly polite waiter who always said 'At your service' whenever he took or served an order, was seen in a moment of inactivity to be swaying a little, eyes half-closed. Somebody at the next table said, 'Early night tonight, then,' but nobody moved.

At this moment a tremendous din started in the side-street to my rear, and within seconds an excited crowd came into view. In their midst, though towering over their heads, stalked a dozen outlandish figures, on six-foot-high stilts to which their legs were bound. They were patched and masked in medieval style, some of them disfigured with beaks and flapping wings like nightmarish birds. There was something compulsive and a little sinister about their vitality as they came on with a pounding of drums and a clash of cymbals, sometimes breaking into a grotesque, stiff-legged dance.

One by one my neighbours straightened in their chairs, rose to their feet as if at the command of a powerful hypnotist, allowed themselves to be drawn away into the street and into the crowd, began to clap and fell into step. The café emptied, and the tired waiter, roused by the rhythm to the point of tapping it out on his tray, caught my eye and came over. 'At your service,' he said, and I asked him what it was all about.

'Clowns,' he said. 'They're limbering up for *Cita en Sevilla* (Appointment in Seville). Starts tomorrow and finishes on 21st June.'

'Don't tell me another fiesta,' I said.

'Not quite that,' the waiter said. 'The Municipality's trying out something new. They didn't want us to be bored, so they're importing six jazz bands, two Czechoslovakian orchestras, visual poetry from Italy (whatever they mean by that), Son et Lumière, an ice-show with a real ice-breaker, musicians from Morocco, and clowns from all over. Half a dozen famous opera singers have been invited; there'll be poetry readings, cante-flamenco, art and photographic shows, and the biggest antiques fair ever, or so they say. That's about the lot.'

'What happens after 21st June?' I asked.

'We haven't been told yet. Be sure they'll think of something.'

'It's a lively town,' I said.

'You're right,' the waiter agreed. 'There's always something to look forward to.'

He dropped into a chair at the next table, putting his head in his hands.

'Sometimes I think they overdo it,' he said.

Genocide

In 1968 Lewis approached The Sunday Times *after he learnt that the Indian Protection Service in Brazil had been complicit in the murderous destruction of Indian communities. The resulting 12,000-word article was the longest that* The Sunday Times *magazine had ever printed.*

As soon as the piece was published, the paper was obliged to hire extra staff to handle the copious correspondence and telephone calls. One of those who contacted the paper was Robin Hanbury-Tenison, who promptly founded Survival International, which campaigns with indigenous peoples worldwide.

On this trip, Lewis travelled on his own. After returning, he made a list of places for a photographer to visit. The Sunday Times *sent out the young Don McCullin, who had made his name in Vietnam and Biafra. Lewis was so impressed by McCullin's photographs that he asked to meet him, the beginning of a long and fruitful friendship.*

A summary of the past and current predicament for Brazil's indigenous population – including continuing assaults – can be found by googling 'brazilian indigenous people survival international'.

Lewis considered this article the most important of his lifetime.

First published in *The Sunday Times*, February 23, 1969

I F YOU HAPPENED TO BE one of those who felt affection for the gentle, backward civilisations – Nagas, Papuans, Mois of Vietnam, Polynesian and Melanesian remnants – the shy primitive peoples, daunted and overshadowed by the juggernaut advance of our ruthless age, then 1968 was a bad year for you.

By the descriptions of all who had seen them, there were no more inoffensive and charming human beings on the planet than the forest Indians of Brazil, and brusquely we were told they had been rushed to the verge of extinction. The tragedy of the Indian in the United States in the last century was being repeated, but it was being compressed into a shorter time. Where a decade ago there had been hundreds of Indians, there were now tens. An American magazine reported with nostalgia on a tribe of which only 135 members had survived. 'They lived as naked as Adam and Eve in the nightfall of an innocent history, catching a few fish, collecting groundnuts, playing their flutes, making love ... waiting for death. We learned that it was due only to the paternal solicitude of the Brazilian Government's Indian Protection Service that they had survived until this day.'

In all such monitory accounts – and there had been many of them – there was a blind spot, a lack of candour, a defect in social responsibility, an evident aversion to pointing to the direction from which doom approached. It seemed that we were expected to suppose that the Indians were simply fading away, killed off by the harsh climate of the times, and we were invited to enquire no further. It was left to the Brazilian Government itself to resolve the mystery, and in March 1968 it did so, with brutal frankness, and with almost no attempt at self-defence. The tribes had been virtually exterminated, not *despite* all the efforts of the Indian Protection Service, but with its *connivance* – often its ardent co-operation.

General Albuquerque Lima, the Brazilian Minister of the Interior, admitted that the Service had been converted into an instrument for the Indians' oppression, and had therefore been dissolved. There was to be a judicial inquiry into the conduct of 134 functionaries. A full newspaper page in small print was required to list the crimes with which these men were charged. Speaking informally, the Attorney General, Senhor Jader Figueiredo, doubted whether ten of the Service's employees out of a total of over a thousand would be fully cleared of guilt.

The official report was calm – phlegmatic almost – all the more effective therefore in its exposure of the atrocity it contained. Pioneers leagued with corrupt politicians had continually usurped Indian lands, destroyed whole tribes in a cruel struggle in which bacteriological warfare had been employed, by issuing clothing impregnated with the virus of smallpox, and poisoning food supplies. Children had been abducted and mass murder gone unpunished. The Government itself was blamed to some extent for the Service's increasing starvation of resources over a period of thirty years. The Service had also had to face 'the disastrous impact of missionary activity'.

Next day the Attorney General met the Press, and was prepared to supply all the details. A commission had spent 58 days visiting Indian Protection Service posts all over the country collecting evidence of abuses and atrocities.

The huge losses sustained by the Indian tribes in this tragic decade were catalogued in part. Of 19,000 Munducurus believed to have existed in the 1930s, only 1,200 were left. The strength of the Guaranis had been reduced from 5,000 to 300. There were 400 Carajas left out of 4,000. Of the Cintas Largas, who had been attacked from the air and driven into the mountains, possibly 500 had survived out of 10,000. The proud and noble nation of the Kadiweus – 'the Indian Cavaliers' – had shrunk to a pitiful scrounging band of about two hundred. Only a few hundred remained of the formidable Chavantes who prowled in the background of Peter Fleming's *Brazilian Journey*, but they had been reduced to mission fodder – the same melancholy fate that had overtaken the Bororos, who helped to change Lévi-Strauss's views on the nature of human evolution. Many tribes were now represented by a single family, a few by one or two individuals. Some, like the Tapaiunas, had disappeared altogether – in this case from a gift of sugar laced with arsenic. It is estimated that only between 50,000 and 100,000 Indians survive today.

Senhor Figueiredo estimated that property worth 62 million dollars had been stolen from the Indians in the past ten years.

He added, 'It is not only through the embezzlement of funds, but by the admission of sexual perversions, murders and all other crimes listed in the penal code against Indians and their property, that one can see that the Indian Protection Service was for years a den of corruption and indiscriminate killings.' The head of the service, Major Luis Neves, was accused of 42 crimes, including collusion in several murders, the illegal sale of lands, and the embezzlement of 300,000 dollars. Senhor Figueiredo informed the newspapermen that the documents containing the evidence collected by the Attorney General weighed 103 kilograms, and amounted to a total of 5,115 pages.

In the following days there were more headlines and more statements by the Ministry:

Rich landowners of the municipality of Pedro Alfonso attacked the tribe of Craos and killed about a hundred.

The worst slaughter took place in Aripuaná, where the Cintas Largas Indians were attacked from the air using sticks of dynamite.

The Maxacalis were given fire-water by the landowners who employed gunmen to shoot them down when they were drunk.

Landowners engaged a notorious *pistoleiro* and his band to massacre the Canelas Indians.

The Nhambiquera Indians were mown down by machine-gun fire.

Two tribes of the Patachós were exterminated by giving them smallpox injections.

In the Ministry of the Interior it was stated yesterday that crimes committed by certain ex-functionaries of the IPS amounted to more than a thousand, ranging from tearing out Indians' finger-nails to allowing them to die without any relief.

To exterminate the tribe Beiços de Pau, Ramis Bucair, Chief of the 6th Inspectorate, explained that an expedition was formed which went up the River Arinos carrying presents and a great quantity of foodstuffs for the Indians. These were mixed with arsenic and formicides.... Next day a great number of the Indians died, and the whites spread the rumour that this was the result of an epidemic.

As ever, the frontiers with Colombia and Peru (scene of the piratical adventures of the old British-registered Peruvian Amazon Company) gave trouble. A minor boom in wild rubber set off by the last war had filled this area with a new generation of men with hearts of flint. In the 1940s one rubber company punished those of their Indian slaves who fell short in their daily collection by the loss of an ear for the first offence, then the loss of the second ear, then death. When chased by Brazilian troops, they simply moved, with all their labour, across the Peruvian border. Today, most of the local landowners are slightly less spectacular in their oppressions. One landowner is alleged to have chained lepers to posts, leaving them to relieve themselves where they stood, without food or almost any water for a week. He was a bad example, but his method of keeping the Ticuna Indians in a state of slavery was the one commonly in use. They were paid half a cruzeiro for a day's labour and then charged three cruzeiros for a piece of soap. Those who attempted to escape were arrested (by the landowner's private police force) as thieves.

Senhora Neves da Costa Vale, a delegate of the Federal Police who investigated this case, and the local conditions in general, found that little had changed since the bad old days. She noted that hundreds of Indians were being enslaved by landowners on both sides of the frontier, and that Colombians and Peruvians hunted for Ticuna Indians up the Brazilian rivers. Semi-civilised Indians, she said, were being carried off for enrolment as bandits in Colombia. The area is known as Solimões, from the local name of the Amazon, and Senhora Neves was shocked by the desperate physical condition of the Indians. Lepers were plentiful, and she confirmed the existence of an island called Armaça, where Indians who were old or sick were concentrated to await death. She said that they were without assistance of any kind.

From all sources it was a tale of disaster. No one knew just how many Indians had survived, because there was no way of counting them in their last mountain and forest strongholds. The most optimistic estimate put the figure at 100,000, but others thought they might be as few as half this number. Nor could more than the roughest estimate be made of the speed of the process of extermination. All accounts suggest that when the Europeans first came on the scene four centuries ago they found a dense and lively population. Fray Gaspar, the diarist of Orellana's expedition in 1542, claims that a force of fifty thousand once attacked their ship. At that time the experts believe that the Indians may have numbered between three and six million. By 1900, the same authorities calculate, there may have been a million left. But in reality, it is all a matter of guesswork.

The first Europeans to set eyes on the Indians of Brazil came ashore from the fleet of Pedro Álvares Cabral in the year 1500 to a reception that enchanted them, and when the ships set sail again they left with reluctance.

Pêro Vaz de Caminha, official clerk to the expedition, sent off a letter to the King that crackled with enthusiasm. It was the fresh-eyed account of a man released from the monotony of the seas to miraculous new experiences that might have been written to any crony back in his hometown. Nude ladies had paraded on the beach splendidly indifferent to the stares of the Portuguese sailors – and Caminha took the King by the elbow to go into their charms at extraordinary length. The Indian girls were fresh from bathing in the river and devoid of body hair. Caminha describes their sexual attractions with minute and sympathetic detail adding that their genitalia would put any Portuguese lady to shame. In those days Europeans rarely washed (a treatise on the avoidance of lousiness was a best seller), so one supposes that the Portuguese were frequently verminous in these regions. Caminha cannot avoid coming back to the subject again before settling to prosaic details of the climate and produce of the newly discovered land. 'Sweet girls,' he says.... 'Like wild birds and animals. Lustrous in a way that so far outshines those in captivity – they could not be cleaner, plumper and more vibrant than they are.'

The Europeans were overwhelmed, too, by the magnificence of the Indians' manners. If they admired any of the necklaces or personal adornments of feathers or shells these were instantly pressed into their hands. In other encounters it was to be the same with golden trinkets, and temporary wives were always to be had for the taking. The bolder of the women came and rubbed themselves against the sailors' legs, showing their fascination at the instant and unmistakable sexual response of the white men.

Such open-handedness was dazzling to these representatives of an inhibited but fanatically acquisitive society. The official clerk filled page after page with a catalogue of Indian virtues. All that was necessary to complete this image of the perfect human society was a knowledge of the true God. And since these people were not circumcised, it followed that they were not Mohammedans or Jews, and that there was nothing to impede their conversion. When the first Mass was said the Indians, with characteristic politeness and tact, knelt beside the Portuguese and, in imitation of their guests, smilingly kissed the crucifixes that were handed to them. As discussion was limited to gestures the Portuguese suspected their missionary labours were incomplete, and when the fleet sailed, two convicts were left behind to attend to the natives' conversion.

It was Caminha's letter that encouraged Voltaire to formulate his theory of the Noble Savage. Here was innocence – here was apparent freedom, even, from the curse of original sin. The Indians, said the first reports, knew of no crimes or punishments. There were no hangmen or torturers among them; no destitute. They treated each other, their children – even their animals – with constant affection. They were to be sacrificed to a process that was beyond the control of these admiring visitors. Spain and Portugal had become parasitic nations who could no longer feed themselves.

The fertile lands at home had been abandoned, the irrigation systems left by the Moors were fallen into decay, the peasants dragged away to fight in endless wars from which they never returned. Economic forces that the newcomers could never have understood were about to transform them into slavers and assassins. The natives gave gracefully, and the invaders took what they offered with grasping hands, and when there was nothing left to give the enslavement and the murder began. The American continent was about to be overwhelmed by what Claude Lévi-Strauss described four hundred years later as 'that monstrous and incomprehensible cataclysm which the development of Western civilisation was for so large and innocent a part of humanity'.

Caminha and his comrades landed at Porto Seguro, about five hundred miles up the coast from the present Rio de Janeiro, and it is no more than a coincidence that a handful of Indians have somehow succeeded in surviving to this day at Itabuna, which is nearby. The continued presence of these Tapachós is something of a mystery, because for four centuries the area has been ravaged by slavers, belligerent pioneers and bandits of all descriptions. The survivors are found in a swarthy, austere landscape, hiding in the ligaments of bare rock, in the crevices of which they have developed an aptitude for concealment; furtive creatures in tropical tatters, scuttling for cover as they are approached. One sees them in patches of wasteland by the roadside or railway track, which they fertilise by their own excrement to grow a few vegetables before moving on. Otherwise they eke out a sub-existence by selling herbal recipes and magic to neurotic whites who visit them in secret, also by a little prostitution and a little theft. They suffer from tuberculosis, venereal disease, ailments of the eye, and from epidemics of measles and influenza, the last two of which adopt particularly lethal forms.

Two of their tribes held on through thick and thin to a little of their original land until ten years ago when a doctor – now alleged to have been sent by the Indian Protection Service of those days – instead of vaccinating them, inoculated them with the virus of smallpox. This operation was totally successful in its aim, and the vacant land was immediately absorbed into the neighbouring white estates.

There are a dozen such dejected encampments along three thousand miles of coastline, and they are the last of the coastal Indians of the kind seen by Caminha, who once appeared from among the trees by their hundreds whenever a ship anchored offshore. The Patachós are officially classified as *integrados*. It is the worst label that can be attached to any Indian, as extinction follows closely on the heels of integration.

The atrocities of the conquistadores described by Bishop Bartolomeo de Las Casas, who was an eyewitness of what must have been the greatest of all wars of extermination, resist the imagination. There is something remote and shadowy about horror on so vast a scale. Numbers begin to mean nothing, as one reads with a sort of detached, unfocused belief of the mass burnings, the flaying, the disembowelling, and the mutilations.

Some twelve million were killed, according to Las Casas, most of them in frightful ways. 'The Almighty seems to have inspired these people with a meekness and softness of humour like that of lambs; and the conquerors who have fallen upon them so fiercely resemble savage tigers, wolves and lions.... I have seen the Spaniards set their fierce and hungry dogs at the Indians to tear them in pieces and devour them.... They set fire to so many towns and villages it is impossible I should recall the number of them.... These things they did without any provocation, purely for the sake of doing mischief.' Wherever they could be reached, in the Caribbean islands, and on the coastal plains, the Indians were exterminated. Those of Brazil were saved from extinction by a tropical rainforest, as big as Europe, and to the south of it, the half million square miles of thicket and swampland – the Mato Grosso – that remained sufficiently mysterious for quite recent explorers, such as Colonel Fawcett, to lose their lives searching for golden cities.

For those who pursued the Indians into the forest there are worse dangers to face than poison-tipped arrows. Jiggers deposited their eggs under their skin; there was a species of fly that fed on the surface of the eye and could produce blindness; bees swarmed to fasten themselves to the traces of mucus in the nostrils and at the corners of the mouth; fire ants could cause temporary paralysis, and worst of all, a tiny beetle sometimes found in the roofs of abandoned huts might drop on the sleeper to administer a single fatal bite.

There were also the common hazards of poisonous snakes, spiders and scorpions in variety, and the rivers contained not only piranhá, electric eels and stingrays, but also a tiny catfish with spiny fins which allegedly wriggled into the human orifices. Above all, the mosquitoes transmitted not only malaria, but the yellow fever endemic in the blood of many of the monkeys. The only non-Indians to penetrate the ultimate recesses of the forest were the black slaves of later invasions, who escaped in great numbers from the sugar estates and mines to form the *quilombas*, the fugitive slave settlements. But these, apart from helping themselves to Indian women, where they found them, followed the rule of live and let live. They merged with the surrounding tribes, and lost their identity.

The processes of murder and enslavement slowed down during the next three centuries, but did so because there were fewer Indians left to murder and enslave. Great expeditions to provide labour for the plantations of Maranhão and Pará depopulated all the easily accessible villages near the main Amazonian waterways, and the loss of life is said to have been greater than that involved in the slave trade with Africa. Those who escaped the plantations often finished in the Jesuit reservations – religious concentration camps where conditions were hardly less severe, and trifling offences were punished with terrible floggings or imprisonment: 'The sword and iron rod are the best kind of preaching,' as the Jesuit missionary José de Anchieta put it.

By the nineteenth century some sort of melancholy stalemate had been reached. Indian slaves were harder to get, and with the increasing rationalisation of supply and the consequent fall in cost of black slaves from West Africa – who in any case stood up to the work more robustly – the price of the local product was undercut. As the Indians became less valuable as a commodity, it became possible to see them through a misty Victorian eye, and at least one novel about them was written, swaddled in sentiment, and in the mood of *The Last of the Mohicans*. A more practical viewpoint reasserted itself at the time of the great rubber boom at the turn of the century, when it was discovered that the harmless and picturesque Indians were better equipped than black slaves to search the forests for rubber trees. While the eyes of the world were averted, all the familiar tortures and excesses were renewed, until with the collapse of the boom and the revival of conscience, the Indian Protection Service was formed.

In the raw, abrasive vulgarity that it displayed in its consumption of easy wealth, the Brazilian rubber boom surpassed anything that had been seen before in the Western world since the days of the Klondike. It was centred on Manaus which had been built where it was at the confluence of two great, navigable rivers, the Amazon and the Rio Negro, for its convenience in launching slaving expeditions. It was a city that had fallen into a decline that matched the wane in interest for its principal commodity.

With the invention of the motorcar and the rubber tyre, and the recognition that the *hevea* tree of the Amazon produced incomparably the best rubber, Manaus was back in business, converted instantly to a tropical Gomorrah. Caruso refused a staggering fee to appear at the opera house, but Madame Patti accepted. There were Babylonian orgies of the period, in which courtesans took semi-public baths in champagne, which was also awarded by the bucketful to winning horses at the races. Men of fashion sent their soiled linen to Europe to be laundered. Ladies had their false teeth set with diamonds, and among exotic importations was a regular shipment of virgins from Poland.

The most dynamic of the great rubber corporations of those days was the British-registered Peruvian Amazon Company, operating in the ill-defined north-western frontier of Brazil, where it could play off the governments of Colombia, Peru and Brazil against each other, all the better to establish its vast, nightmarish empire of exploitation and death.

A young American engineer, Walter Hardenburg, carried accidentally in a fit of wanderlust over the company's frontier, was immediately seized and imprisoned for a few days during which time he was given a chance to see the kind of thing that went on. Several thousand Huitoto Indians had been enslaved and at the post where Hardenburg was held, El Encanto (Enchantment), he saw the rubber tappers bringing back their collection of latex at the end of the day. Their bodies were covered with great raised weals from the overseers' tapir-hide whips, and Hardenburg noticed that the Indians who had managed to collect their quota of rubber danced with joy, whereas those who had failed to do so seemed terror-stricken, although he was not present to witness their punishment. Later he learned that repeated deficiencies in collection could mean a sentence of a hundred lashes, from which it took as much as six months to recover.

An element of competition was present when it came to killing Indians. On one occasion one hundred and fifty hopelessly inefficient workers were rounded up and slashed to pieces by *macheteiros* employing a grisly local expertise, which included the *corte do bananeiro*, a backward and forward swing of the blade which removed two heads at one blow, and the *corte maior*, which sliced a body into two or more parts before it could fall to the ground. High feast days, too, were celebrated by sporting events when a few of the more active – and therefore more valuable – tappers might be sacrificed to make an occasion. They were blindfolded and encouraged to do their best to escape while the overseers and their guests potted at them with their rifles.

Barbadian British subjects were recruited by the Peruvian Amazon Company as the hunters of wild Indians, being sent on numerous expeditions into those areas where the company proposed to establish new rubber trails. These were paid on the basis of piecework and were obliged to collect the heads of their victims, and return with them as proof of their claims to payment. Stud farms existed in the area where selected Indian girls would breed the slave-labour of the future when the wild Indian had been wiped out. Some rubber companies have been suspected, too, of not stopping short of cannibalism, and there were strong rumours of camps in which ailing and unsatisfactory workers were used to supply the tappers' meat.

The worldwide scandal of the Peruvian Amazon Company, exposed by Sir Roger Casement, coincided with the collapse of the rubber boom caused by the competition of the new Malayan plantations, and a crisis of conscience was sharpened by the threat of economic disaster. The instant bankruptcy of Manaus was attended by spectacular happenings. Sources of cash suddenly dried up, and the surplus population of cardsharpers, adventurers and whores poured into the river steamers in the rush to escape to the coast. They paid for the passages with such possessions as diamond cufflinks and solitaire rings. Merchant princes with their fortunes tied up in unsaleable rubber committed suicide. The celebrated electric street cars – first of their kind in Latin America – came suddenly to a halt as the power was cut off, and were set on fire by their enraged passengers. A few racehorses found themselves between the shafts of converted bullock carts. The opera house closed, never to open again.

When the Brazilians had got used to the idea that their rubber income was substantially at an end, they began to examine the matter of its cost in human lives in the light of the fact, now generally known, that the Peruvian Amazon Company alone had murdered nearly thirty thousand Indians. Brazil was now Indian-conscious again and its legislators reminded each other of the principles so nobly enunciated by José Bonifacio in 1823, and embodied in the constitution: 'We must never forget,' Bonifacio said, 'that we are usurpers, in this land, but also that we are Christians.'

It was a mood responsible for the determination that nothing of this kind should ever happen again, and an Indian Protection Service – unique and extraordinary in its altruism in the Americas – was founded in 1910 under the leadership of Marshall Rondon, himself an Indian, and therefore, it was supposed, exceptionally qualified to be able to interpret the Indian's needs.

Rondon's solution was to integrate the Indian into the mainstream of Brazilian life – to educate him, to change his faith, to break his habit of nomadism, to change the colour of his skin by inter-marriage, to draw him away from the forests and into the cities, to turn him into a wage-earner and a voter. He spent the last years of his life trying to do this, but just before his death came a great change of heart. He no longer believed that integration was to be desired. It had all been, he said now, a tragic mistake.

The conclusion of all those who have lived among and studied the Indian beyond the reach of Western civilisation is that he is the perfect human product of his environment – from which it should follow that he cannot be removed without calamitous results. Ensconced in the forest in which his ancestors have lived for thousands of years, he is as much a component of it as the tapir and the jaguar: self-sufficient, the artificer of all his requirements, integrated with his surroundings, deeply conscious of his place in the living patterns of the visible and invisible universe.

It is admitted now that the average Indian Protection Service official recruited to deal with this complicated but satisfactory human being was all too often venal, ignorant and witless. It was inevitable that he should call to his aid the missionaries who were in Brazil by the thousand, and were backed by resources that he himself lacked. But the missionary record was not an impressive one, and even those incomparable colonisers of the faith, the Jesuits, had little to show but failure.

In the early days they had put their luckless converts into long white robes, segregated the sexes, and set them to 'godly labours', lightened by the chanting of psalms in Latin, as well as mind-developing exercises in mnemonics, and speculative discussions on such topics as the number of angels able to perch on the point of a pin. It was offered as a foretaste of the delights of the Christian heaven, complete with its absence of marrying or giving in marriage. Many of the converts died of melancholy. After a while demoralisation spread to the fathers themselves and some of them went off the rails to the extent of dabbling in the slave trade. When these missionary settlements were finally overrun by the bloodthirsty pioneers and frontiersman from São Paulo, death can hardly have been more than a happy release for the listless and bewildered Indian flock.

When the Indian Protection Service was formed the missionaries of the various Catholic orders were rapidly being outnumbered by nonconformists, mostly from the United States. These were a very different order of man, no longer armed only with hellfire and damnation, but with up-to-date techniques of salesmanship in their approach to the problems of conversion. By 1968 the *Jornal de Brasil* could state:

> In reality, those in command of these Indian Protection posts are North American missionaries – they are in all the posts – and they disfigure the original Indian culture and enforce the acceptance of Protestantism.

Whereas the Catholics for all their disastrous mistakes, had mostly led simple, often austere lives, the nonconformists seemed to see themselves as the representatives of the more ebullient and materialistic brand of the faith. They made a point of installing themselves, wherever they went, in large, well-built stone houses, inevitably equipped with an electric generator and every modern labour-saving device. Some of them even had their own planes. If there were roads they had a car or two, and when they travelled by river they preferred a launch with an outboard engine to the native canoe habitually used by the Catholic fathers.

As soon as Indians were attracted to the neighbourhood, a mission store might be opened, and then the first short step towards the ultimate goal of conversion be taken by the explanation of the value and uses of money, and how with it the Indian could obtain all those goods which it was hoped would become necessary to him. The missionaries are absolutely candid and self-congratulatory about their methods. To hold the Indian, wants must be created and then continually expanded – wants that in such remote parts only the missionary can supply. A greed for unessential trifles must be inculcated and fostered.

The Portuguese verb employed to describe this process is *conquistar* and it is applied with differentiation to subjection by force or guile. What normally happens is that presents – usually of food – are left where the 'uncivilised' Indians can find them. Great patience is called for. It may be years before the tribesmen are won over by repeated overtures, but when it happens the end is in sight. All that remains is to encourage them to move their village into the mission area, and let things take their natural course.

In nine cases out of ten the local landowner has been waiting for the Indians to make such a move – he may have been alerted by the missionary himself – and as soon as it happens he is ready to occupy the tribal land. The Indians are now trapped. They cannot go back, but at the time it seems unimportant, because for a little longer the missionary continues to feed them, although now the matter of conversion will be broached. This usually presents slight difficulty: natural Indian politeness – and in this case gratitude – accomplishes the rest. Whether the Indian understands what it is all about is another matter. He will be asked to go through what he may regard with great sympathy as a rain-making ceremony, as water is splashed about, and formulae repeated in an unknown language. Beyond that it is likely to be a case of let well alone. Any missionary will tell you that an Indian has no capacity for abstract thought. How can he comprehend the mystery and universality of God when the nearest to a deity his own traditions have to offer may be a common tribal ancestor seen as a jaguar or an alligator?

From now on the orders and the prohibitions will flow thick and fast. The innocence of nudity is first to be destroyed, and the Indian who has never worn anything but a beautifully made and decorated penis-sheath to suppress unexpected erections, must now clothe himself from the mission's store of cast-offs to the instant detriment of his health. He becomes subject to skin diseases, and since in practice clothes once put on are never taken off again, pneumonia is the frequent outcome of allowing clothing to dry on the body after a rainstorm.

The man who has hitherto lived by practising the skills of the hunter and horticulturist – the Indians are devoted and incomparable gardeners of their kind – now finds himself, broom or shovel in hand as an odd-job man about the mission compound. He shrinks visibly within his miserable, dirty clothing, his face becomes puckered and wizened, his body more disease-ridden, his mind more apathetic. There is a terrible testimony to the process in the Brazilian Ministry of Agriculture's handbook on Indians, in which one is photographed genial and smiling on the first day of his arrival from the jungle, and then the same man who by this time appears to be crazy with grief is shown again, ten years later. 'His expression makes comment unnecessary,' the caption says. 'Ninety percent of his people have died of influenza and measles. Little did he imagine the fate that awaited them when they sought their first contact with the whites.'

There is a ring about these stories of enticement down the path to extinction, of the cruel fairytale of children trapped by the witch in the house made of gingerbread and barley sugar. But even the slow decay, the living death of the missionaries' compound was not the worst that could happen. What could be far more terrible would be the decision of the *fazendeiro* – as so often happened – to recruit the labour of the Indians whose lands he had invaded, and who are left to starve.

Extract from the atrocity commission's report:

> In his evidence Senhor Jordao Aires said that eight years previously the six hundred Ticuna Indians were brought by Fray Jeremias to his estate. The missionary succeeded in convincing them that the end of the world was about to take place, and Belem was the only place where they would be safe ... Senhor Aires confirmed that when the Indians disobeyed his orders his private police chained them hand and foot. Federal Police Delegate Neves said that some of the Indians thus chained were lepers who had lost their fingers.

Officially it is the Indian Protection Service and 134 of its agents that are on trial, but from all these reports the features of a more sinister personality soon emerge, the *fazendeiro* – the great landowner – and in his shadow the IPS agent shrinks to a subservient figure, too often corrupted by bribes.

One would have wished to find an English equivalent for this Portuguese word *fazendeiro*, but there is none. Titles such as landowner or estate owner, which call to mind nothing harsher than the mild despotism of the English class system, will not do. The *fazendeiro* by European standards is huge in anachronistic power, often the lord of a tropical fief as large as an English county, protected from central authority's interference by vast distances, traditions of submission, and the absolute silence of his vassals. All the lands he holds – much of which may not even have been explored – have been taken by him or by his ancestors from the Indians, or have been bought from others who have obtained them in this way. In most cases his great fortress-like house, the *fazenda*, has been built by the labour of the Indian slaves, who have been imprisoned when necessary in its dungeons. In the past a *fazendeiro* could only survive by his domination of a ferocious environment, and although in these days he will probably have had a university education, he may still sleep with a loaded rifle beside his bed. Lonely *fazendas* are still occasionally attacked by wild Indians (i.e. Indians with a grievance against the whites), by gold prospectors turned bandit, by downright professional bandits themselves, or by their own mutinous slaves. The *fazendeiro* defends himself by a bodyguard enrolled from the toughest of his workers – in the backwoods many of them are fugitives from justice.

It has often been hard by ordinary Christian standards for the *fazendeiro* to be a good man, only too easy for him to degenerate into a Gilles de Raïs, or some murderous and unpredictable Ivan the Terrible of the Amazon forests. It can be Eisenstein's *Thunder Over Mexico* complete with the horses galloping over men buried up to their necks – or worse. Some of the stories told about the great houses of Brazil of the last century in their days of respectable slavery and Roman licence bring the imagination to a halt: a male slave accused of some petty crime castrated and burned alive ... a pretty young girl's teeth ordered by her jealous mistress to be drawn, and her breasts amputated, to be on the safe side ... another, found pregnant, thrown alive into the kitchen furnace.

An extract from the report by the President of last year's inquiry commission into atrocities against the Indians corrects the complacent viewpoint that we live in milder days.

> In the 7th Inspectorate, Paraná, Indians were tortured by grinding the bones of their feet in the angle of two wooden stakes, driven into the ground. Wives took turns with their husbands in applying this torture.

It is alleged, as well, in this investigation, that there were cases of an Indian's naked body being smeared with honey before leaving him to be bitten to death by ants.

Why all this pointless cruelty? What is it that causes men and women probably of extreme respectability in their everyday lives to torture for the sake of torturing? Montaigne believed that cruelty is the revenge of the weak man for his weakness, a sort of sickly parody of valour. 'The killing after a victory is usually done by the rabble and baggage officials.'

It is the beginning of the rainy season, and from an altitude of 2,000 feet the forest smokes here and there as if under sporadic bombardment, while the sun sucks up the vapour from the local downpour.

The Mato Grosso seen from the air is supposed to offer a scene of monotonous green, but this is not always so. At this moment, for example, a pitch-black swamp lapped by ivory sands presents itself. It is obscured by shifting feathers of cloud, which part again to show a Cheddar Gorge in lugubrious reds. The forest returns, pitted with lakes which appear to contain not water but brilliant chemical solutions: copper sulphate, gentian violet. The air taxi settles wobbling to a scrubbed patch of earth where vultures flutter like black rags.

All these small towns in this meagre earth are the same. A street of clapboard, tapering off to mud and palm thatch at each end; a general store, a hotel, Laramie-style with men asleep on the veranda; a scarecrow horse with bones about to burst through the hide, tied up in a square yard of shade; hairy pigs; aromatic dust blown up by the hot breeze.

Life is in slow motion and on a small scale. The store sells cigarettes, meticulously bisected if necessary with a razor blade, ladlefuls of mandioca flour, little piles of entrails for soup, purgative pills a half-inch in diameter, and handsomely tooled gun holsters. The customers enter not so much to buy but to be there, wandering through the paperchains of dusty dried fish hanging from the ceiling. They are Indians, but so de-racialised by the climate of boredom and their grubby cotton clothing, that they could be Eskimos or Vietnamese. They have the expression of men gazing, narrow-eyed, into crystal balls, and they speak in childish voices of great sweetness. Like Indians everywhere, the smallest intake of alcohol produces an instant deadly change.

The only entertainment the town offers is a cartomancer, operating largely on a barter basis. He tells fortunes in a negative but realistic way, concerned not so much with good luck, but the avoidance of bad. All the children's eyes are rimmed with torpid, hardly moving flies. The *fazenda*, some miles away, has absorbed everything; owns the whole town, even the main street itself.

This is a place where cruelty is supposed to have happened, but the surface of things has been patched and renovated and the aroma of atrocity has dispersed. Everything can now be explained away as examples of extreme exaggeration, or the malice of political enemies, and all the witnesses for the defence have been mustered. Finally, the everyday violences of a violent country are quoted to remind one that this is not Europe.

Senhor Fulano lives with his family in three rooms in one of the few brick-built houses. His position is ambiguous. An ex-Indian Protection Service agent, he has been cleared of financial malpractices, and hopes shortly for employment in the new Foundation. He has an Abyssinian face with melancholy, faintly disdainful eyes, a high Nilotic forehead, and a delicate Semite nose. He is proud of the fact that his father was half Negro, half Jewish; a trader who captured in marriage a robust girl from one of the Indian tribes.

'Not all *fazendeiros* are bad,' Fulano says. 'Far from it. On the contrary, the majority are good men. People are jealous of their success, and they are on the lookout for a way to damage them.

'In the case you mention the man was a thief and a troublemaker. As a punishment he was locked in the shed, nothing more. He was drunk, you understand, and he set fire to the shed himself. He died in the fire, yes, but the doctor certified accidental death. There was no case for a police inquiry. In thirty years' of service I have only seen one instance of violence – if you wish to call it violence. The Indians were drunk with *cachaça* again, and they attacked the post.

They were given a chance by firing over their heads, but it didn't stop them. They were mad with liquor. What could we do? There's no blood on my hands.' He holds them up as if a confirmation. They are small and well cared for with pale, pinkish palms. His wife rattles about out of sight in the scullery of their tiny flat. A picture of the President hangs on the wall, and another of Fulano's little girl dressed for her first communion, and there is no evidence in the cheap, ugly furniture that Senhor Fulano has been able to feather his nest to any useful extent.

He joined the service out of a sense of vocation, he says. 'We were all young and idealistic. They paid us less than they paid a post man, but nobody gave any thought to that. We were going to dedicate our lives to the service of our less fortunate fellow men. If anyone happened to live in Rio de Janeiro, the Minister himself would see him when he was posted, shaking hands with him and wish him good luck. I happened to be a country boy, but my friends hired a band to see me off to the station. Everybody insisted in giving me a present. I had so many lace handkerchiefs I could have opened a shop. There was a lot of prestige in being in the service in those days.'

There are three whitish, glossy pockmarks in the slope of each cheek under the sad, Amharic eyes, and it is difficult not to watch them. He shakes his head. 'No one would believe the conditions some of us lived under. They used to show you photographs of the kind of place where you'd be working: a house with a veranda, the school and the dispensary. When I went to my first post I wept like a child when I saw it. The journey took a month and in the meanwhile the man I was supposed to be assisting had died of the smallpox. I remember the first thing I saw was a dead Indian in the water where they tied up the boat. I'd hit a measles epidemic. Half the roof of the house had caved in. There never had been a school, and there wasn't a bottle of aspirin in the place. When the sun went down the mosquitoes were so thick, they were on your skin like fur.'

He finds an album of press-cuttings in which are recorded the meagre occasions of his life. A picture shows him in dark suit and stiff collar receiving a certificate and the congratulations of a politician for his work as a civiliser. In another he is shown posing at the side of Miss Pernambuco 1952, and in another he is a paternal presence at a ceremony when a newly pacified tribe are to put on their first clothing. There are 'before' and 'after' pictures of the tribal women, first naked and then in jumpers and skirts, not only changed but facially unrecognisable from one moment to the next, as if some malignant spell had been laid upon them as they wriggled into the shapeless garments. The few cuttings, which I scan through out of politeness, speak of Senhor Fulano as the pattern of self-abnegation, and the words *servicio* and *devoçao* constantly reappear. 'My trifling pay was only one hundred cruzeiros a month,' he says, 'and it was sometimes up to six months overdue. In just the first year I caught measles, jaundice and malaria three times. If it hadn't been for the *fazendeiro*, I'd have died. He looked after me like a father. He was a man of the greatest possible principles, and among many other benefactions he gave 100,000 cruzeiros to a church in Salvador. I see now that his son has been formally charged with invading Indian lands. All I can say to that is, what the Indians would do without him, I don't know.'

Fulano is nothing if not loyal. '*Fazendeiros* are no different from anyone else,' he says. 'They try to make out they're monsters these days. You mustn't believe all you read.'

It was certain that no one would be found now in this town to contradict him.

For a half-century rubber had been the great destroyer of the Indian, and then suddenly it changed to speculation in land. Rumour spread of huge mineral resources awaiting exploitation in the million square miles that were inaccessible until recently – and the great speculative rush was on. Nowhere, however remote, however sketchily mapped, was secure from the surveyors, who were sent out to measure out the claims of the politicians, the real-estate companies and the *fazendeiros*. Back in São Paulo, the headquarters of the land boom, the *grileiro* – specialist in shady land deals – went into secret partnerships with his friend in the Government, who was well positioned to see that the deals went through. A great deal of this apparently empty land was only empty to the extent that it contained no white settlements, and the map-makers had not yet put in the rivers and the mountains. There might well be Indians there – nobody knew until it had been explored – but this possibility introduced only a slight inconvenience. In theory the undisturbed possession of all land occupied by Indians is guaranteed to them by the Brazilian constitution, but if it can be shown that Indian land has been abandoned it reverts to the Government, after which it can be sold in the ordinary way. The *grileiro*'s task is to discover or manufacture evidence that such land is no longer in occupation – a problem, if sincerely confronted, complicated by the fact that most Indians are semi-nomadic, cultivating crops in one area during the period of the summer rains, then moving elsewhere to hunt and fish during the dry winter season.

A short cut to the solution of the problem is simply to drive the Indians out. Other *grileiros* quite simply ignore its existence, offering land to the gullible by map reference, sight unseen, and hoping to be able to settle the legal difficulties by political manipulations at some later date.

The *grileiro* with his manoeuvrings behind the scenes was kept under some control while President João Goulart was in power, and it finally became clear to the big-scale land speculators that they were going to get nowhere until they got a new President. Goulart, although a rich landowner himself, held the opinion that Brazil would never occupy the place in the Western Hemisphere to which its colossal size and resources entitled it, while it limped along in its feudalistic way with an 86 percent illiteracy figure and the land in the hands of an infinitesimally small minority, many of which made no effort to develop it in any way. The remedy he proposed was to redistribute 3 percent of privately owned land, but also – what was far more serious – he announced the resuscitation of an old law permitting the Government to nationalise land up to six miles in depth on each side of the national means of communication – roads, railways and canals.

This would have been a death blow to the speculators, who hoped to resell their land at many times the price they had paid, as soon as it was made accessible by the building of roads. One such firm had advertised 100,000 acres of land for sale in the English Press. The land was offered in 100-acre minimum lots at £5 an acre. An initial purchase of land had already been sold, the company announced, 'mainly to investment houses and trusts, insurance companies and a number of syndicates'. A charter flight would be arranged for buyers from Manchester, Birmingham, Glasgow, Edinburgh and Liverpool, and representatives of Kenya farmers who had already bought 50,000 acres. 'There is little hope,' said the promotion literature, 'of any return from the purchase of the land for a few years yet.'

But in 1964 the speculative prospects brightened enormously when a *coup d'état* was staged to depose the troublesome Goulart, and the land rush could go ahead. A promotional assault was launched on the United States market with lavishly produced and cunningly worded brochures offering glamour as well as profit, being phrased in the poetic style of American car advertisements. Amazon Adventure Estates were offered, and there were allusions to monkeys and macaws and the occult glitter of gems in the banks of mighty rivers sailed by the ships of Orellana. They had some success. Several film stars took a gamble in the Mato Grosso. In April 1968, in fact, a Brazilian deputy, Haroldo Veloso, revealed that most of the area of the mouth of the Amazon had passed into the hands of foreigners. He mentioned that Prince Rainier of Monaco had bought land in the Mato Grosso twelve times larger than the principality. Someone, presumably stabbing with a pencil at a map, had picked up the highest mountain in Brazil – the Pico de Nieblina – for a song, although it would have taken a properly equipped expedition a matter of weeks to reach it.

This was doomsday for the tribes who had been pacified and settled in areas where they could be conveniently dealt with. Down in the plains on the frontiers with Paraguay it was the end of the road for the Kadiweus. In 1865 in the war against Paraguay they had taken their spears and ridden naked, barebacked, but impeccably painted – a fantastic Charge of the Light Brigade, at the head of the Brazilian army – to rout the cavalry of the psychopathic Paraguayan dictator Solano Lopez. For their aid in the war the Emperor Pedro II had received the principal chief, clad for the occasion in a loincloth sewn with precious stones, and granted the Kadiweu nation in perpetuity two million acres of the borderland. Here these Spartans of the West were reduced now to two hundred survivors, working as the cowhands of *fazendeiros* who had taken all their lands.

It was also doomsday for Lévi-Strauss's Bororos. The great anthropologist had lived for several years among them in the 1930s, and they had led him to the conclusions of 'structural anthropology', including the proposition that 'a primitive people is not a backward or retarded people, indeed it may possess a genius for invention or action that leaves the achievements of civilised people far behind'. He had said of the Bororos, 'few people are so profoundly religious ... few possess a metaphysical system of such complexity. Their spiritual beliefs and everyday activities are inextricably mixed.' They had been living for some years now far from the complicated villages where Lévi-Strauss studied them, in the Teresa Cristina reserve in the South Mato Grosso, given them 'in perpetuity', as ever, in tribute to the memory of the great Marshall Rondon, who had been part-Bororo himself.

Life in the reserve was far from happy for the Bororos. They were hunters, and fishermen, and in their way excellent agriculturists, but the reserve was small, and there was no game left and the rivers in the area had been illegally fished-out by commercial firms operating on a big scale, and there was no room to practise cultivation in the old-fashioned semi-nomadic way. The Government had tried to turn them into cattle-raisers, but they knew nothing of cattle. Many of the cows were quietly sold off by agents of the Indian Protection Service, who pocketed the money. Others – as the Bororos had no idea of building corrals – wandered out of the reservation before being impounded by neighbouring *fazendeiros*. The Indians ate the few cows that remained before they could die of disease or starvation, after which they were reduced to the normal diet of hard times – lizards, locusts and snakes – plus an occasional handout of food from one of the missions.

They also suffered from the great emptiness and aimlessness of the Indian whose traditional culture has been destroyed. The missionaries, upon whom they were wretchedly dependent, forbade dancing, singing or smoking, and while they accepted with inbred stoicism this attack on the principle of pleasure, there was a fourth prohibition against which they continually rebelled, but in vain.

The Indians are obsessed by their relationship with the dead, and by the condition of the souls of the dead in the afterlife – a concern reflected in the manner of the ancient Egyptians by the most elaborate funerary rites: orgies of grief and intoxication, sometimes lasting for days. The Bororos, seemingly unable to part with their dead, bury them twice, and the custom is at the emotional basis of their lives. In the first instance – as if in hope of some miraculous revival – the body is placed in a temporary grave, in the centre of the village, and covered with branches. When decomposition is advanced, the flesh is removed from the bones, which are painted and lovingly adorned with feathers, after which final burial takes place in the depths of the forest. The outlawing of this custom by an American missionary reduced the Bororos to despair, but the missionary was able to persuade the local police to enforce the ban, and the party of half-starved tribesmen who dragged themselves two hundred miles on foot to the State capital and presented themselves, weeping, to the *commissario* were turned away.

Final catastrophe followed the devolution by the Federal Government of certain of its powers – particularly those relating to the ownership and sale of land – to the Legislative Assembly of the Mato Grosso State. This at once invoked a law by which land that, after a certain time limit, had not been legally measured and demarcated, reverted to the Government. It was a legal device which saddled Indians, many of whom did not even realise that they were living in Brazil, with the responsibility of employing lawyers to look after their interests. It had been employed once before, and with additional refinements of trickery, in an attempt to snatch away the last of the land of the unfortunate Kadiweus. On this occasion it seems that only two copies of the official publication recording the enactment were available, one of which had been lodged in the State archives, and the other taken the same day to the reserve by the persons proposing to share the land between them.

Hardly less haste was shown in the occupation of the Teresa Cristina reserve. It was a muddled, untidy operation, and it finally turned out that considerably more land had been sold on paper than the actual area of the reserve. This was before the final demoralisation and collapse of the Indian Protection Service. Local officials not only challenged the legality of the sale but called in vain for State troops to be sent to repel an invasion of *fazendeiros,* who were supported by their private armies carrying sub-machineguns.

The state of affairs that had come to pass at Teresa Cristina only five years later, in 1968, is depicted in the testimony of a Bororo Indian girl.

> There were two *fazendas,* one called Teresa, where the Indians worked as slaves. They took me from my mother when I was a child. Afterwards I heard that they hung my mother up all night.... She was very ill and I wanted to see her before she died.... When I got back they thrashed me with a raw-hide whip.... They prostituted the Indian girls.... One day the IPS agent called an old carpenter and told him to make an oven for the farmhouse. When the carpenter had finished the agent asked him what he wanted for doing the job. The carpenter said he wanted an Indian girl, and the agent took him to the school and told him to choose one. No one saw or heard any more of her.... Not even the children escaped. From two years of age they worked under the whip.... There was a mill for crushing the cane, and to save the horses they used four children to turn the mill. They forced the Indian Ottaviano to beat his own mother.... The Indians were used for target practice.

Thus were the Indians disarmed, betrayed, and hustled down the path towards final extinction. Yet in the heart of the Mato Grosso and the Amazon forests, there were tribes that still held out. Classified by the Government manual on Indians as *isolados*, they are described as those that possess the greatest physical vigour. Nobody knows how many such tribes there are. There may be three hundred or more with a total population of fifty thousand, including tiny, self-contained and apparently indestructible nations having their own completely separate language, organisation and customs. Some of these people are giants with herculean limbs, armed with immense longbows of the kind an archer at Crécy might have used. A few groups are ethnically mysterious with blue eyes and fairish hair, provokers of wild theories among Amazonian travellers, that there is one tribe supposed by some to have migrated to these forests some two thousand years ago from the island of Hokkaido in Japan. One common factor unites them all; a brilliant fitness for survival – until now. For 400 years they have avoided the slavers and lived through the epidemics. They have armed themselves with constant alertness. They have been ready to embrace a new tactical nomadism. They have made distrust the greatest of their virtues. Above all, the chieftains have had the intelligence and the strength to reject those deadly offerings left outside their villages by which the whites seek first to buy their friendship, then take away their freedom.

The Cintas Largas were one such tribe living in magnificent if precarious isolation in the upper reaches of the Aripuaná River. There were about five hundred of them, occupying several villages.

They used stone axes, tipped their arrows with curare, caught small fish, played four-feet long flutes made from gigantic bamboos, and celebrated two great annual feasts: one of the initiation of young girls at puberty, and the other of the dead. At both of these they were said to use some unknown herbal concoction to produce ritual drunkenness. They were in a region still dependent for its meagre revenues on wild rubber, and this exposed them to routine attacks by rubber tappers, against whom they had learned to defend themselves. Their tragedy was that deposits of rare metals were being found in the area. What these metals were, it was not clear. Some sort of a security blackout had been imposed, only fitfully penetrated by vague news reports of the activities of American and European companies, and of the smuggling of planeloads of the said rare metals back to the United States.

David St Clair in his book *The Mighty Mighty Amazon* mentions the existence of companies who specialised in dealing with tribes when their presence came to be considered a nuisance, attacking their villages with famished dogs, and shooting down everyone who tried to escape. Such expeditions depended for their success on the assistance of a navigable river which would carry the attacking party to within striking distance of the village or villages to be destroyed. The Beiços de Pau had been reached in this way and dealt with by the gifts of foodstuffs mixed with poisons, but the two inches on the small-scale map of Brazil separating these two neighbouring tribes contained unexplored mountain ranges, and the single river ran in the wrong direction. The Cintas Largas, then, remained for the time being out of reach. In 1962, a missionary, John Dornstander, had reached and made an attempt to pacify them but he had given them up as a bad job.

The plans for disposing of the Cintas Largas were laid in Aripuaná. This small festering tropical version of Dodge City in 1860 has the face and physique of all such Latin-American hellholes, populated by hopeless men who remain there simply because for one reason or other, they cannot leave. A row of wooden huts on stilts stands in the hard sunshine down by the river. Swollen-bellied children squat to delouse each other; dogs eat excrement; vultures limp and balance on the edge of a ditch full of black sewage; the driver of an oxcart urges on the animal wreckage of hide and bones by jabbing with a stick under its tail. Everyone carries a gun. *Cachaça* offers oblivion at a shilling a pint, but boredom rots the mind. There are two classes: those who impose suffering, and the utterly servile. In this case nine-tenths of the working population are rubber tappers, and most of them fugitives from justice.

A cheap and sometimes effective ruse – besides being the quite normal procedure where a tribe's villages are beyond reach – is to bribe other Indians to attack them; and this was tried in the first instance with the Cintas Largas. The Kayabis, neighbours both of the Cintas Largas and the Beiços de Pau, had been dispersed when the State of Mato Grosso sold their land to various commercial enterprises, part of the tribe migrating to a distant range of mountains, while a small group that had split off remained in this Aripuaná area, where it lived in destitution. This group took the food and guns that they were offered in down-payment, and then decamped in the opposite direction and no more was seen of them.

Later a *garimpa* – an organised body of diamond prospectors – appeared in the neighbourhood. They were all in very bad shape through malnutritional disorders. They had attacked an Indian village and had been beaten off and then ambushed, and several of them were wounded. The intention had been to capture at least one woman, not only for sexual uses, but as a source of supply of the fresh female urine believed to be a certain cure for the infected sores from which the *garimpeiros* habitually suffer, and which are caused by the stingrays abounding in the rivers in which they work. *Garimpeiros* are organised under a captain who supplies their food and equipment, and to whom they are bound to sell their diamonds – under pain of being abandoned in the forest to die of starvation. Like the rubber tappers – who are their traditional enemies – they are mostly wanted by the police. The feud existing between these two types of desperado is based on the rubber tappers' habit of stalking and shooting the lonely *garimpeiro*, in the hope that he may be found with a diamond or two. In this case emissaries arranged a truce, and the *garimpeiros* were brought into town, and given food; a company doctor patched up the wounded men. Common action against the Cintas Largas was then proposed, and the captain fell in with the suggestion and agreed to detach six men for this purpose as soon as everyone was fully rested. In the condition in which he found himself, he may have been ready to agree to anything, but by the time the *garimpeiros* had put on a little flesh and their wounds had cleared up, there was an abrupt cooling in the climate of amity. Aripuaná was not a big enough town to contain two such trigger-happy personalities as the *garimpa* captain and the overseer of the rubber tappers. For a while the poverty-stricken rubber tappers put up with it, while the affluent *garimpeiros* swaggered in the bars, and monopolised the town's prostitutes. Then, inevitably, the *entente cordiale* foundered in gunplay.

In 1963 a series of expeditions were now organised under the leadership of Francisco de Brito, general overseer of the rubber extraction firm of Arruda Junqueira of Juina-Mirim near Aripuaná, on the river Juruana.

De Brito was a legendary monster who kept order among the ruffians he commanded by a .45 automatic and a five-foot tapir-hide whip. He made cruel fun of the Indians, and when one was captured he was taken on what was known as 'the visit to the dentist', being ordered to 'open wide' whereupon De Brito drew a pistol and shot him through his mouth. There was a lively competition among the rubber men for the title of champion Indian killer, and although this was claimed by De Brito, local opinion was that his score was bettered by one of his underlings who specialised in casual sniping from the riverbanks.

The expeditions mounted by De Brito were successful in clearing the Cintas Largas from an area, insignificant by Brazilian standards, although about half as big as England's southern counties; but there remained a large village considered inaccessible on foot or by canoe, and it was decided to attack this by plane. At this stage it is evident that a better type of brain began to interest itself in these operations, and whoever planned the air-attack was clearly at some pains to find out all he could about the customs of the Cintas Largas.

It was seen as essential to produce the maximum number of casualties in one single, devastating attack, at a time when as many Indians as possible would be present in the village, and an expert was found to advise that this could best be done at the annual feast of the *Quarup*. This great ceremony lasts for a day and a night, and under one name or another it is conducted by almost all the Indian tribes whose culture has not been destroyed. The *Quarup* is a theatrical representation of the legends of creation interwoven with those of the tribe itself, both a mystery play and a family reunion attended by not only the living but the ancestral spirits. These appear as dancers in masquerade, to be consulted on immediate problems, to comfort the mourners, to testify that not even death can disrupt the unity of the tribe.

A Cessna light plane used for ordinary commercial services was hired for the attack, and its normal pilot replaced by an adventurer of mixed Italian-Japanese birth. It was loaded with sticks of dynamite – 'bananas' they are called in Brazil – and took off from a jungle airstrip near Aripuaná. The Cessna arrived over the village at about midday. The Indians had been preparing themselves all night by prayer and singing, and now they were all gathered in the open space in the village's centre. On the first run, packets of sugar were dropped to calm the fears of those who had scattered and run for shelter at the sight of the plane. They had opened the packets and were tasting the sugar ten minutes later when it returned to carry out the attack. No one has ever been able to find out how many Indians were killed, because the bodies were buried in the bank of the river and the village deserted.

But even this solution proved not to be final. Survivors had been spotted from the air and were reported to be building fresh settlements in the upper reaches of the Aripuaná, and once again De Brito got together an overland force.

They were to be led, in canoes, by one Chico, a De Brito underling. The full story of what happened was described by a member of the force, Ataide Pereira, who, troubled by his conscience and also by the fact that he had never been paid the fifteen dollars promised him for his bloody deeds, went to confess them to a Padre Edgar Smith, a Jesuit priest, who took his statement on a tape recorder and then handed the tape to the Indian Protection Service.

'We went by launch up the Juruana,' Ataide says. 'There were six of us, men of experience, commanded by Chico, who used to shove his tommy-gun in your direction whenever he gave you an order!' (Chico, it was to turn out, was no mere average sadist of the Brazilian Badlands. For this kind of Latin American – and they have been the executioners of so many revolutions – the ultimate excitement lies in the maniac use of the machete on the victims, and it was to use this machete that Chico had gone on this expedition.) 'It took a good many days upstream to the Serra do Norte. After that we lost ourselves in the woods, although Chico had brought a Japanese compass with us. In the end the plane found us. It was the same plane they used to massacre the Indians, and they threw us down some provisions and ammunition. After that we went on for five days. Then we ran out of food again. We came across an Indian village that had been wiped out by a gang led by a gunman called Tenente, and we dug up some of the Indians' mandioca for food and caught a few small fish. By this time we were fed up and some of us wanted to go back, but Chico said he'd kill anybody who tried to desert. It was another five days after that before we saw any smoke. Even then the Cintas Largas were days away. We were all pretty scared of each other. In this kind of place people shoot each other and get shot, you might say without knowing why.

'When they drill a hole in you, they have this habit of sticking an Indian arrow in the wound, to put the blame on the Indians.'

This expedition breathed in the air of fear. Ataide reports that there were diamonds and gold in all the rivers, and the shadow of the *garimpeiro* stalked them from behind every rock and tree. A violent death would claim most of these men sooner or later. Premature middle age brought on by endless fever, malnutrition, exhaustion, hopelessness and drink overtook the rubber tappers in their late twenties, and few lived to see their thirtieth birthday. An infection turning to gangrene or blood-poisoning would carry them off; or they would die in an ugly fashion, paralysed, blind and mad from some obscure tropical disease; or they would simply kill each other in a sudden neurotic outburst of hate provoked by nothing in particular – for a bet, or in a brawl over some sickly prostitute picked up at a village dance.

Hacking their way through this sunless forest a month or more's march from the dreadful barracks that was their home, they were dependent for survival on the psychopathic Chico and his Japanese compass. It was the beginning of the rainy season when, after a morning of choking heat, sudden storms would drench them every afternoon. They were plagued with freshly hatched insects, worst of all the myriads of almost invisible *piums* that burrow into the skin to gorge themselves with blood, and against which the only defence is a coating of grime on every exposed part of the body. Some of the men were blistered from the burning sap squirted on them from the lianas they were chopping

'We were handpicked for the job,' Ataide says, with a lacklustre attempt at *esprit de corps*, 'as quiet as any Indian party when it came to slipping in and out of trees. When we got to Cintas Largas country there were no more fires and no talking. As soon as we spotted their village we made a stop for the night. We got up before dawn, then we dragged ourselves yard by yard through the underbrush till we were in range, and after that we waited for the sun to come up.

'As soon as it was light the Indians all came out and started to work on some huts they were building. Chico had given me the job of seeking out the chief and killing him. I noticed there was one of these Indians who wasn't doing any work. All he did was to lean on a rock and boss the others about, and this gave me the idea he must be the man we were after. I told Chico and he said, "Take care of him, and leave the rest to me," and I got him in the chest with the first shot. I was supposed to be the marksman of the team, and although I only have an ancient carbine, I can safely say I never miss. Chico gave the chief a burst with his tommy-gun to make sure, and after that he let the rest of them have it ... all the other fellows had to do was to finish off anyone who showed signs of life.

'What I'm coming to now is brutal, and I was all against it. There was a young Indian girl they didn't shoot, with a kid of about five in one hand, yelling his head off. Chico started after her and I told him to hold it, and he said, "All these bastards have to be knocked off." I said, "Look, you can't do that – what are the padres going to say about it when you get back?" He just wouldn't listen. He shot the kid through the head with his .45, then he grabbed hold of the woman – who by the way was very pretty. "Be reasonable," I said. "Why do you have to kill her?" In my view, apart from anything else, it was a waste. "What's wrong with giving her to the boys?" I said. "They haven't set eyes on a woman for six weeks. Or failing that we could take her back with us and make a present of her to De Brito. There's no harm in keeping in with him." All he said was, "If any man wants a woman he can go and look for her in the forest."

'We all thought he'd gone off his head, and we were pretty scared of him. He tied the Indian girl up and hung her head downwards from a tree, legs apart, and chopped her in half right down the middle with his machete. Almost with a single stroke I'd say. The village was like a slaughterhouse. He calmed down after he'd cut the woman up, then told us to burn down all the huts and throw the bodies into the river. After that we grabbed our things and started back. We kept going until nightfall and we took care to cover our tracks. If the Indians had found us it wouldn't have been much use trying to kid them that we were just ordinary backwoodsmen. It took us six weeks to find the Cintas Largas, and about a week to get back. I want to say now that personally I've nothing against Indians. Chico found some minerals and took them back to keep the company pleased. The fact is the Indians are sitting on valuable land and doing nothing with it. They've got a way of finding the best plantation land and there's all these valuable minerals about too. They have to be persuaded to go, and if all else fails, well then, it has to be force.'

De Brito, the man who organised this expedition, was to die within a year of it in the most horrific circumstances. When he found cause for complaint in one of his men, he would normally tie him up and thrash him until the blood ran down and squelched into the man's boots, but in an aggravated case he would have one of his henchmen use the whip while he raped the culprit's wife as the punishment was being inflicted. An Italian called Cavalcanti, who tried to attack the overseer after receiving the more serious punishment, was promptly shot dead and his body burned. A revolt of the rubber tappers followed in which nine men were killed. De Brito when cornered was like Rasputin, very difficult to disable, and absorbed several bullets and a thrust in the stomach with a machete before he went down. After this he was stripped, the bowels plugged back with a tampon of straw, then dragged alive into the open and left 'for the ants'.

How many Indian hunts of the kind mounted against the Cintas Largas must have gone on unnoticed in the past, condemned at worst as a necessary evil? Ataide speaks of them as if they were commonplace, and the likelihood is confirmed by a statement, made to the police inspector of the 3rd Divisional Area of Cuiabá Salgado who investigated the case, by a Padre Valdemar Veber. The Padre said, 'It is not the first time that the firm of Arruda Junqueira has committed crimes against the Indians. A number of expeditions have been organised in the past. This firm acts as a cover for other undertakings who are interested in acquiring land, or who plan to exploit the rich mineral deposits existing in this area.'

When one considers the miasmic climate of subjection in which these remote rubber baronies operate, in which the voice raised in protest can be instantly suffocated, and as many false witnesses as required created at the lifting of a finger, it seems extraordinary that police action could ever have been contemplated against Arruda Junqueira. It appears even more so when one surveys the sparse judicial resources of the area.

Denunciations by the hundred, of the kind made by Atiade, lie forgotten in police files, simply because the police have learned not to waste their strength in attempting the impossible. Nine major crimes out of ten probably never come to light. The problem of the disposal of the body – so powerful a deterrent to murder – does not exist where it can be thrown into the nearest stream, where – if a cayman does not dispose of it – the piranhás can reduce it to a clean skeleton in a matter of minutes.

In the case of the brazen and contemptuous tenth, where a man murders his victim in public view, and makes not the slightest attempt to hide the crime, he knows he is under the powerful protection of distance and inaccessibility. Aripuaná is 600 miles from Cuiabá, the capital and seat of justice of Mato Grosso, and it can be reached only by irregular planes. Moreover at the time Inspector Salgado began his investigation, about a thousand criminal cases were awaiting trial in Cuiabá, where, since the tiny local lock-up can accommodate only some fifty persons (all ages and sexes are kept together), most criminals manage to remain at liberty awaiting their trial, which may be long delayed.

Salgado's task was immediately complicated by factors unrelated to the normal frustrations of geography and communications. Ataide, principal witness and self-confessed murderer, was now the owner of a sweet stall on the streets of Cuiabá, and could be picked up at any time, but other essential witnesses were beginning to disappear. Two of the members of Chico's expedition had managed to drown themselves 'while on fishing trips'. The pilot of the plane used in the attack on the Cintas Largas was reported to have been killed in a plane crash. De Brito had of course been murdered in the rubber tappers' revolt, and even Padre Smith, who had taped Ataide's confession, could not be found.

Despite the series of contretemps, Salgado completed the police's case against Antonio Junqueira and Sebastião Arruda exactly three years after his investigations had begun, and the documents were sent to the judge. Under Brazilian law, however, the next procedure is the formal charge, the *denuncia*, which must be made by the public prosecutor, and it now became evident that the case might never surmount this hurdle. In all such countries as Brazil where a middle class is only just emerging, the landed aristocracy and the heads of great commercial firms are almost impregnably protected from the consequences of misdemeanour by dynastic marriages, interlocking interests, and the mutual security pacts of men with powerful political friends. This is by no means an exclusively Latin American phenomenon, even, and is often prevalent in Mediterranean Europe.

In this case the public prosecutor, Sr Luis Vidal da Fonseca, promptly objected that the case could not be tried in Cuiabá because

Aripuaná came under the jurisdiction, he said, of Diamantino. The papers were therefore sent to Diamantino where the judge immediately sent them back to Cuiabá. The question being referred to the supreme judiciary, it was ruled that the trial should take place in Cuiabá. So far only a month had been lost.

Fonseca now claimed exemption from officiating on the grounds that he was lawyer to the firm of Arruda and Junqueira. A second public prosecutor refused to be saddled with this embarrassing obligation, and the judge of the Cuiabá assize agreed with him and turned down Fonseca's application. Fonseca then applied to the supreme court again for an annulment of the local decision. The application was refused. By now nine months had been used up in manoeuvres of this kind, and it was April 1967.

At this point an attempt was made to settle these difficulties, to the satisfaction of all concerned, by the appointment of a substitute public prosecutor – who immediately claimed exemption on the grounds of his wife's somewhat remote relationship with Sebastião Arruda. The plea was accepted and another public prosecutor found, who declined to officiate, basing his refusal on the legal invalidity of Fonseca's objection. All papers were therefore returned to Fonseca.

In September 1967 a fourth substitute public prosecutor was appointed who, instead of taking action, sent the papers to the Attorney General who confirmed the original decision that Fonseca, who had moved away, was competent to act. This was followed by an endless bandying of legal quibbles and the appearance and departure of a succession of substitute prosecutors until March 1968 when the Attorney General was goaded to a protest: 'Since August 1966 the papers relating to this case have been shuffled about in an endless game of farcical excuses and pretexts, to the grave detriment of the prestige of justice.' Thus encouraged, the eighth or ninth substitute public prosecutor took action, and made a formal charge against the murderers of the Cintas Largas nearly all of whom were by now, after five years, either dead or not to be found. The names of Antonio Junqueira and Sebastião Arruda were omitted from the *denuncia* 'as their assent to the massacre of the Indians has never been established'. At this, the police attempted to take the law into their own hands by ordering the two men's preventive arrest. This could not be carried out, because they had gone into hiding.

One reads the history of the four years' legal battle against the firm of Arruda and Junqueira, and the imagination reels at the thought of what lies in store for the champions of justice for the Indians – the practised and methodical wasting of time, the pleas for exemption, the demands for retrials, the appeals and the counter-appeals, while the months run into years, and the years into decades, and the Indian slowly vanishes from the earth.

And when, if ever, after all the lawsuits are settled, a little land is wrested back from the great banks, the corporations, the *fazendeiros*, the timber and mining concessionaires that now hold it – still what is to be done? Can the mission hanger-on, miraculously refurbished in the body and spirit, return once again to the free life of the *isolado*? Does any remedy exist for the Indian, who, when the great day comes for the repossession of his land, finds the forest gone, and in its place a ruined plain, choked with scrub? Can a happy, viable, self-sufficient people be reassembled from those few broken human parts?

Manhunt

This journey to Paraguay in 1974 took place six years after Lewis published his famous article, 'Genocide'. 'Manhunt' was Lewis's second trip with the renowned photographer Don McCullin, and a strong bond had formed between them. In an interview with Lewis's biographer, McCullin later said: 'I have argued with a lot of journalists in my time, and the fault was often not theirs. I could be bad-tempered and erratic... There was one writer, however, who always brought out the best in me. His name was Norman Lewis, and in a way I became his disciple.' In 1982 McCullin was quoted as saying, 'I don't think I've been happier on any job with any writer. You get a private education with him, if you keep your mouth shut and soak up his wisdom.'

But their relationship could be described as symbiotic. In an interview with Pico Iyer, published in Lewis's biography, Lewis said 'I learnt from Donald how beautiful the ordinary can be, where previously the only things that attracted me really were non-ordinary things. But travelling with him I would see him suddenly spellbound by the beauty of rain drizzling down the mountainside, where I would normally pass it by... Trained by Donald, I've developed more observation in these matters.' Lewis would refer to McCullin as 'the Leonardo of photographers'.

In some respects the predicament of some of the Aché/ Guayaki Indians may have improved since this article, and a hopeful sign came in 2008 when an Aché – who had once been enslaved – was appointed the Minister in charge of indigenous affairs. But many of the remaining Aché still need to battle against illegal loggers, speculators, and landless peasants. And some are still being lured to indebtedness and semi-slavery hundreds of miles from their communities.

The horrifying 33-page report that inspired Lewis to go to Paraguay can be found by googling 'Mark Münzel IWGIA The Aché genocide continues'. (Lewis has used the word Guayaki to describe the tribe, but they have also been known as the Aché, now considered the more respectful name.)

In 1988, Lewis published an entire book, The Missionaries, examining the behaviour of these Christian fundamentalist missions.

First published in *The Sunday Times*, January 26, 1975

IN 1972, DR MARK MÜNZEL, an anthropologist working in Paraguay, reported wholesale enslavement, torture and massacre of the Guayaki Indians, among whom he was conducting his field studies. The Guayaki, who once occupied the whole of the forests of eastern Paraguay, have in recent years been reduced to a few bands roaming an area as large, perhaps, as Wales. They are of unique ethnic interest in that in a high proportion of cases they possess fair skin – for which reason they are sometimes known as 'white Indians'. They live as hunters and gatherers, and are notable poets, composers of epics and laments of extraordinary beauty.

Like all the forest Indians of South America the Guayaki have always suffered the persecution of settlers, ranchers and agriculturists, but until recent years have been able to survive by withdrawing farther and farther into the depths of the forest. With the forest's gradual destruction and its replacement by ranching and farmland, their position has become increasingly desperate, and their hunger greater. These lovers of nature in all its forms – who actually embrace and talk to trees – are non-aggressive, and there is no recorded instance of a Guayaki having drawn his bow against a settler without provocation; but here and there hunger, caused by the removal of game, has caused them to kill a cow, with instant and terrible reprisals. By 1972 it seemed that official policy called for their elimination, or 'sedenterisation' by forcible removal from the forest into a small reserve; the Colonia Nacionál Guayaki.

This operation was attended by atrocious circumstances. Professional Indian killers were employed to carry out the raids into the forest. In many cases adult Indians were simply shot on sight, and the fate of their children was to be sold as slaves to farmers all over eastern Paraguay – a fact which has been confirmed by the accounts of numerous travellers in the area. Reports reached the international press, including that of the UK, that so great was the glut of child slaves that their market price had fallen as low as five dollars. Later the figure fell to one dollar 50 cents. At this time Dr Münzel and his wife happened to be working on the collection and translation of Guayaki poems in the neighbourhood of Cecilio Baez, where the reservation had been established. He was already familiar with Jesús Pereira, the camp's administrator, an ageing manhunter with a criminal record, now transformed into a government official. Pereira still drew his gun on slight provocation, and his method of disciplining recalcitrant Guayakis at the camp was to cram them into a wooden frame called the 'tronco', in which the victim, unable to sit down or stand up, was left as long as necessary in full sun. He was a notorious sexual pervert, attracted to very young girls. Münzel was at the camp on one occasion when a batch of captured Indians were brought in, and he says that Pereira offered him an immature girl of about 11, presumably to keep him quiet.

Thereafter, Dr Münzel visited the camp on several occasions. He noticed extraordinary variations in the number of its inhabitants. Although Indians were constantly being brought in, the number present on the reservation never exceeded two hundred. By the end of July 1972 there were sixty fresh graves to be counted, and Münzel calculated that seventy-five Guayakis had disappeared since March of that year. He also recorded a great disparity in the sexes of the captives. There were hardly any girls between the age of five and puberty. Female slaves in this age-group attracted the best prices on the market, and Münzel was forced to assume that a clandestine trade in slaves was being carried on. All the evidence – not only that of Münzel but of Paraguayan intellectuals and leaders of the Paraguayan Church – leads to the view that this was a camp through which the remaining Indian population of eastern Paraguay was doomed to pass into servitude or oblivion.

For obvious reasons Münzel's visit to the camp soon came to an end, after which a curtain of secrecy descended on its operations. For all this the manhunters seem to have felt no shame when by some accident the general public happened to witness them at their work – nor in fact did the general public evince any signs of the moral outrage one would have expected. 'I was on the bus to Asunción when it made the usual stop at Arroyo Guasú. We heard that *señuelos* had just brought in a large number of Guayakis, and all the passengers got down to see them.' They found the Guayakis, guarded by the *señuelos*, bathing naked in the river. A passing car stopped. 'A woman and her two daughters got down. A daughter went down to the river's bank and took away a child that had been feeding at the breast of its mother, then returned with it to the car. The woman made no attempt to stop her, or even cried out. She seemed petrified.' The writer of this account, a Paraguayan zoologist, Dr Luigi Miraglia, rushed after the abductress and took the child back. One wonders how many more such pathetic human souvenirs were taken when the hunters so often came in with their prey.

The term *señuelo* calls for explanation. *Señuelos* are 'tame' Guayakis turned hunter. Wild Guayakis cannot be captured by a white man in what, to him, is an impenetrable forest, but they can easily be captured by their own kind, armed with the belief shared by Indians of both conditions that the whites are jaguars in human form, and that when a Guayaki is captured by a jaguar, he too becomes a jaguar and is compelled to capture more of his own people. When the free Guayakis find themselves face to face with the 'jaguars' they throw down their bows, offer no resistance, make no attempt to escape. With their capture, they lose their humanity. The magic power of their chiefs is lost; the ceremonies are forgotten, the musical instruments thrown aside. Their only purpose in life now is to hunt as jaguars themselves. The *señuelo*'s immediate reward is the wives of the free men he captures.

In the summer of 1972 the Roman Catholic church of Paraguay stated its grave concern over these events, and announced that it had informed the Holy See, while the Paraguayan anthropologists – Padre Bartolomé Meliá, and Dr Chase Sardi – and the zoologist Dr Luigi Miraglia made a public declaration of genocide. In the resulting scandal Jesús Pereira was relieved of his functions and served a short term in prison. The administration of the sinister Colonia Nacionál Guayaki was then transferred to Mr Jack (Santiago) Stolz, of the New Tribes Mission – a North American Protestant missionary sect which, in dealing with the problems of conversion, combines a streamlined business approach with religious fundamentalism. The mission has its own air service and radio transmitters, and three of its four centres in Paraguay are provided with airstrips.

A few months after Mr Stolz's takeover there were reported to be only some 20 Indians left on the reservation. However, in February 1973 a German army officer, who succeeded in visiting the camp in the guise of a tourist, found a group of fifteen or twenty 'obviously just arrived, in a desperate state of mind, just sitting around passively and staring at the ground'. A North American working on the reservation told him that these had just been brought in by none other than the old convicted manhunter, Jesús Pereira, who had caught a whole band and afterwards divided up the captives, some for the reservation, some for sale, and some for a farm he was now running with slave labour. The Jesuit Father Meliá writing in June 1973 to the German firm Farbwerke Hoechst, one of the original sponsors of the reservation, said that in the first nine months of the Stolz incumbency some 120 Guayakis had disappeared. It had been established that on April 2, 1973, more Indians had been brought into the reservation from the Itakyry region, but at this time no visits were allowed and therefore no investigations possible. At this time, Mark Münzel says in a report published by International Work Group for Indigenous Affairs (Denmark), 'there are signs that hunger was a problem on the reservation after the "arrival" of new Indians'. A letter from a Paraguayan contact said, 'There was a Guayaki who, in order to be able to buy something to eat, sold his son to some settlers for 80 guaranies (25p).'

Hunger is also mentioned in a letter written on May 1, 1973, by the Paraguayan rancher Mr Arnaldo Kant to Mr Nelido Rios, at that time assistant to the administrator of the reservation.

'Yesterday Mr Jack (Santiago) Stolz, administrator of the Colonia Guayaki was here ... He threatened to report me because I had that group of Guayakis you had gathered for me. I explained to him that I had them on your request, and only to prevent them from being used as slaves ... I was struck by the fear that this man (Jack Stolz) inspires in these Indians: when they noticed he was there (to return them to the reservation), they started to run away into the forest. The women wept, telling me they did not want to return to the camp because there they were given no food ... The administrator claimed payment for the work the Guayakis had done cleaning up around their houses, and I gave him the sum of 2500 Gs., as proved by the enclosed receipt...'

The receipt, given at Cecilio Baez on April 30, 1973, is 'for labour performed by a group of Guayakis'. According to Mr Stolz, he wanted the money only in order to pay it to the Indians later on.

It should be understood that by this time the forest sheltering the Guayaki had been cleared to leave the reservation surrounded to a great depth by ranches and farms, and that no Indians had been seen anywhere in the vicinity since the great manhunts of 1972. In the unlikely event of any Guayaki leaving the forest to enter the reservation of his own free will, he would have had to travel for many miles across these cleared areas, now the property of settlers said to be prepared to shoot Indians on sight. Despite this the population at Cecilio Baez showed a sudden increase, reaching 110 by June 1973. At this time, Mark Münzel, back in Frankfurt, received a letter from a local contact to say that: 'The New Tribes missionaries are now hunting by motor vehicles for Guayaki in the region of Igatimi (one hundred miles from Cecilio Baez) in order to reintegrate them into the reservation.' Nevertheless a visitor to the reservation on August 23, 1973, counted only 25 Indians – who were considerably outnumbered by the missionaries and their families. There was an upswing in population again by September 17 when (by Münzel's account), 'according to the North American missionary ... a band of 46 Guayaki were brought to the reservation on a truck, "by the decision of God" and with the help of the Native Affairs Department of the Ministry of Defence, and of local authorities from the region of Laurel, Department of Alto Paraná'. Thus, by the end of September, there should have been some 70 Indians on the reservation, but in January 1974, when it was visited by *New York Times* correspondent Jonathan Kandell, less than 50 were counted. The reservation continued to devour Indians.

In March 1974 the International League for the Rights of Man, joined by the Inter-American Association for Democracy and Freedom, charged the Government of Paraguay with complicity in the enslavement and genocide of the Guayaki Indians in violation of the United Nations Charter, the Genocide Convention and the Universal Declaration of Human Rights.

In a protest to the United Nations Secretary-General, documented by four annexes, eyewitness accounts and photographs, the organisations stipulated the following violations leading to the 'wholesale disappearance of a group of human beings', the Guayaki ethnic group: (1) enslavement, torture and killing of the Guayaki Indians in reservations in eastern Paraguay; (2) withholding of food and medicine from them resulting in their death by starvation and disease; (3) massacre of their members outside the reservations by hunters and slave traders with the toleration and even encouragement of members of the government and with the aid of the armed forces; (4) splitting up of families and selling into slavery of children, in particular girls for prostitution; and (5) denial and destruction of Guayaki cultural traditions, including use of their language, traditional music and religious practices.

On March 8, Senator Abourezk, supported by 44 other senators, took to the US Senate floor 'to denounce genocidal activities still rampant in Paraguay'. Revealing that he had a copy of a receipt for work done by slaves from the Colonia Nacionál Guayaki, he went on: 'While on the reservation the Indian slaves are discouraged from using their own language, and music is expressly forbidden. The death rate from diseases of malnutrition and sheer lack of will to survive is one of the highest in the world.' The senator called for the cutting off of aid to Paraguay. He concluded: 'A government which is bent on the mass extermination of part of its people does not deserve our aid any more than a convicted and professed killer deserves a welfare check.'

Following Senator Abourezk's speech, the US Ambassador to Paraguay was recalled and the senator and his supporters were privately admonished by the ambassador, who reminded them that the abiding friendship of Paraguay was indispensable in the framework of American hemispheric defence. The US Press remained strangely unresponsive to the news from Paraguay. Professor Richard Arens, Counsel to the League for the Rights of Man, says of this episode: 'A careful survey of the national media ... left us solely with an impression of a consciously or unconsciously determined news block out.'

In Paraguay, however, repercussions were rapid. On April 28, 1974, the Department of Missions of the Paraguayan Episcopal Conference sent a letter to the Asunción daily newspaper *La Tribuna*, containing the following passage: 'Our secretariat has in its possession documentation of cases of massacre, cases which, moreover, have been partly published in your very paper.' The letter was signed by the Bishop of the Chaco region, Mgr Alejo Ovelar, and by Father Bartolomé Meliá, respectively president and secretary of the organisation.

A second letter to *La Tribuna* on May 8, 1974, said:

'The Department of Missions of the Paraguayan Episcopal Conference has denounced and denounces, basing its denunciations on concrete data which has been duly investigated, the existence of cases of genocide ... [It] desires there be a sweeping investigation, especially into the situation of certain indigenous groups of

Paraguay, who are especially threatened in their ethnic survival...'

Next day the Paraguayan Minister of Defence, General Marcial Samaniego, called a conference to discuss these allegations. The minister did not attempt to deny that crimes against the Indians had taken place. He stressed however that there was no *intention* of destroying the Guayaki, and thus, by definition, genocide was excluded. 'Although there are victims and victimiser, there is not the third element necessary to establish the crime of genocide – that is "intent". Therefore, as there is no "intent", one cannot speak of "genocide".'

Thereafter whatever news breached the silence has been discouraging. Latest reports from Paraguay suggested that the New Tribes missionaries pursued their aims with undiminished zeal, and Mr Jack Stolz continued to 'attract' Indians – as the euphemism puts it – with methods only too similar to those employed by his predecessor Mr Jesús Pereira. Pereira was still around, too, the master now of some fifty slaves, and engaged in continual incursions into the forest on his own account. A new Guayaki group had been located in Amambay, in the far North East, and an expedition planned against it. Colonel Infanzón, Director of the Native Affairs Department, responsible in its time for the administrations of both Mr Pereira and Mr Stolz, had said: 'In spite of all our critics, we shall go on fighting and working, because we do not wish to leave unfinished the work we have started.'

In October 1974, Donald McCullin and I went to Paraguay for *The Sunday Times* to try to find out what was happening.

The opinion of a contact in Asunción was that we should not make formal application to the Paraguayan Ministry of Defence to visit Cecilio Baez, because permission might well be denied, and even if granted, we should be sacrificing the element of surprise. Professor Chase Sardi, who had accompanied the correspondent of the *New York Times* in January, described their mission as a waste of time, because of advance notice given to the camp's administration, who were suitably prepared. Donald and I discussed the advice but felt obliged to reject it, because there was no certainty that having mounted an expedition to reach Cecilio Baez – in a country in which we supposed that every move made by a foreigner is under observation – we should be admitted to the reservation. It was conceivable, moreover, that any such abortive effort might be followed by expulsion from the country.

We therefore made application for the visit and were duly summoned into the presence of Colonel T. Infanzón. It was hard to believe that the colonel, a man of mild appearance and courteous manner, could have been the subject of the serious allegations relating to the procuring of young Indian women. He seemed uneasy at our request to visit the reservation, explaining his reluctance by the circumstance that a French couple who had been there on what had been described to him as a scientific mission had filmed Guayakis engaged in sexual intercourse. Their film had been shown in Panama. We assured the Colonel that we had no film cameras with us, and he seemed relieved. A trip might be possible, he said, but we should have to be accompanied by an official from his department. He would also have to obtain the permission of his superior, the minister.

A wait of some days followed, which we filled in with visits to country towns in the neighbourhood of Asunción. Many of these, once early Jesuit settlements, were of great charm. The grandiose churches of the 17th century remain – a superb example being that of Yaguarón, a building of such external severity as to be hardly recognisable as a church, yet which astonishes with the extravagant baroque of its interior, decorated by Indian artists.

The delay provided time for interesting interviews, notably one with the Bishop of the Chaco region, Mgr Alejo Ovelar. The bishop spoke of growing concern felt by churchmen and intellectuals throughout Latin America over the activities of certain missionary sects 'who seem indifferent to the spiritual – let alone material – welfare of the primitive peoples among whom they work, but subject them instead to commercial exploitation'. As a specific example he said that North American missionaries had set themselves up as middlemen in the Chaco region compelling Indians such as the Moros, who lived by hunting and trapping, to sell their skins through the mission. Traders who attempted to deal directly with the Moros had actually been threatened with violence. 'These missionaries,' said the bishop, 'are also implicated in the grave crime of ethnocide.'

In due course, Colonel Infanzón announced his decision. We could go to Cecilio Baez if we were prepared to give an undertaking that any article written about Paraguay would contain no evaluation of the situation of the Indians in that country. Since we proposed to avoid evaluations and deal in facts, not about 'the Indians of Paraguay' as a whole but about the Guayaki, it seemed possible to agree to this. The permit was then given, together with a letter of introduction to Mr Jack Stolz, 'American Missionary in charge at the Colonia Nacionál Guayaki, Cecilio Baez'. The proviso that we should have to be accompanied by an official from the Ministry of Defence had evidently been forgotten. Next morning at six, we set off.

The weather in Paraguay at this season of the year is subject to dramatic variation. Rainfall is at its highest, a sweltering summer is only weeks away, and hurricanes occur. We were being driven by a Paraguayan friend in his Citroen 'Deux Chevaux', and he had warned us that we could reach the reservation only if the rain held off. There are only two paved highways in Paraguay; one from Asunción, to Foz do Iguaçu on the Brazilian frontier, and the other, only about 30 miles in length, to the town of Paraguari, south-east of the capital. All the others are dirt roads, which are closed to traffic by steel barriers as soon as it rains, when the surfaces are instantly transformed into slimy red mud, and makeshift bridges are sometimes carried away. To reach Cecilio Baez we should cover about 120 miles of the paved highway to Caaguazú and there take a left turning along a dirt road, leading eventually through fifty miles of forest to the reservation. This road was reported to be in exceedingly bad condition.

At Coronel Oviedo, thirty miles short of Caaguazú, clouds were building up ahead, and road police told us that it was raining in eastern Paraguay, and that all the roads were closed. We therefore turned off southwards in the hope of reaching our friend's house in Caazapá. At Villarrica the road was barred, but after delicate and protracted negotiations with the road police, the barrier was unlocked and we were waved on into a landscape veiled already in pearly rain.

In this area all the Arcadian charm, the style and the swagger of South America had survived unscathed. The streets of small towns like Caacayí (named after the call of a bird) had turned to grassland cropped by cows. Diurnal bats fluttered from the windows of great sepulchral mansions emptied by so many wars and revolutions. Aloof horsemen went thudding past under their wide hats, a palm always upraised in greeting. Enormous blue butterflies floated by – never molested by the local Indians, who believe them to be the 'ears of God'. The distances dispensed frugal harp music, and the melodious hoot of the bellbird. Eagles were settled in the flowering trees by the roadside, and once we spotted a Model T Ford that had brought a couple to market. After pushing the Citroen through – what seemed to us – miles of slippery mud, we finally reached Caazapá at sundown. If there was any place to escape from the world, this was it.

In living memory there were Guayaki in the woods round Caazapá, but it is a quarter of a century since the last of them were cut up to bait jaguar traps, or otherwise killed or enslaved. At San Juan Nepomuceno, twenty miles away, an episode took place in 1949 which proved too much even for the stomachs of the local whites. An Indian hunter called Pichín López managed to round up the last of the Guayakis in the area, and having hacked to death the aged and enfeebled, carried off the viable survivors to San Juan, where naked and in chains, they were exposed in the town square for public sale. Pichín, denounced by the local priest, had to leave the country. It was the first time that Indian killing had provoked such disapproval. Pichín's lieutenant at that epoch was none other than Jesús Pereira – twenty years later the government's administrator of the Colonia Guayaki Nacionál at Cecilio Baez.

The present calm of towns such as San Juan and Caazapá conceal undercurrents of extraordinary violence. In 1947 the political boss of Caazapá, one Matilde Villalba – himself a notable Indian killer in his time – was ambushed and shot down with six of his sixteen sons. Because the local cemetery at that time was full to overflowing with victims of the recent short but incredibly bloody civil war, a charitable neighbour had to lend space in his family vault to accommodate the bodies. Guerrilla fighters captured in the vicinity round about the time of the Che Guevara fiasco in Bolivia were taken straight up and pitched from planes. Remembering this sombre period our friend said, 'Thank God we now have peace and tranquillity.' No sooner had this utterance been made than we noticed several people running to get a better view of something that was happening in the next street. 'It is nothing,' we were assured, 'probably two men have challenged each other to a duel.' As it happened the excitement was caused by an informally staged bullfight, but confrontations – High Noon style – are said to remain common.

Next morning we left Caazapá early and drove all day over officially closed roads and collapsing bridges to reach Caaguazú by evening. The town was under water, and having found beds in a lodging house we splashed through the floods to the nearest cantina for a meal. Here, by the purest chance, a boy of about eight fetching and carrying in the restaurant was pointed out as a slave. In such cases the child will be passed off as 'adopted', and given the family name, although he will remain in a subservient position in the family for the rest of his days. The boy admitted to me to being an Indian – the family had rather absurdly denied that he was one – and he was clearly delighted by the rare experience of being for once the object of attention.

Incessant rain compelled us next day to return to Asunción, and this, as it turned out later, may have been a good thing. Three days later, a young Englishman employed on a scientific project in Paraguay, offered to take us in his Land-Rover to Cecilio Baez, and once again we set out at dawn. The reservation, reached through narrow and constantly branching tracks inside the partially cleared forest was hard to find, but eventually we came into a large clearing at the end of which several huts clustered round a building which, from its size and style, was clearly the missionary's house.

Several Indians in near rags were mooching about, and a white man in mechanic's overalls tinkered with a piece of machinery. This proved to be Mr Jack Stolz, who received us with marked coolness. I presented Colonel Infanzón's letter, and without lowering his guard, Mr Stolz said that he had expected us three days before. From this it was clear that Colonel Infanzón had been in touch with Cecilio Baez by radio. I explained that we had been held up by the weather, but it was clear that the missionary remained suspicious. He questioned us much in the way that Colonel Infanzón had done, and the occasion seemed a delicate one. I formed the opinion that Mr Stolz would have to be regarded to all intents and purposes as a functionary of the government, empowered to turn us away from the reservation if he thought fit. We gave non-committal replies but a positive assurance that we had no intention of making indecent films of the Indians. At this point it turned out that Mr Stolz had never heard of Colonel Infanzón's French couple. Mrs Stolz now came out of the house. Although I would have taken her husband for a top sergeant in an American combat unit rather than a missionary, she was what one thought of as a typical missionary's wife; resolutely smiling, and calm in what might have seemed to her a crisis. She invited us into the house, which was pleasantly furnished without being luxurious, and devoid of the labour-saving gadgetry commonly found in such missionary establishments. Donald McCullin then found an excuse to wander off and start photographing, while I began to attempt to prise the missionary out of a defensive silence.

I asked Mr Stolz how many Indians were on the reservation, and he said there were three hundred. Most of them were out working on the land, he said. When I asked if I could see them at work, he said they were on their way back to the camp and would be coming in soon. This seemed reasonable enough. It was now nearly noon. In Paraguay the working day starts soon after dawn, and the whole country knocks off for a siesta at midday. Work then restarts at three and continues until sunset. I asked Mr Stolz what the Indians were paid, and after a little hesitation he said they received 100 guaranies (about 33p) a day. 'They have no sense of trade or money,' he said.

In view of the camp's large increase in population since the visit of the *New York Times*' correspondent in January, I thought it reasonable to enquire where the new arrivals had come from. Mr Stolz seemed evasive. They just came in from time to time, he said. 'In the night.' The last group had arrived some weeks previously. He agreed that no 'wild' Guayaki were to be found any longer in the vicinity, and that those recently arrived had come from a long way away. 'What made them come?' I asked, and Mr Stolz said: 'Maybe they heard this was a good place to be in.' He confirmed that there were many enslaved Indian children in the neighbourhood. He mentioned four that had been taken two years previously at Kuruzu some twenty kilometres away, but had not survived. One of these had died from a gun-butt blow received at the time of his capture, and the others shortly afterwards from measles, a disease against which forest Indians have built up no immunity.

'It's the smart thing to own a Guayaki round here,' he said. 'I guess it's a kind of status symbol.'

Donald was anxious to photograph Guayakis playing their musical instruments; their flutes and above all a species of one-stringed fiddle with which a range of about three notes is obtained simply by bending the neck, and thus varying the tension of the string. Mr Stolz said flatly that there were no musical instruments of any kind on the reservation. Did the Indians perform any traditional ceremonies? I asked. No, he said, none. Were there any chiefs? No. Any medicine-men? Absolutely not. The only thing the Guayakis ever seemed to do was to sing, he said. The words of their songs were 'not too interesting'. The men were always trying to build themselves up as great hunters, and the women sang those terrible groaning, belly-aching songs. They blamed everything on their ancestors.

At this point I decided to ask Mr Stolz to explain the function of the mission. He replied that it was to bring salvation to those who were in a state of sin. This was to be done, eventually, by baptism after the converts had accepted Christ in their hearts and by 'admission through the mouth'. He thought it was a good thing that I should write down his replies to these questions on topics of faith and conversion and this I carefully did.

How many Guayakis had he baptised? I asked, and Mr Stolz replied, 'None. Before I can bring them to Christ I must first understand what they believe.'

'And do you?' I asked him.

Mr Stolz said, 'Vaguely.' He had a problem with the language, he added, but at least he knew that they believed in three gods. The tiger (jaguar), the alligator, and the grandfather. 'This makes things difficult. When we talk of God's son they think of a tiger's son. It's hard to get across the idea they can be redeemed from sin by a tiger's son nailed to a cross. None of these Indians can make the admission because they do not know what to admit.'

I asked Mr Stolz what progress, if any, had been made towards conversion, and he replied that most of them at least realised that they were living in a state of sin, particularly in sexual matters.

As by Mr Stolz's own estimate it would be many years before any of the Guayakis were brought by his efforts into the fold of Christianity, it was supposed that many in the meanwhile would die, while a few, at least, would probably remain beyond reach of the missionary effort. What in the view of the New Tribes mission, I asked, would be their spiritual fate?

Mr Stolz was on firm ground now. 'There is no salvation,' he said, 'for those who cannot be reached. The Book tells us that there are only two places in the hereafter: heaven and hell. Hell is where those who cannot be reached will spend eternity.'

It seemed to me unreasonable that divine retribution should be visited on the Guayakis because Mr Stolz had been unable to learn their language, but the missionary shrugged his shoulders. Such things were beyond his jurisdiction, he suggested. Soon after this he excused himself to return to stripping down an electrical generator some part of which had to be got away to Asunción that day, and I only saw him briefly again before we left.

I now joined Donald at his photography, noticing by this time that several young missionaries, not in evidence before, had come on the scene. There were about twelve small huts in the immediate area of the mission-house. These averaged some 15ft square and it was difficult to imagine how as many as three hundred Indians could have been sheltered in them. We saw about thirty-five Guayakis in all, about half of them the possessors of skin of the extraordinary waxen whiteness for which anthropologists have been able to offer no explanation. All of them were extremely mongoloid in appearance, and many could have passed for Eskimos. At an approximate estimate about fifteen of these were nubile young women, or mothers with babies. There were a half-dozen boys between eight and twelve years of age, and two young girls in this age-bracket, all with the distended stomachs and decayed teeth suggestive of malnutrition. The rest of the visible population was made up of adult males.

All the Guayakis, except for two or three men wearing pseudo-military uniforms, were in dirty cast-offs. The camp stank of excrement, and there appeared to be no sanitary arrangements. We noted that the adult males had access to bows and arrows with which they showed off their skill to the missionaries. The fact that these men were not at work, and that the missionaries fraternised with them in an affectionate manner suggested at least the possibility that they were privileged, possibly as camp 'trusties', or even *señuelos*. In common with all other visitors to the reservation we observed an extreme disproportion in the sexes of its population. If Mr Stolz had told us correctly there were three hundred Indians, and – presuming women were not compelled to work with their menfolk on the farms – men outnumbered women by about twenty to one. We also found it strange that we never saw any old men.

About half the Indian adults were lying on the ground in their huts in what seemed a condition of total apathy, giving no evidence of awareness of our presence as we came and went. We saw no signs of food anywhere in the huts – no scraps or leftovers.

Some of the Indians managed to keep the pets from which they are never parted when at liberty. In one hut we found three tame coatis, in another a fox cub, in another a baby vulture. A bunch of little boys, with distended stomachs under their filthy shirts, came running up to stroke our hands and caress our fingers (the Guayaki may be the most affectionate and outgoing of the Indian races). Some of the boys showed us their tame lizards; one had a hen perched on his shoulder, and another a hawk.

Having finished his photography in the central area of the camp, Donald strolled off towards two huts on the outskirts, followed by Mr Stolz's son and a smiling young missionary, who assured him that there was nothing more to be seen. In one hut he found two old Indian ladies in the last stages of emaciation, and clearly on the verge of death. They lay on the ground apparently having been abandoned to their fate. This was a scene of the kind one associates with the ultimate disasters of Ethiopia and Bangladesh. In the second hut lay a third woman, also in a desperate condition, and Mr Stolz's son, who was carrying Donald's tripod, explained that she had been shot in the side while being brought in.

It was clear that behind the evasions and the resentful silences of Cecilio Baez a great deal remained to be explained before the charges made by the International League for the Rights of Man could be brushed aside. But it was also clear that attempts to probe further, at this time, into what went on behind the scenes would have been futile. Paraguay, the firmest of the Latin American dictatorships, is not a country where it is recommended to put too many inconvenient questions to persons entrusted by the government. We should have been very unhappy indeed, for example, to have been obliged to surrender Donald's photographs when leaving the country.

But as it turned out, a few more of the facts were let slip by a visiting New Tribes missionary from the Chaco, who asked us when we left – after several hours' fruitless waiting for the Guayakis to come in from the farms – to give him a lift back to Asunción. He had been stranded by the rains for four days at Cecilio Baez, and, convinced that God had sent us in answer to his prayers, in his euphoria and relief he threw caution to the winds.

Mr Stolz had told us that the Indians received 100 guaranies a day (spendable at the mission's store), but the Chaco missionary dismissed this as absurd. The farmers they worked for promised them 200 guaranies (66p) a week, but usually fobbed them off in the end with an old shirt, or a worn-out pair of shorts. The Chaco missionary also succeeded in letting the cat out of the bag in the matter of Mr Stolz's activities as an Indian catcher. He had been out several times recently 'to make contact' he said, and once, indeed, had been narrowly missed by a Guayaki arrow.

It would be impossible without an investigation to substantiate the allegations that have been made that Mr Stolz has engaged in traditional manhunts using *señuelos*, and no investigator is ever likely to be permitted at the camp. But clearly the Indian population of Cecilio Baez has much increased in the past few months, and it is hard to believe that free Indians would wish to make a journey of at least a hundred miles from the remote forests where they still exist, in order to deliver themselves up to a condition hardly distinguishable from slavery. In fact their reluctance to be 'attracted' by Mr Stolz is very clear from his colleague's account of the arrow that narrowly missed him.

By Mr Stolz's own admission the New Tribes mission at Cecilio Baez performs no religious function. What then is its purpose? It is hard not to agree with the view of Dr Mark Münzel in IWGIA Document No. 17, published in Copenhagen in August 1974, that: 'The reservation has the function of a transitional "taming" camp: the proud and "wild" Indians of the forest would not be immediately willing to work in the white man's fields; but they are willing once they have passed through the reservation, because they see no other solution, or because they are so instructed by the missionaries.' It has seemed strange to outside observation that the countries of Latin America should tolerate and even favour the presence of such North American Protestant missions dispensing – when religious instruction is given at all – a version of Christianity which must be repellent to their own Catholic beliefs. The reason can only be that they are regarded by governments, intent at all costs on the 'development' of natural resources, as efficient in the performance of actions that no other organisations are qualified by philosophy, temperament and – above all – by tradition, to undertake.

I do not believe that Mr Stolz would be particularly concerned to defend himself from inclusion in the category of those missionaries who, by the verdict of Bishop Alejo Ovelar, 'are implicated in the grave crime of ethnocide', because he would see nothing wrong in the destruction of the racial identity of Indians for which he feels little but contempt.

'We believe,' says the printed doctrinal statement of the New Tribes mission, in 'the unending punishment of the unsaved.'

The White Promised Land

Lewis visited Bolivia in October 1977. He was accompanied by the photographer Colin Jones, whom The Sunday Times *had dubbed 'the George Orwell of British Photography.' Jones was another of the eminent photographers who fell under Lewis's spell, comparing him favourably to Bruce Chatwin and Paul Theroux, with whom Jones had suffered fractious relationships.*

Jones told Lewis's biographer, Julian Evans, 'Whenever I went on an assignment with Norman I never knew exactly what we were going to do. He had no definite way of doing things until he hit the ground.' He remembered this journey with Lewis as being an especially dangerous one. 'We were told not to do it and were warned off several times.'

The Summer Institute of Linguistics, described in the article, has since been expelled from Brazil, Mexico and Peru, but still has staff in more than eighty nations.

First published in *The Observer*, March 5, 1978

D R GUIDO STRAUSS, Bolivian Under Secretary for Immigration, caused a stir throughout Latin America last year [1977] when he announced his government's intention to encourage the entry into Bolivia 'of large and important numbers of white immigrants ... especially from Namibia, Rhodesia and South Africa'.

His statement was published by Bolivia's leading newspaper, *Presencia*, which later revealed that 150,000 whites would be accommodated and that the scheme had been financed by a 150-million-dollar credit to Bolivia offered by the Federal German Republic.

Censorship in Bolivia fosters eagerness on the part of the Press to study and reflect the governmental viewpoint, but in this instance *Presencia* seems to have been in a quandary as to the official line. A vigorous denial by one department of the plan's very existence coincided with a wealth of confirmatory detail poured forth by another. 'At the very time,' wailed the newspaper, 'when the Institute of Colonisation affirmed that they had no knowledge of such a project ... Under Secretary Guido Strauss said that the immigration plan was a matter of top priority.'

Confusion was worse confounded by publication of extracts from a confidential letter written by Strauss to his minister, General Juan Lechín Suarez. The letter was full of precise figures and facts. The white settlers would be admitted in stages, taking possibly as long as six years, although 30,000 families could be admitted in the first year if the financial arrangements were settled by then. The exact areas to be taken over by the newcomers were listed and the amount of land they would receive (800,000 hectares). Strauss noted that the cost of purchasing this land – which was to be given to settlers – and of building roads was 250 million dollars, and that this sum had been funded.

The letter breathed the sentiments of humanity, and warned of the holocaust that awaited South African whites once black majority rule became a fact. Mr Sean McBride, High Commissioner for Namibia at the UN, was quoted as having said that Namibian whites would have to abandon the country. 'There is no doubt that the factors of a catastrophe are imminent,' Dr Strauss wrote. Motives of national self-interest were also touched upon. Bolivia's economically under-developed, under-populated areas cried out for the drive and the technical skills of the energetic South Africans. He also made the claim that Britain, the US and France between them were ready to put up 2,000 million dollars to indemnify white Rhodesians, 'who would be unable to resist the process of Africanisation'.

Adverse reaction to this colonisation was to be expected, and was led by the Catholic Church, the only body in Bolivia prepared to stand up to the dictatorship. A conference of religious leaders was held last July and a declaration followed listing numerous objections to the plan. What clearly disturbed the Church was the prospect of apartheid in Bolivia. *Presencia* was permitted to publish the criticisms. I quote from only two paragraphs:

> The South African immigrants, with their violently racial mentality, condemned even in their own countries, could import the principles of apartheid into those under-populated areas where they would form compact groups. Bolivia, as the South Africans write so often in their newspapers, is the richest of the Latin American countries, requiring only an advanced technology for the exploitation of its raw materials … 'three-quarters of its population are illiterate natives'. This is a point of view echoed by the contemptuous remarks of some of our own authorities who say, 'The Indians cost more to keep than animals. They have to be fed and work less.'

The well-meaning objections of the Church, and of so many liberal Bolivians, are as naive as they are creditable, since in some ways apartheid already exists in a purer and more extreme form in Bolivia than the version professed by the racists of South Africa. This, a visitor to the country quickly discovers.

La Paz, at 12,400 feet, is the highest capital in the world. The plane lands on the edge of a plateau high above it, engines screaming in reverse and wing-flaps clawing at the thin air. A few hundred yards from the runway's end, the abyss awaits. The city lying below is crammed into a monstrous crater. Here, in this hole in the earth, its original Spanish builders, who went there to mine gold, huddled out of reach of the terrible wind that whines like a persistent beggar at every turn. From the almost infernal vision beneath, one turns back to that of the flat world of the Altiplano, almost all of it over 13,000 feet.

This is the homeland of the Aymara Indians, whose grandiose civilisation preceded that of the Incas. They have been forcibly Christianised, and enslaved over four centuries, and they are still tremendously exploited. But somehow they have survived. Now, with the Church turned benign, they are no longer compelled to carry priests in chairs on their backs, or scourged for persisting in their ancient worship of Pachamama, the mother-goddess, and Tio, the Devil, who is also, appropriately, god of the tin mines. In their honour the Aymara sacrifice innumerable llama foetuses and get drunk whenever they can. Their distrust of all whites has become instinctive and can be belligerent. At best one is ignored, and at worst hustled away. Whites who insist on fraternising with Aymaras in their state of holy drunkenness may even find themselves attacked.

Bolivia is a poor country; its per capita income of about £200 a year putting it at the bottom of the league of South American nations. Its adult literacy is about thirty percent. Oil revenues have brought about some increase in prosperity in recent years, but this has been diverted to a small sector of the population and largely spent on non-essential consumer goods. Far from bringing comfort to the peasant majority, the new prosperity has in fact done the reverse, for while the prices of agricultural produce have been rigorously held down, almost everything else is imported and subject to the inflationary process.

National poverty and under-development formed the theme of Under Secretary Guido Strauss's argument in favour of mass immigration of whites from South African countries when I interviewed him in La Paz. He was communicative and direct, a man with a reputation for not mincing words. Occasionally he is indiscreet, and is widely quoted as having said in public, 'They [the white immigrants] will certainly find our Indians no more stupid or lazy than their own blacks.' He confirmed to me with enthusiasm all the details of the project so far published, whether or not they had been denied elsewhere, and added to these a little fresh data.

Bolivia, Dr Strauss explained, a country twice the size of Spain, had a population of five million. Such people as it had were crowded into the semi-barren Altiplano and a number of upland valleys, leaving the vast and rich territory of the eastern provinces virtually unpopulated. Through lack of development of its agricultural wealth, the country was even obliged to import food. When, therefore, discreet international moves had been set afoot to discover possible areas of resettlement for whites who it was believed would sooner or later be forced out of Rhodesia, Namibia and South Africa, Bolivia had recognised that its acceptance of such refugees might provide a partial solution to this problem.

Dr Strauss said that approaches had been made (through the German Federal Government) not only to Bolivia, but to Argentina, Brazil, Uruguay and Venezuela. Brazil and Venezuela had agreed to accept a limited number of technicians.

Only Bolivia had been ready to take on immigrants of all classes en bloc. Dr Strauss said that any white settler would be given, free, a minimum of fifty hectares of first-class agricultural land, and would also receive social, technical and economic assistance. Those who wished to engage in ranching would receive 'very much more', together with ample low-cost labour.

Dr Strauss handed me a copy of the Bolivian immigration laws, which also stressed Bolivia's demographic predicament, and the inducements offered to immigrants who could contribute to its solution. Settlers from all countries would be welcomed with open arms, but, he said, a special and natural sympathy predisposed Bolivians in favour of persons of European origin, who shared with them a common heritage of culture and religion.

Asked how many immigrants from the South African countries had already arrived, Dr Strauss said that some 'spontaneous' immigration had taken place. He believed that this trickle would soon become a flood, an inundation which could be expected as soon as black majority rule becomes a fact. The infrastructure, including the building of roads in areas where the colonialists were to be settled, was complete. Bolivia for them, Dr Strauss said, is a promised land.

One question remained. Forty-one Indian tribes, with a total population of about 120,000, are recorded as living in the nominal emptiness of Eastern Bolivia. Some of them occupy precisely those areas shown on Dr Strauss's map as designated for development. What was to become of them?

This was a question to which Dr Strauss did not feel competent to reply. If I wanted to know anything about Indians, and the nation's plans for them, he suggested that I should go and talk to the Summer Institute of Linguistics, the largest of the group of North American evangelical missionaries working in Bolivia. This I did.

The mildness of Mr Victor Halterman's personality came as a surprise after learning something of his formidable reputation as head of the Summer Institute of Linguistics. With the exception of Under Secretary Strauss, who was naturally obliged to keep his thoughts to himself, I never met a Bolivian who did not regard the Summer Institute of Linguistics as a base for operations of the CIA in Bolivia; possibly in South America itself.

Mr Halterman's reticence and modesty were reflected in his bare office in a ramshackle building. Shoved into a corner at the back of the cheap furniture stood a splendid object of carved wood and macaw's feathers, an Indian god, said the missionary, that had been joyously surrendered to him by some of his converts. The other decoration was a coloured photograph of a Chácobo Indian wearing handsome nose-tusks, and a long gown of bark.

The presence of these reminders of the Indians' traditional past came as a surprise because, in the mood of the Pilgrim Fathers, most missionaries frown on all such things, banning personal adornments of all kinds (unless produced in a modern factory) as well as outlawing musical instruments, and jollifications of any kind in missionary compounds. Mr Halterman was more liberal in his outlook. Indians might dress up as they pleased, and even sing and dance, but only in a 'folkloric' spirit, in other words as long as such activities were stripped of any possibility of a hidden 'superstitious significance'.

Among the innumerable North American religious bodies devoted to the spiritual advancement of South America are three main missionary groups: the New Tribes Mission, the South American Mission, and the Summer Institute of Linguistics, all of whom concern themselves with the capture of Indian souls. Of these the SIL is possibly the richest and most powerful, with thirteen active posts throughout the country. Not only does it have the government's support, but one learns with surprise that it comes under the Ministry of Culture and Education, of which Mr Halterman is an official.

It may be in acknowledgement of this official co-operation that the biblical text that features most prominently in the SIL's well-produced promotional literature is Romans 13:1, offered in Spanish and eight Indian translations. The Institute's text is at variance with both that of the English Revised Version of the Bible, and its Spanish equivalent. 'Let every soul be subject unto the higher powers' becomes 'Obey your legal superiors, because God has given them command', while the SIL quite remarkably re-translates 'the powers that be are ordained by God' as 'There is no government on earth that God has not permitted to come to power.' (Could General Banzer, who seized control of the country in 1971, have had a hand in this linguistic exercise?)

Mr Halterman agreed that the SIL, as well as the two other evangelical missions, were religious fundamentalists, and therefore ready with a tooth-and-nail defence of every line of the Holy Writ, including the world's literal creation in six days, and Eve's origin as a rib from Adam's side.

Fundamentalists also believe that all the non-Christians of this world, including those who have never heard of the existence of the Christian faith, are doomed to spend eternity in hell. As the printed doctrinal statement of the New Tribes Mission – with whose theology Mr Halterman said he was in complete agreement

– puts it: 'We believe in the unending punishment of the unsaved.' It is this belief that inspires so many missionaries to save souls at all costs, often with disregard for the converts' welfare in this world.

'We have a very limited medical programme,' Mr Halterman said, and one could be sure he meant what he said. It is this indifference to anything but the act of conversion that explains the almost incredible experience reported by the German anthropologist Jürgen Riester in an encounter with a missionary who in 1962 had been entrusted by the Bolivian government with the pacification of the Ayoreo Indians.

'The missionary allowed more than 150 Ayoreos to die in cold blood, after establishing contact with them. The Indians were dying of a respiratory disease accompanied by high fever, and the missionary held back medicine, using the following argument: "In any case they won't allow themselves to be converted. If I baptise them just before they die, they'll go straight to heaven."'

Mr Halterman agreed that a certain number of Indians remained at large in the forest areas designated for future occupation by the white immigrants. It was a matter for regret, he thought, and he seemed to blame himself and his brother missionaries for incompetence in this matter. The Indians could be dangerous, he said, mentioning that only two days before, a member of an oil exploration team had been shot to death by arrows. However, there were still souls to be harvested, and he described with quiet relish the methods used to entice the occasional surviving Indian group from its natural environment so that this could be accomplished.

'When we learn of the presence of an uncontacted group,' said the missionary, 'we move into the area, build a strong shelter – say of logs – and cut paths radiating from it into the forest. We leave gifts along these paths – knives, axes, mirrors, the kind of things that Indians can't resist – and sometimes they leave gifts in exchange. After a while the relationship develops. Maybe they are mistrustful at first, but finally they stop running away when we show, and we all get together and make friends.'

But the trail of gifts leads inevitably to the mission compound, and here, often at the end of a long journey, far from the Indian's sources of food, his fish, his game, it comes abruptly to an end.

'We have to break their dependency on us next,' Mr Halterman said. 'Naturally they want to go on receiving all those desirable things we've been giving them, and sometimes it comes as a surprise when we explain that from now on, if they want to possess them, they must work for money. We don't employ them, but we can usually fix them up with something to do on local farms. They settle down to it when they realise that there's no going back.'

'Something to do' on a local farm is only too often indistinguishable from slavery. Mr Halterman, whether he knows it or not, is the first human link in the chain of a process that eventually reduces the Indians to the lamentable condition of all those we saw in Bolivia. There are many hundreds of missionaries like Mr Halterman all over South America, striving with zeal and devotion to save souls whose bodies are condemned to grinding labour in an alien culture.

While the North American missionaries have become – often officially – the servants of such right-wing military dictatorships as that of Bolivia, opposition in the name of Human Rights is frequently organised by Catholic priests and members of some religious orders. Their efforts, although at best hardly more than an attempt to alleviate harshness and cultivate compassion, involve them in some risk, and shortly before we arrived in La Paz gunmen murdered two assistants of a priest who had been troublesome. We were told that such gunmen could be hired to assassinate someone of small importance for as low a fee as one hundred pesos (rather less than £3).

A number of our informants were churchmen of a staunchly liberal kind, but it is not possible to mention them by name, or even identify the organisations to which they belong. One of them described the current Banzer regime as a confident and therefore fairly mild form of Fascism. Unlike Chile, which had lost all self-respect, it valued its good name. When, in January 1974, the peasants in Cochabamba showed too spirited a resistance to its authority, it had not been above sending in planes and tanks and killing a hundred or so, but for the moment it had fallen like a digesting crocodile into a kind of watchful inactivity.

In the meanwhile, the Church had cautiously involved itself in the formation of peasant groups that could bypass the fraudulent government-rigged trades unions. He expected that eventually the crocodile would show that it was far from asleep. Future repression, he thought, was certain.

It was a priest who introduced us to some of the facts about the abundant labour promised by Dr Strauss to the new white immigrants, and he took us to see cane-cutters at work on an estate near Santa Cruz. Some forty thousand migrant workers are brought in to deal with the cotton harvesting and the cane-cutting. These are all Indians, the majority from the Altiplano. Of the two groups, the cotton pickers seem marginally the worse off, being housed in dreadful barracoons in which they sleep packed in rows, sexes mixed, thirty or forty to a hut.

The working day is from just before dawn until dark. Altiplano Indians, accustomed to the cold, clear air of the high plateau, suffered dreadfully, the priest said, in the heat of the tropics, and also from the incessant attacks of insects, unknown in the highlands, but which made life unbearable to them here. It was difficult to estimate how much a worker earned, but, allowing for loss of working time through bad weather, the priest estimated that this might average out at fifty pence a day. But this was far from being the take-home wage. Various deductions had to be made, including the contractor's cut.

All the estates employed agents who scoured the country in their search for suitable labour. The plantation owners paid them fifty pence to two pounds fifty for every man, plus a percentage deducted from the worker's pay.

There were other drawbacks. The migrant worker would be forced to buy his supplies from the estate's stores, where prices could be three or four times those normally charged. Jürgen Riester noted in his work on the Indians of Eastern Bolivia that a kilo of salt might be charged up in this way at ten times its market price, and a bottle of rum at twenty times normal cost. Thus the average daily wage could be reduced from fifty pence to twenty pence, or less. Worst of all, said the priest, almost all migrant workers were debt-slaves, and the debts they had been induced to incur went on mounting up every year, so that they were bound for life to a particular employer; the children, who would inherit the debt, would be bound to him, too.

The cane-cutters we visited were on a sugar estate about twenty miles from the city. They were Chiriguano Indians, from Abapó Izozóg – one of the two principal areas designated for the new white settlements. These men worked a fifteen-hour day, starting at 3 a.m., by moonlight or the light of kerosene flares, except on Sunday, when thirteen hours were worked. The two free hours on Sunday were dedicated to a visit by lorry to buy supplies at the estate owner's shop in Montero, the nearest village. They were paid about fifty pence a day. Although their contracts stipulated that water, firewood and medicines would be provided free, there was no wood, two inches of a muddy brown liquid in the bottom of one only of the two wells, and the only medicine given was aspirin, used impartially in the treatment of enteritis, tuberculosis (from which many of them suffer), and snake-bite. Every woman over the age of eighteen had lost her front teeth as a result of poor nutrition, leaving the gums blue and hideously swollen.

A contractor hung about, keeping his eye on us; a sleek and smirking young man in a big sombrero, a digital watch strapped on his wrist and a transistor to his ear. Part of his duties would be to keep a look-out for cane-cutters that were obviously not long for this world, so that they could be shipped back to their villages where they could die out of sight. All these cane-cutters were debt-slaves. It should be stressed that the estate was not specially singled out for this investigation, but was chosen at random, largely because it was easily reached from the main road. It was almost certainly no better nor any worse than the rest.

The current fight championed by the Church on such estates is for the elimination of the contractor and the abolition of a system by which twenty percent of wages are withheld until the end of the harvest to prevent desertion. Even when, despite this precaution, workers do cut and run, they may be brought back by the police as absconding debtors. Little objection is seen from above to controlling labour by brute force: a mentality inherited from the days before 1962, when estates were bought and sold *with* their workers.

In 1972, at the time of the cotton boom and the trebling of cotton prices on the world market, workers were forcibly prevented from leaving the cotton fields. Troops were sent to the Altiplano to recruit labour, while in Santa Cruz schools were closed so that the children might be free to help out.

The following news-clip from the *Excelsior*, dated 23rd June 1977, gives some idea of labour conditions that can still exist in odd corners.

Slave camp denounced in Bolivia

La Paz, 22nd June. The unusual case of a slave camp's existence was denounced here today. The denunciation was received in the labour office of the town of Oruro against the owners of the Sacacasa estate alleging this to be a slave camp. Apart from harsh treatment received by both adults and children, they are forced to work from 6 a.m. to 6 p.m. for a daily wage of 10 pesos [about twenty-five pence]. The slaves are threatened with firearms and brutal floggings to compel them to submit to this exploitation.

The migrant workers I have described are classified as Indians integrated into the national society. They have, in some cases, been in painful contact with white people for several centuries. They dress as whites, are nominally Catholics, and often no longer speak an Indian language. The Chiriguano cane-cutters were descendants of those who survived the wholesale exterminations of the great rubber boom of the nineteenth century, when those Indians who were dragged from their villages to become rubber tappers had no more than a two-year expectation of life, and could expect punishment, which might include the amputation of a limb for failure to produce the expected quota of rubber. There exists a class of even cheaper and more defenceless labour. These are 'non-integrated' Indians, who have only recently been driven or enticed from the jungle. They are at the bottom of the pyramid of enslavement.

On 16th October 1977, two days before our arrival in Santa Cruz, *Presencia* published an account of the kind of misadventure that can befall a forest Indian – in this case one of a band of refugees who escaped from the mission compound – who happens to follow a road and arrive at the end of it, dazzled, bewildered, and quite unable to make himself understood, in the streets of a boom city.

This Indian, an Ayoreo named Cañe, was washing his clothes in the River Piraí, a few yards from the Santa Cruz main railway station, when he heard screams coming from a parked car in which two men were attacking a girl. Cañe ran to the girl's aid, and the two men drove off, but they soon returned with a police car. In this, after a thorough beating, Cañe was taken to the police station where, being unable to give any account of himself in Spanish, and in the absence of an interpreter, one of the policemen simply drew his gun and shot him through the head. The bullet entered the right side of the head, low down, behind the ear, and exited, astonishingly, without damage to the brain.

The unusual aspect of Cañe's story, apart from the miraculous escape which got him into the newspapers, was that he was taken to hospital. In Latin America it is unusual for an ambulance ever to be sent for an Indian.

This happened on 9th October, and on 20th October, learning that Cañe and the rest of his fugitive groups were still to be seen on a piece of wasteland outside the Brazil Station, we found an interpreter and went in a taxi to talk to him. The taxi-driver had some reason to know just where the Ayoreos were to be found, because he mentioned that the Ayoreo women had been driven by starvation to prostitute themselves for five pesos (thirteen pence) per visit. He himself had had intercourse with one of them several days before, copulation having taken place at dusk, in the open, by the side of the well-illuminated and busy road. He now waited with anxiety the possible appearance of dread symptoms.

We found approximately twenty Ayoreos on wasteland by the side of the new dual carriageway. Cañe was among them, and we examined the still raw wound in the back of his head. He told us that a number of ribs had been broken in the beating he had been given, and that an attempt had been made to break both his wrists, using a device kept at the police station for that purpose.

Cañe, a magnificently strong young man, had put up such resistance to this that the policeman had had to give up his efforts, and then in frustration had drawn his gun and shot him instead.

The Ayoreos are the proudest of the tribes of Bolivia, making a fetish of manly strength and courage, particularly as demonstrated in their hunting of the jaguar. To acquire status in the tribe and marry well, an Ayoreo must be prepared to tackle a jaguar at close quarters, in such a way that the maximum amount of scarring is left by the encounter on his limbs and his torso.

For these Ayoreos the days of hunting the jaguar in the Gran Chaco were at an end. They had gone through the mission, been deprived of their skills and been taught the power of money. As a last resort, since food had to be bought, they sold their women. Cañe remembered being taken by a missionary as a boy from the Chaco. Since then he had slaved for farmers, being paid with an occasional cast-off garment or a little rice. In the end, he and his companions could stand the life no longer, and had just wandered away following the road through the jungle, and then a railway track until they reached Santa Cruz.

We learned that the mission from which the Ayoreos had decamped was a South American mission station in the jungle some twenty kilometres from the village of Pailón. Deciding to see for ourselves what were the conditions that could have caused this apparently hopeless headlong flight into nowhere, we visited the mission on 22nd October, in the company of three Germans, one of whom spoke Ayoreo.

A Quiet Evening

The scene, when we arrived in the camp, was a depressingly familiar one: the swollen bellies, inflated pulpy flesh, toothless gums and chronic sores of malnutrition, the slow, listless movements, the eyes emptied by apathy. Here, 275 Ayoreos, a substantial proportion of the survivors of the tribe, had been rounded up with their jaguar-scarred chief, who presented himself, grotesque in his dignity, wearing a motorcycle crash-helmet. We inspected a deep cleft in his forehead where he had attempted to commit suicide, using an axe.

The only signs of food we saw was a bone completely covered by a black furry layer of putrefaction, being passed round to be gnawed, as well as a cooked tortoise being shared among a group. With our arrival a commotion began, led by some weeping women, and we soon learned the reason. Here in the tropics, at the height of the dry season, the water supply had been cut off by the missionary in punishment for some offence. The Indians, several of them ill, and with sick children in the camp, had been without water for two days.

We saw the missionary, Mr Depue, a lean shaven-headed North American of somewhat austere presence, who confirmed that he had ordered a collective punishment he believed most likely to be effective to deal with the case in which two or three children had broken into a store and stolen petrol. There was to be no more water until the culprits were found, and brought into his compound, there to be publicly thrashed.

The situation was a difficult one because, as Mr Depue explained, in all the years he had spent as a missionary, he had never heard of a single instance of an Indian punishing a child, which was to say that the conception of corrective chastisement seemed to be beyond their grasp. Mr Depue spoke of this aversion to punishment as of some genetic defect inherited by the whole race. It had now come to be a trial of strength, and he could only hope that the deadlock would soon be resolved. He took up an 'it-hurts-me-as-much-as-it-hurts-them' attitude, assuring us that he had decided to share the general discomfort by ordering the water supply to the mission house to be cut off as well. It occurred to us that he might have prudently arranged for a reserve, because we happened to arrive when the missionary and his family were at lunch, and both water and soft drinks appeared to be in reasonable supply.

Mr Depue happened to have read the newspaper's account of Cañe's misfortunes, and remembered that he himself had 'brought him in', during a pacification drive in the Chaco. Three or four youngsters, including Cañe, had become separated in the panic from their tribe. 'I kept out of sight and sent Ayoreo-speaking Indians to offer them a better life, and to persuade them to come in, which they did.'

It was by chance on our way back from this expedition that we saw our first *criada* – a Chiquitano Indian girl who had been 'adopted' by a white family, and happened to serve us in her foster-parents' bar. The *criada* system is far from being exclusively Bolivian, and exists under different names in backward rural areas in most Latin American countries where there are groups of depressed and exploited Indians.

In the hope that they will receive some education, and an economically brighter future, Indians give their children away to white families. The little Indian – usually a girl – becomes a Cinderella that no prince will ever discover. She will be put to unpaid drudgery from the age of four or five, be traditionally available for the sexual needs of the sons of the family, and will not be able to marry, although she will be allowed to have children, who in their turn will become *criadas*.

Our German friends knew this girl well, but having lived for some years in Bolivia, and become accustomed to its institutions, they were not horrified, as we were, that such barely disguised forms of slavery could exist.

A *criada*, they informed us, could be lent or given away. They had no rights of any kind: they were the rural Untouchables of Latin America, whose existence went unnoticed. There was no saying how many there were in Bolivia, they said. In some parts of the eastern provinces almost every farm kept one.

A Quiet Evening

Santa Cruz is a boom town, a little dizzy with quick profits, and displaying its wealth as best it can. It has a new Holiday Inn, full of American oilmen in baseball caps, and possesses no fewer than four ring-roads. Among its leading citizens are Germans, the most successful and affluent of the foreigners in Bolivia. There are about 300,000 of them, and it is said that President Banzer, who is a descendant of one of the older German families, came to power through a military coup financed by the German colony.

The powerful Dr Strauss, in control of immigration, comes of German forebears, and Teutonic surnames are scattered liberally through the lists of directors of the country's leading enterprises. Near Abapó Izozóg, immediately adjoining the nominally empty area (save for a few thousand Indians) that Dr Strauss proposes to people with Rhodesians, South-West-African whites of German descent and refugees from South Africa, there are Germans who occupy a colony about the size of Holland. They have founded other vast colonies at Ascención de Guarayos in the centre of the country, and at Rurrenaháque, in the north, where they have sunk fortunes into building roads and costly irrigation schemes.

A high percentage of German immigrants arriving since the war had remained loyal to Nazi political philosophy, and recently neo-Nazi groups had also emerged. The Bolivian government appears indifferent to this phenomenon, and has pushed its neutrality to the lengths of resisting the extradition of at least one war criminal. Neo-Nazi journals imported from Germany, and such militaristic publications as *Soldaten-Zeitung*, find avid readers.

Our own brief experience of martial nostalgia was to be warned on arrival at our hotel of an impending dinner for some three hundred Germans, to raise funds for the German school (the best in Eastern Bolivia). With some embarrassment it was hinted that many of the guests were ex, or actual, Nazis and that we might find some of their old wartime drinking songs offensive. We listened to the clamour but kept out of the way. Most extraordinary of all was the assurance that German Jews in Bolivia had sunk their differences with their old Aryan persecutors, and now fraternised at such gatherings, joining to chorus the 'Horst Wessel' along with the rest.

Our final conclusion was that Dr Strauss's justification for the plan to bring in whites from the South African countries – i.e. economic necessity – was not quite the whole story. Bolivia is potentially one of the richest of South American countries, since apart from other largely untapped timber resources in the east, aerial surveys have indicated that the Amazon Basin it shares with Brazil contains one of the most valuable and diverse mineral profiles in the world. It is also among the weakest of these countries, with a population of five million, many of whom are illiterate subjugated Indians, contributing little to the country's muscle. With these human resources it must be ready to defend the thousand miles of frontier it also shares with Brazil, which has more than a hundred million.

At the moment, Brazil is fully engaged in gobbling up its own resources, but it can be imagined that sooner or later it might turn its eyes westwards with a renewed appetite and in a mood for expansion. Brazilian roads have either been built or resurfaced by Army engineers to take the heaviest tanks. One such road points to the heart of Bolivia through Corumbá, after which, crossing the frontier, it dwindles to dirt. Also, through foreclosures, Brazilian banks already own much land along Bolivia's eastern borders.

Almost imperceptibly, Bolivia has been a country bleeding to death. In a series of wars fought and lost over the past century it has seen its territory whittled away by victorious neighbours: first the nitrate-rich Atacama Desert and the port of Antofagasta to Chile; then the Acre territory to Brazil; then three-quarters of the Chaco to Paraguay. Always these losses have been the result of its failure to fill 'empty' spaces occupied by the Indians, who do not count.

Seeing into the future with Dr Strauss's eyes one might be tempted to agree with him that, from his viewpoint, and that of the Bolivian government, the problem of filling this territorial vacuum is urgent. Strauss has to have his immigrants, and sooner or later he will probably get them, as the intransigence of the South African whites stokes up fuel for the fire tomorrow.

Together with the powerful German-Dutch minority already in place, these newcomers could transform Bolivia into a strong, white-dominated, ultra-right, anti-Communist state in the heart of Latin America. This vigorous transformation would discourage the future covetousness of neighbouring states, and it would delight the United States by laying forever the ghost of Che Guevara – himself once attracted to empty spaces in Bolivia.

The Tribe that Crucified Christ

In 1982, five years after his visit to Bolivia, Lewis happened to see a BBC film about the Panare Indians of southern Venezuela, which renewed his preoccupation with the fate of South America's indigenous tribes. He contacted Paul Henley, the film's consultant, and early the following year set off with the photographer Don McCullin. He soon discovered that the New Tribes Mission had already entrenched itself in the area. It continued to operate in Venezuela until Hugo Chavez ordered its expulsion in 2005, accusing the mission of imperialist infiltration, destruction of indigenous culture and connections with the CIA.

In 2017, the New Tribes Mission changed its name to Ethnos 360 'because we're positioning to reach a changing world'. The mission reaffirmed its founder's principle, laid down in 1943: 'By unflinching determination we hazard our lives and gamble all for Christ until we have reached the last tribe regardless of where that tribe might be.' According to its website, it still retains more than two thousand missionaries in more than twenty nations.

First published in *The Sunday Times*, May 15, 1983

OUR FIRST VIEW OF THE PANARE was at the village of Guanama, sweltering at the end of the track from an unfinished dirt road that faltered southwards through Venezuela in the general direction of the Amazon. A half-dozen Panare males soon came out of a communal roundhouse, moving springily like ballet dancers, with an offering of hot mango juice. They were good examples of a people described as incredibly impervious to Western influence, dressed therefore in no more than scrupulously woven loincloths, and armlets of blue and white beads. A long history of nomadism had shaped them to look unlike their nearest white neighbours who, having spent their lives on horseback or in cars, are often somewhat misshapen and inclined to fat. The Panare who, if put to it, could run and walk fifty miles a day across the savannah were lean, lithe and supple, coming close in their bodily proportions to the classic ideal.

Guanama was spruce and trim with everything in place, a little like an anthropological model. Its roundhouses were masterpieces of stone-age architecture, built for all weather, and marvellously cool under their deep fringing of thatch. It was a quiet place as Panare villages are wont to be. The dogs remained silent and respectful, children didn't cry, and the adults, back from hunting or work in their gardens, slipped into their hammocks after greeting us, to resume soft-voiced discussions. Only one thing seemed out of place in this calm and confiding atmosphere – the new barbed-wire fence, a symbolic intrusion of an alien culture.

We had made a point of this visit to Guanama after a report of extraordinary happenings by Maria Eugenia Villalón, who had gone there while employed in a census of the Indian population. A year before she had been in Guanama to record Panare songs, and now, returning for the census, she proposed to entertain the villagers by playing these back to them. No sooner had the tape recorder been switched on than the Indians leaped to their feet in a state of panic, running in all directions, their hands clasped over their ears. The machine was switched off and the commotion subsided. The Panare explained that what they had been compelled to listen to was the voice of the Devil speaking through their mouths. Now they had found Jesus, and henceforward would sing nothing but hymns. They lined up to oblige with one of these, a Panare version of 'Weary of earth and laden with my sin...', the first line repeated ad infinitum to Mexican guitars and the rattle of maracas. It was clear to Señora Villalón that the New Tribes Mission had moved in. Members of this organisation are the standard-bearers in Venezuela of the new computerised, airborne evangelism that insists not only on conversion, but on the demolition of all those ceremonies and beliefs by which an indigenous culture is defined.

The question now was how far the missionary labours had progressed. Evangelists rush to cover the unclothed human form, and the spectacle of Indians dressed in shapeless and often grubby Western cast-offs is frequently a glum reminder of their presence. Apart from the barbed-wire fence Guanama was free from the ugliness too often associated with the disruption of belief. We preferred not to abandon hope, and Paul Henley, the British anthropologist who was with us, who had lived, off and on, among the Panare since 1975 and speaks their language fluently, now put the fatal question. 'When is your initiation ceremony to be held?' The reply was a depressing one, confirming our worst fears. 'There will be no ceremony. God is against it. We have turned our backs on all these things.'

It was indeed a breach with the past. The *Katayinto*, the great male initiation ceremony, held in the dry season when food is most plentiful, is for the Panare the culmination of the annual cycle, and to them the equivalent of all the religious and secular feasts of the West rolled into one. Weeks of food gathering and general preparations are necessary, and the festival itself, involving dramatic episodes and three major dances and a number of minor ones, may stretch over six weeks. It ends with the boys' investiture with the loincloths signifying their attainment of adult status. To the Panare the loincloth represents what the turban does to the Sikh, and to destroy the *Katayinto* ceremony is to remove the cornerstone and expunge the future of a culture believed by those who have studied it to have developed over thousands of years. No one understands this better than the missionaries, for whom all such ceremonies, and the wearing of the loincloth itself, shackles the Indian – as they see it – to the heathen past. From time-to-time *Brown Gold*, house magazine of the New Tribes Mission, prints a jubilant notice of a tribe that has been persuaded to change loincloths for trousers, evidence that it is at last on a road from which there is no turning back. Thus in Tanjung Maju: 'The first time we entered the village they were wearing loincloths and very primitive ... see how they have grown in the Lord.'

The New Tribes Mission, now continuing its implacable advance in those parts of the world where 'un-contacted' tribal people remain to be swept into the evangelical net, was founded in 1941 in El Chico, California. It now has some 1,500 missionaries working with 125 tribes in 16 countries. In South America (which it has divided up with its missionary rival, the Summer Institute of Linguistics) it is represented in Venezuela, Bolivia, Brazil and Paraguay, where it has rolled over the Catholic opposition. The Catholic Fathers, sometimes reproaching their flock with desertion, are discouraged by the reply, 'You have no aeroplanes. You are not in touch with God by radio.' To the outsider both fundamentalist missions are identical in their aim and the methods employed, but the New Tribes Mission criticises the Summer Institute of Linguistics as 'too liberal'. Both view Catholics with distaste, and consider their converts as hardly better placed in the salvation stakes than outright pagans.

Mission finances, according to its prospectus, depend upon public donations. These do not necessarily take the form of cash. Survival International (1980) reports an offering of 2,500 hectares of land by the Government of Paraguay, and in 1975 a missionary spoke to me of 'a heck of a piece of land given to the Mission by a company in the Paraguayan Gran Chaco engaged in the extraction of tannin. They figured we could help with the Indians.' The Mission does not hold itself aloof from engaging in commerce, acting frequently as middleman in the supply of goods to the Indians or the resale of their artefacts. Survival International mentions that they are in the fur trade in Paraguay, dealing in jaguar skins which fetch high prices since the jaguar elsewhere is an internationally protected animal.

Impressive technical equipment and abundant funds give the New Tribes Mission more than a head start in the race for souls. Thereafter its work is carried forward by the zeal of its 'born-again' fundamentalist missionaries recruited from those areas of the United States where Darwin is excluded from the school curriculum, while fossils are explained away as Devil's devices implanted in the rocks to cause confusion among the servants of God. The reappearance of the witches of Salem would cause no great surprise. The Mission proclaims with fervour and enthusiasm the imminent Second Coming of Christ and the destruction of this world, and its doctrinal statement includes the belief in the 'unending punishment of the unsaved', thus committing to the flames of Hell all adherents of Judaism, Hinduism, Buddhism and Islam, besides several thousand minor religious faiths that had evolved before the coming of Christ.

Two thousand tribes remain to be contacted, all of them under a threat of everlasting fire, so conversion is a task of utmost urgency. It is this sense of time being so short that tends to outweigh all considerations of the convert's welfare in this life, provided that his soul is saved for eternity. 'He saves the souls of men,' runs the New Tribes Mission doctrine, 'not that they might continue to live in the world, but that they might live forever with Him, in the world to come.'

Military dictatorships are the natural supporters of the New Tribes Mission, with whom they share similar views. Les Pederson, a director, illustrates this identity of outlook in his autobiography *Poisoned Arrows*: 'The President of the Republic of Paraguay, Don Alfredo Stroessner ... assured me of his appreciation of what we are doing among the indigenous peoples of the country.' Elsewhere efforts to get rid of the missionaries have been frequent, strenuous, and sometimes crowned with temporary success. The clamour has been loudest in Venezuela where a united front of jurists, anthropologists and churchmen have accused the Mission of infringement of the Indians' human rights, including coercion and forcible conversion. Public opinion has led to the creation of two Congressional investigations in Venezuela. The latter of these opened in 1979, and remained in session for some two years, filling the Venezuelan press with bizarre accounts of missionary activities.

Naval Captain Mariño Blanco, charged with keeping an eye on the doings of foreigners in the country's remote regions, spoke at the latter Congressional hearing about scientific espionage. Noting that the missionaries inevitably installed themselves in areas known to contain strategic minerals such as cobalt and uranium, the captain claimed to have proved that they were in the pay of American multinationals, naming two of them as Westinghouse and General Dynamics. He noted that the Mission had been in trouble in Colombia, suffering expulsion for 'damage to national interests and for having assisted illicit explorations carried out by transnational companies in areas likely to contain deposits of strategic materials'. The captain had found missionary baggage labelled 'combustible materials' to contain military uniforms and 'other articles' – this being taken by the press to refer to geiger counters. The uniforms were explained away by the missionaries as intended to impress the Indians. Captain Blanco said that the head of the New Tribes Mission had tried to bribe him. He gave his opinion that the missionaries' involvement with the Indians was only a cover for their other activities.

A Ye'cuana Indian, Simeón Jiménez, speaking defective Spanish with much eloquence, appeared at the hearing to describe the prohibitions imposed upon his people as soon as the missionaries had taken hold. They included the drinking of fermented juices, dancing, singing, the use of musical instruments, tribal medicines and tobacco, and the tribal custom of arranging marriages within the framework of kinship groups.

Jiménez stressed the psychological terror inflicted on the Ye'cuanas to force conversions. In particular he cited the appearance of a comet, described by the chief missionary in the area as heralding the end of the world. The missionary had gathered the Ye'cuanas together and given them three days, on pain of suffering a fiery extinction, to break with their wicked past. They were later warned by the same men of a communist plot to drive the missionaries out of the country, saying that if this were to happen US Airforce planes would be sent to bomb Ye'cuana villages.

I was unable to see Simeón himself and listen to an account of their traumatic experience from his own lips, because he was seven days away by canoe in the Orinoco jungle. Instead I called on his wife, Dr Nelly Arvelo, a distinguished anthropologist who had set a seal on her approval of the lifestyle of primitive hunter-gatherers by marrying one. She confirmed all her husband had had to say, including an incident when Simeón's aged grandmother had come to him in tears, imploring him to give up his struggle before they were all reduced to ashes.

Terror apart, Dr Arvelo said, the missionaries had worked out a new kind of punishment for those who resisted conversion. 'Indians,' she said, 'like to do everything together. They share everything, particularly their food. They're very close to each other. The missionaries understood this so they worked out that the best way to punish those who didn't want to be converted was by isolation. As soon as they had a strong following in a village they would order the converts to have nothing more to do with those who held out. No one, not even their own parents, was allowed to talk to them, and they were obliged to eat apart from the rest. It was the worst punishment an Indian could imagine, and often it worked.'

Following the Congressional Hearings the press had delved into the Mission's history, noting that in Paraguay they had been involved in manhunts carried out against the Aché Indians, and in more manhunts, enforced relocation and enslavement of 'wild' Ayoreos (Survival International, 1980). It was further noted that a description of such an armed manhunt, when Indian fugitives were taken as slaves, had actually appeared in a Mission publication. A group of foreign anthropologists, three of them British, wrote a letter to a Caracas newspaper calling for the Mission's expulsion, resulting in two of the American signatories being immediately summoned to their embassy for an ambassadorial rebuke. According to Captain

Blanco there was at least one other intervention by the US Embassy in support of the New Tribes Mission. 'I ordered the arrest of two American engineers named Ward and Curry, who were carrying out (illegal) scientific investigations ... Later it was proved that James Bou (head of the New Tribes Mission in Venezuela) had organised their journey ... Mr Bou telephoned the US Embassy, and the Counsellor of the Embassy then called me, asking me to release the two men.'

The feelings of the Venezuelans as a whole was summed up by the Apostolic Vicar of Puerto Ayacucho, the Amazonian capital, who said: 'These people have created a terrible confusion in the Indian's mind. They have no conception of Indian culture. When you forbid the Indian to dance, drink his *yarake* or eat the ashes of his dead ones, you destroy his culture. One doesn't spread God's message by terror. The New Tribes Mission relies on force, and if the native allows himself to be converted he does so not out of conviction, but fear.'

The methods used by the New Tribes Mission to deal with the Ye'cuana seemed to have proved successful: a high percentage of the tribe – perhaps as many as 75 percent – had been induced to accept conversion and to renounce their old customs. Attention was now focused on the Panare, who had been least receptive of all Venezuela's twenty Indian tribes to the evangelical message.

In April 1972, a Mr and Mrs Price of the New Tribes Mission had carried out an aerial survey of the Panare region and decided on establishing a mission in Venezuela's Colorado valley where an easily accessible Indian settlement had been observed. A jeep was sent to the spot, where they were well received. 'The Lord provided us with a Panare guide, without whom we would not have known where to go.' Although they had been told before that the Panare never worked for anyone, such was the native hospitality that 'the Indians seemed willing to have us come to live there and to build a house for us ... the Panare fellows pitched in and worked really hard'. Clearly there was satisfactory human material for the missionary labours here, and only a small note of disapproval obtrudes. 'On the other side of the clearing could be seen a large, hollowed-out log in which they had their drink, made of mashed corn, sugar cane and sweet potato. The tracks where they had danced were still visible.'

Thereafter progress towards salvation went at a snail's pace. The Indians were helpful and friendly in every way, but they had had contact with missionaries – Jesuits and Franciscans – in the past, and had clearly not enjoyed the experience. Five years after the Lord had 'impressed upon the hearts' of the original three missionaries to settle where they did, the Panare continued to lead their same old easy-going lives, to drink and to dance, to share their food and do as little work as they had to. They remained eager recipients of trade goods, using the missionaries' iron tools to increase the size of the communal houses that the missionaries so much disliked, and of their gardens where far too much of the produce went into the preparation of liquor. In matters relating to the acceptance of the new faith they remained as wary and unreceptive as ever.

Henry Corradini, a Venezuelan anthropologist who had worked with the Panare for a number of years and speaks their language, began an investigation of books of scriptural stories translated by the Mission into the Panare language, which he suspected might have embodied manipulations of the holy text.

Two books based on what purported to be stories from the Bible were soon available in translation, the first *Learning about God* (1975), the second *The Panare Learn About the Devil* (1976). The creation of these had presented certain linguistic problems, solved in the end in a resolute fashion. Difficulties arose from the fact that, like many other Indian languages, there are no equivalents in Panare for many words held as basic to the concepts of the Christian religion. There are none, for example, for sin, punishment or redemption. God cannot be thanked or praised, only congratulated. Above all Panare lacks any word for guilt.

This above all was a situation that had to be rectified. A way had to be found to manufacture the sense of guilt upon which repentance and salvation depended, and the missionary translators may have decided that the best way of tackling this was by re-editing the scriptures in such a way as to implicate the Panare in Christ's death. Henry Corradini soon discovered that the New Tribes Mission's version of the Crucifixion as arranged for Indian consumption was at striking variance with that of the Bible. Gone were the Romans, the Last Supper, the trial and Pontius Pilate turning away to wash his hands. He read on:

> *The Panare killed Jesus Christ*
> *because they were wicked*
> *Let's kill Jesus Christ*
> *said the Panare.*
> *The Panare seized Jesus Christ.*
> *The Panare killed in this way.*
> *They laid a cross on the ground.*
> *They fastened his hands and his feet*
> *against the wooden beams, with nails.*
> *They raised him straight up, nailed.*
> *The man died like that, nailed.*
> *Thus the Panare killed Jesus Christ.*

If this could not create feelings of guilt, nothing could. Now there was talk of God's vengeance for the dreadful deed.

> *God will burn you all,*
> *burn all the animals, burn also the earth,*
> *the heavens, absolutely everything.*
> *He will burn also the Panare themselves.*
> *God will exterminate the Panare*
> *by throwing them on to the fire.*
> *It is a huge fire.*
> *I'm going to hurl the Panare into the fire*
> *said God.*

The comet had come and gone but the frightening memory of it remained. God had relented once but might not a second time.

God is good.
'Do you want to be roasted in the fire?'
asks God.
'Do you have something to pay me with
so that I won't roast you in the fire?
What is it you're going to pay me with?'

The nature of the payment demanded is a foregone conclusion; unquestioning submission to the missionaries' demands, the abandonment of their traditional life and their customs, their culture. Ethnocide is the price exacted by the ferocious deity the missionaries have created. The pressure proved too much even for the well-tried nerves of the Panare, and within months results began to come in. The following, headed 'Panare Breakthrough', is quoted from *Brown Gold*, dated 1977: '...I finished stressing the need for each one to ask God for the payment of their own sins.... A few hours later Achen (a Panare woman) came by the house, she said, "I asked God like this: I want my payment for my sin (sic). I don't want to burn in the big fire. I love Jesus."

'...Here we had sat for almost a year teaching one believer and nothing else happening, and then all of a sudden, WOW!'

Venezuela's Colorado valley, where it had all started, came as a surprise. It gave a feeling of being in the Orient rather than the West – a landscape sketched in briefly by a Chinese artist, red earth with angular trees set among immense black boulders, backed by a recession of low hills afloat in the mist. Communal houses showed among the trees down by the river like delicately woven straw hats, and we could see the Panare women moving about, walking with quick, strutting steps, and wearing nothing but G-strings, tassels and beads. The course of the river was marked by a tight border of forest, full of noisy birds and great dark, blundering butterflies. In this arcadian setting the missionary building, solid and rectangular at the head of the airstrip, seemed austere and aloof.

Paul Henley presented us to the 32 adult men and women of the extended family who had adopted him. We had brought gifts for them all, and in accordance with egalitarian principles each man received an identical nylon fishing line, and each woman a garishly decorated enamel bowl. In addition we handed over a sack of rice in return for our share in communal meals we might be invited to join. We were then directed to hang our hammocks in an empty house at the highest point of the village, recommended as being relatively free from mosquitoes. It was a traditional thatched construction, well-swept and free in Panare style from litter of any kind. Following a perfunctory inspection to make sure that there were no rattlesnakes ensconced, we installed ourselves. Soon after, Panare of all ages and both sexes began their visits, examining and commenting in soft, clucking monosyllables on our persons and our equipment, dropping into unoccupied hammocks, and just standing about in companionable groups long after darkness had fallen, clearly trying to make us feel at home.

Next day the news, as in Guanama, turned out to be discouraging. Two years before when the *Katayinto* ceremony was last held, it had been truncated by the omission of its most dramatic component: a piece of theatre involving the ritual appearance at the height of the dancing by a group of strangers who behave in a hostile and menacing manner, but who are finally pacified and induced to join in the general merriment. This episode seemed to symbolise the young initiates' necessity for arriving at a pacific arrangement with the threatening outside world. The Panare said that they had been obliged to cut it out 'because God did not like it'.

In the following year, 1982, there had been no initiation ceremony at all, and Paul had assumed that this had been no more than a postponement. Now we were to hear that again God had raised objections, and that the *Katayinto* would not take place once more, although it 'might' be held next year. It seemed likely that the missionaries' strategy was to encourage indefinite postponement. The Mission had been careful to keep a low profile while the Congressional investigation was going on. Now there were signs it was moving to the counterattack. In February 1982, Elizabeth Stucky, one of the missionaries at Colorado, wrote in *Brown Gold*: 'On the surface it seems as though they (the Panare) have the least interest in spiritual things.' She defends current Mission strategy, anticipating the possibility of the American evangelists' eventual expulsion. 'Santos Casanova is one of the six men who Maurice was teaching ... and who in turn teaches his own people. His group is the largest in the valley who meet together, numbering 100.' This suggests that about one half of the Panare of Colorado have been evangelised, and if it is true, the *Katayinto* is at an end. Maria Villalón described a native evangelist, trained perhaps by Mrs Stucky, at work in a remote Panare community they had visited by helicopter for the census. 'The village children were made to kneel down in a row. No one could understand what was going on, nor could the Panare evangelist make them understand. In the end he said, "every time I say the word Jesus, you must bang your head on the ground", and this they did.'

In the past it had been possible to organise what the Panare call a 'For nothing', a watered-down version of the *Katayinto*, devoid of any ritual significance – certain to have called down the missionaries' ban. The Panare stage a 'For nothing' whenever they can, purely because they like to drink and dance, and they can normally be induced to go through a full repertoire of dances if provided with a sack of sugar with which to brew the very mild, sweet beer obtainable from only three days' fermentation. In preparation for this, a day or so is spent in cutting down a tree and hollowing out from it the 'canoe' to contain the beer – in itself a traditional community exercise in which everyone takes part, and it is seen as contributing to the fun. We asked if a 'For nothing' could be arranged, but there was always a doubt at the back of the mind. The first sign of fermentation can be detected in a warm climate in any sweetened juice only hours after it has been exposed to the air, and we had heard of native 'deacons' keeping a stern watch to see that all such drinks were jettisoned as soon as the first bubbles appeared on the surface.

While this proposal was under consideration we settled down to give the Panare the chance to get to know us, and familiarise ourselves with the village scene.

Missionary propaganda has taken a new turn recently, assuring us that peoples not reached by their message have a miserable time in this world as well as being doomed to perdition in the next. 'In the Panare way of life before the Gospel was shared with them, everything was bad. It was their way of life to expect the worst. Misfortunes hung over their heads. Constant fears were always in their hearts. This ever-present fear seems to be the very pulsation of life itself.' Thus Mrs Linda Myers, writing about our hosts shortly before our visit.

All that we saw from our visit, or the enquiries we made, presented a strikingly less dismal picture. We had previously noticed that the Indians' physique was superior to that of the local whites, and now it seemed likely that they enjoyed better health in general. A number of families had produced six or more children, all of whom seemed lively. The Panare claim that before introduced diseases such as influenza, measles and malaria took their toll, they suffered from hardly any illnesses. Their mental health appeared equally robust. The close-knit communal life of the Panare protects them from most of the pressures familiar in our society, and the crime-rate is nil.

The missionaries supply tools and consumer goods to the Panare which have to be paid for in cash. Aspirin and penicillin are now driving out effective remedies derived from local plants, and Western medicines cost money. Sales promotions, sometimes divinely backed, can seem unnecessary; one in particular infuriated Henry Corradini, who had now joined us. 'God wants us to use soap. He wants us to eliminate unpleasant odours; to wash under the armpits, and round the anal area.' Corradini said, 'The Indians are never out of the water. Without exception they're the cleanest people in the world. How dare these gringos tell them they stink?'

Cash for these purchases has to be found, so the Panare make decorative baskets which they sell to the local whites. It was part of an evangelical manoeuvre to settle the Indians in the vicinity of the missions, wean them away from the barter system, persuade them to buy more and more inessential goods, converting them in this way into wage-earners working a 48-hour week. It has been calculated that with all their household, horticultural and other chores, the Panare work on average only three hours a day. The missionaries' initiation of basket-weaving to rescue the Panare from the evil effects of this sloth has misfired, for the weaving is easily done while lying in a hammock, while in a state of almost trance-like Panare reverie.

In other directions the irresistible bait of trade-goods, of fishhooks, hunting-knives, axes and aluminium bowls has done its work, for the old nomadic expeditions in search of fresh hunting grounds have become fewer and fewer, and this in turn has wiped out stocks of game and fish that are in the vicinity. This being the case, the Panare are always on the lookout for someone with transport who may be cajoled into giving them a lift on a hunting trip to an area which can no longer be reached on foot in a single day.

We took six Indians in the back of the Toyota Land Cruiser deep into the endless park of the savannah in search of mangoes. The fruit-bearing trees could be picked out in the little spinneys dotted about the grassland by the almost artificial brilliance of their foliage among the delicate lavenders and greys of savannah trees. When the Indians spotted them they jumped down, cut bamboos and stirred the branches to dislodge fruit, touching off explosions of toucans and parakeets which streaked away squawking into the sky.

An inclination to keep on the best possible terms with the Panare, with the hope of the 'For nothing' in mind, compelled us to agree with the suggestion they next put up, which was a major fishing expedition which would involve poisoning a stretch of river. This would be done by the use of *enerima*, a liana growing in the mountains which is pounded up and added to the water. The idea can be disturbing to Westerners, but sporting restraints are meaningless in the context of primitive food-gathering realities, where no one kills for the fun of it, but simply to eat. Fishing for pleasure is unknown in the outback of such countries as Venezuela, and insofar as the town-dweller eats fish at all, it is frozen and imported. Consequently the rivers remain stocked, probably to capacity, and when the flow virtually ceases in the dry season, pools form in which stranded fish are confined in an ever-shrinking volume of water, where they are preyed upon by fishing eagles and otters. In this season alone, the Indian uses his poison. His evolution has made a conservationist of him, and in fact the 'poison' seems to do no environmental damage.

It took a day to find and cut the *enerima* and next morning we set out for the Tortuga River, a tributary of the Orinoco, at its nearest point about thirty miles away. The Toyota was crammed, as before, with Panare, but a large number had set out before us on veteran bicycles purchased through the missionaries, with the message 'Christ is coming' painted on the mudguards. On these, pedalling furiously across the savannah, their arrival coincided roughly with ours.

The pool chosen was some 100 yards long by 20 in width. Shoals of kingfishers as big as starlings were splashing into the water when we arrived. Some fifty Panare lined both banks while the poison was being pounded up and put into baskets which were rinsed into the water.

Within five minutes of a milk whiteness spreading into the pool a great subaqueous commotion began, a spinning catherine wheel of tin-plate reflections just beneath the surface, from which a big fish sometimes spun away then shot off in a straight line, dorsal fin cutting the water, making for the shallows. Occasionally one broke surface, launched itself into the air, thumped down on the bank, then propelled itself in a series of leaps a dozen feet across dry land. The Panare waited for the fish to slow down then speared them phlegmatically and without obvious effort, striking home with their barbed lances at thirty feet or more, and always clean through the head.

Fishing, stone-age style, was sensationally productive. In less than two hours several hundred fish had been taken, among them 25 pounders, and the total weight of the catch was in the neighbourhood of a ton. A few remained in the pool twisting and turning beyond easy reach, and the Panare said that these would recover in about four hours. *Enerima* seems to be a nerve poison of a sort, for it has no effect upon edibility. The fish were cleaned on the spot, and the first caracara – a spruce and elegant hawk that stands in here for the vulture – dropped from the sky to attend to the clearing up. The only problem remaining was to get the fish back to the village, where it would immediately be smoked on the many frames already prepared, after which it could be kept some weeks before consumption. It was a highly successful occasion. And the Panare showed pleasure in their usual restrained way. One convert triumphantly produced a tract from the folds of his loincloth – although clearly muddled as to the nature of its message – headed, 'Has life nothing better to offer than this?'

The missionaries, with whom it might have been enlightening, if not useful, to discuss the matter of a 'For nothing', and of a reported ban on photography, were not at home to callers during our first two days in Colorado, and on the third day a plane came and carried them away. Thereafter the mission remained empty but there was little doubt that evangelical interests were entrusted to their trainee 'deacons' who would report on all happenings.

We had never felt over-optimistic about the 'For nothing' and were resigned now on being told that it could not be arranged after all. The excuse given was that a number of essential participants were about to leave on a trip to the mountains to collect tonka beans, for sale to the whites who used them to add fragrance to tobacco.

Following this setback there was nothing further to keep us at Colorado, and we set out on our return to Caracas. On the way we made a side trip to a diamond-mining camp, attracted there by its name, Tiro Loco (Crazy Shot), and by the news of a recent settlement on its outskirts by Panare who had come down from the forests to taste what was to be had of the joys of civilisation in the form of trade goods.

Tiro Loco prided itself on being tough. It was straight out of Chaplin's *Gold Rush*, a shanty town built on a stratum of crushed beer cans, full of hatchet-faced villains in big hats and spectacular whores. In Tiro Loco you could actually see the swing-doors of a bar fly open and an unwanted customer pitched through them headfirst into the street.

The mild, calm Panare newcomers had built their roundhouses (the best examples we had seen) on a hillock above this dynamic scene, and the hard men of our times and the peaceful ones representing the distant past had got together to establish an easy-going and mutually satisfactory relationship. Food for the miners, apart from what the Panare had to offer, had to be flown in at great cost, most of it not worth eating when it arrived. The Panare grew excellent vegetables which they were very happy to offer in exchange for gardening tools from the store, or made up easily enough by the miners.

This Panare village was one of the few as yet unreached by the New Tribes Mission, and here the Indians lived happily under the protection and patronage of as hard-bitten a selection of humanity as it would be possible to find. The miners supplied them with all they needed, with no strings attached. In Tiro Loco the Panare could drink, dance, paint themselves and perform their ceremonies to their hearts' content.

The Congressional committee investigating the New Tribes Mission failed to reach any positive conclusion, nor outrageously was its report ever made public. Inevitably the born-again Christian general had described the Mission as a geopolitical necessity, by which he meant it was useful for Indians in remote jungle areas to be under the control of people who were so far politically to the right that they classified all their opponents as communists, including archaeologists, journalists, army officers, and the Apostolic Vicar of Puerto Ayacucho. The Congressional committee found that charges of espionage were unproven. Meanwhile the Mission hardly bothered to defend itself against charges of ethnocide, since in its doctrinal statement, and its literature, it made it abundantly clear that – presented under another name – ethnocide was precisely its goal.

A Letter from Belize

In 1955 Norman Lewis visited British Honduras, now Belize, immediately before his Guatemala visit (described in 'A Quiet Evening in Huehuetenango'). Britain had taken full control of the territory in 1862 after decades of conflict with the Spanish, declaring it a Crown Colony, and naming it British Honduras. In 1964, nine years after Lewis's visit, the nation was granted self-government, and in 1973 it renamed itself Belize.

Full independence was granted in 1981. Belize remains a member of the Commonwealth, with the British monarch as its titular head of state. It is the only nation in Central America whose official language is English.

First published in the *New Yorker*, October 7, 1955

SOMEONE IN MERIDA said that a good way to go to British Honduras was from Chetumal in south-east Mexico by a plane known in those parts as 'El Insecto', which did the twice-weekly run. My informant pointed out that this route was cheaper and more direct than going via Guatemala, as well as giving anyone the chance to get away from the insipidities of air travel with the big international lines. I agreed with him and went down to Chetumal on a veteran DC-3 that was the last surviving plane of a small tattered fleet.

For every person travelling from Chetumal's airport, there were seven people to see them off, and passing emigration became a purely family affair. I found 'El Insecto', which was a four-seated Cessna, in a field full of yellow daisies, and then helped the pilot to pull it out onto the runway. When it took off he leaned across me to make sure that the door was properly shut. There were a few cosy rattles in the cabin, of the kind that most cars develop after some years of honourable service. These added to the pleasantly casual feeling of the trip. Duplicate controls wavered a foot or two from the tip of my nose, and the pilot cautioned me against taking hold of them to steady myself in an air pocket. 'These small planes take more flying than an airliner,' he said. But apart from fiddling with the throttle lever, probably out of pure habit, and an occasional dab at the joystick, he did nothing to influence our course as we wobbled on through the air currents.

As the Cessna flew at about 2,000 feet, the details below were clear enough. Even birds were visible. A pair of flamingos parted company like a torn flag, and a collection of white maggots, that were egrets, were eating into the margins of a pool. We were following the coastline, a mile or two inland, with the horizons wrapped up in turbans of cumulus cloud, and a few white thorns of fishing-boats' sails sticking up through the sea's surface. Approaching British Honduras, swamps began to lap through the dull, dusty green of the jungle. They were gaudy with stagnation; sulphurous yellows, vitriolic greens and inky blues stirred together like badly mixed dyes in a vat. The pilot pointed out some insignificant humps and thickenings in the forest's texture. These were Mayan remains; root-shattered pyramids and temples. Around them would lie the undisturbed tombs, the skeletons in their jade ornaments. The pilot estimated that only ten percent of these sites had suffered any interference.

The airport at British Honduras was negatively satisfying. There were no machines selling anything, playing anything, or changing money. Nor were there any curios, soft drinks or best-sellers in sight. Under a notice imparting uninteresting information about the colony's industries, a nurse waited, ready to pop a thermometer into the mouth of each incoming passenger. The atmosphere was one of somnolent rectitude. A customs officer, as severely aloof as a voodoo priest, ignored my luggage, which was taken over by a laconic taxi-driver, who opened the door of his car with a spanner and nodded to me to get in. We drove off at a startling pace down a palmetto-fringed road, by a river that was full of slowly moving, very green water. Presently the road crossed the river over an iron bridge, and the driver stopped the car. Winding down the window he put out his head and peered down with silent concentration at the water. Although he made no comment, I subsequently learned that he was probably admiring a thirty-foot-long sawfish, which lived on the riverbed at this spot, and was claimed locally to be the largest of its species recorded anywhere in the world.

From a view of its outskirts British Honduras promised to live up to the romantic picture I had formed of it in my imagination. There were the wraiths of old English thatched cottages (a class of structure pleasantly known here as 'trash'), complete with rose gardens with half the palings missing from the fences. Some of their black occupants were to be seen shambling about aimlessly, and others had fallen asleep in the attitudes of victims in murder plots. Pigeons and vultures huddled amicably about the roofs. Notices on gates that hung askew from single rusty hinges warned the world at large to beware of non-existent dogs.

Disillusionment came a few minutes later when we pulled up at the hotel. Here it was that I realised that what information I had succeeded in collecting about British Honduras before leaving England was out of date. According to an account published in the most recent book dealing with this part of the world, the single hotel had possessed all the seedy glamour one might have looked for in such a remote and reputedly neglected colonial possession. But I had arrived eighteen months too late. Newcomers are now conducted, without option, to a resplendent construction of the kind for which basic responsibility must rest with Frank Lloyd Wright – a svelte confection of pinkish ferroconcrete, artfully simple, and doubtlessly earthquake-resistant. As the Fort George turned out to serve good strong English tea, as the waiter didn't expect to be tipped after each meal, and as you could leave your shoes outside the bedroom door

A Quiet Evening

to be cleaned without their being stolen, there were – even from the first – no possible grounds for complaint. But it soon became clear that besides these considerable virtues the Fort George had many secondary attractions which peeped out shyly as the days went by. Little by little the rich, homely, slightly dotty savour of British Honduras seeped through its protective walls to reach me. I began to take a collector's pride in such small frustrations as the impossibility of getting a double whisky served in one glass. Two single whiskies always came. Also, the architectural pretensions were much relieved by such pleasing touches as the showcases in the vestibule which displayed, along with a fine Mayan incense-burner in the form of a grotesque head, a few pink antlers of coral, odd-shaped roots, horns carved into absurd birds, and a detachable pocket made of pink shells, recommended as 'a chic addition to the cocktail frock'.

Part of the Fort George's charm arose from the fact that the staff, who spoke among themselves a kind of creole dialect, sometimes had difficulty in understanding a guest's requirements. This went with a certain weakness in internal liaisons, and from the operation of these two factors arose many delightfully surrealistic incidents. At any hour of the night, for example, one might be awakened by a maid bearing a raw potato on a silver tray, or be presented with four small whiskies, a bottle of aspirins and a picture postcard of the main façade of the capital's fish market, dated 1904. The Fort George, incidentally, must be one of the very few hotels in the world where the manager is prepared to supply to order, and without supplementing the all-in charge, such local delicacies as roast armadillo, tapir or paca – the last-mentioned being a large edible rodent, in appearance something between a rabbit and a pig, whose flesh costs more per pound than any other variety offered for sale in the market. Of these exotic specialities I was only able to try the paca, and can report that, as usual in the case of such rare and sought-after meats, the flavour was delicate to the point of non-existence. The fascination of life at the Fort George grew steadily. It was a place where any beginner could have gone to get his basic

346

training in watching the world go by. Many an hour I spent there, over a cold beer and the free plateful of lobster that always came with it, listening to the slap of the pelicans as they hit the water, while doves the size of sparrows fidgeted through the flowering bushes all round; and the rich Syrian – part of the human furniture of such places – drove his yellow Cadillac endlessly up and down the deserted hundred yards of the Marine Parade.

Among the many self-deprecatory reports sponsored by the citizens of British Honduras is one that their town was built upon a foundation of mahogany chips and rum bottles. True enough the mahogany, which is the principal source of the colony's income, is everywhere. It is a quarter of the price of the cheapest pitch-pine sold anywhere else, and everything from river barges to kitchen tables are made from it. Local taste, however, which has become contemptuous of a too familiar beauty, prefers to conceal the wood, where possible, beneath a layer of fibreglass, or patterned linoleum. As for the rum, it costs thirty-five cents a bottle, tastes of ether, and is seriously recommended by local people as an application for dogs suffering from the mange. It is drunk strictly within British licensing hours, which take no account of tropical thirst, and plays its essential part in the rhythm of sin and atonement in the lives of a people with a nonconformist tradition and too much time on their hands.

Although mostly of almost pure African stock, the citizens of British Honduras have succeeded in creating a pattern of society – if due allowance is made for economic limitations – modelled with remarkable fidelity upon that of their colonial overlords. From their vociferous nonconformity, as well as the curiously Welsh accent underlying the local creole, it is tempting to theorise that the lower-grade colonials they came most in contact with hailed from the Principality, and therefore it is sometimes possible here to imagine oneself in a district of Cardiff settled by people from Africa. The evangelism of the chronically depressed area flourishes. Around every corner there is a chapel; commercial enterprises give themselves such titles as The Holy Redeemer Credit Union; and one is constantly confronted by angry notices urging repentance and the adoption of the Good Life. Even the prophetic books are unable to supply enough warning texts to satisfy the Honduran appetite for admonition. An eating-house, which advertises the excellence of its cow-heel, observes enigmatically at the foot of its list of plats du jour, 'The soul, like the body, lives on what it feeds.' Not, by the way, that one Englishman in fifty thousand had ever tasted cow-heel – a variety of soup which as far as I know is indigenous to the neighbourhood of Liverpool. This was only one of a number of intriguing gastronomic survivals: 'savoury duck' – a rude but vigorous forefather of the hamburger, once eaten in Birmingham; 'spotted dick' – rolled suet-pudding containing raisins; 'toad-in-the-hole' – sausages baked in batter: both the latter dishes once a feature of popular eating-houses all over England, but now usually disregarded.

One constantly stumbles upon relics of provincial Britain preserved in the embalming fluid of the Honduran way of life, but often what has been taken over from the mother country is strikingly unsuitable in its new surroundings. The minor industries, for instance, such as boat-building, are carried on in enormous wooden sheds, the roofs of which are supported by the most complicated system of interlacing beams and girders I have ever seen. One thinks immediately of hurricanes, but on second thoughts it is clear that all this reinforcement would be valueless against the lateral thrust of a high wind. It turns out that such buildings were copied from originals put up by Scottish immigrants: the roofs were originally designed to withstand the snow-loads imposed by the severest northern storms.

Many of the Scotsmen themselves lie buried in the city's two cemeteries, both of which are located in neat formal gardens between wide roads, just where in Latin America the living would have taken their nightly promenade. According to inscriptions, many of the dead were sea-captains. They came here to die of fever, or were sometimes murdered, and in this case the inscription supplies an affirmation of the victim's hope of immortality, including the exact time of the tragedy, but alas no further detail. The tombstones serve conveniently for the drying of the washing of the neighbours on both sides of the road. The cemetery is not at all a bad place to be buried, for those who were confident of the body's resurrection – surrounded by the white houses, and the lemon-striped telegraph poles, with the constant bustle and chatter of bright-eyed crows in the trees above, and the eternal British-Sunday-afternoon strumming of a piano in a chapel just down the road.

Death took these captains by surprise. It was never old age or a wasting sickness, but always the mosquito or the dagger that struck them down. No Britisher ever wanted to lay his bones anywhere but in the graveyard of his own parish church in the home country. In this lies the key to all the unsoundable differences between the Spanish and the British colonies. The Spaniard took Spain with him. The Briton was always an exile, living a provisional and makeshift existence, even creating for himself a symbol of impermanence in his ramshackle wooden house.

One of the first things that strikes the newcomer to British Honduras, who has seen anything of life in the West Indies, is the mysterious absence of anything that might come under the heading of Having a Good Time. There are no calypsos, no ash-can orchestras, no jungle drums, no half-frantic voodoo devotees gyrating round some picturesque mountebank. The Hondurans sacrifice no cocks to the old African gods, and feuds are settled by interminable lawsuits or swift machete blows, but in either case without recourse to the black magic of the obeahman. This in some ways is a pity. Because timber extraction, the main occupation, ceases with the wet season, people are left with several months to fill in, without the faintest idea of what to do with themselves, apart from chapel-going, playing dominoes, and suffering the afflictions of love. This highly un-African existence, with its complete ineptitude for self-entertainment, is probably the result of certain historical factors. The colony was founded by an English buccaneer called Wallace – Belize is believed to be a corruption of his name – who turned from piracy to the more dependable profits of logwood extraction. The slave-owning Wallace and his successors were very few in number. They were exposed to frequent attacks by the warlike Indians of southern Yucatan, and to the constant threat of action by the Spanish, who never recognised the legality of their settlement. The interlopers could only hope to defend themselves, and to keep their foothold, by arming their slaves. These slaves might have been more likely to shoot their masters in the back, if their servitude had been as oppressive as in other Caribbean colonies. In Jamaica for instance, the slave-owners produced a formidable breed of mastiff which they trained not only to track down, but to savage, black runaways, and these dogs were also in great demand in the French and Dutch colonies.

One supposes that the atrocious treatment meted out to the blacks encouraged the slaves to conserve all that was African in their lives, and to unite them in their hate for all that was white. However, the black inhabitants of British Honduras, with their musketry drill, their smallholdings and their Sunday holidays, would have been encouraged to turn their backs on their African past and to struggle ever onwards and upwards towards the resplendent human ideal of the suburban Englishman.

The test of this culture *malgré-soi* came on September 10th, 1798, when a Spanish flotilla commanded by Field-Marshall Arthur O'Neil, Captain-General of Yucatan, appeared off the coast. The field-marshal was carrying orders to liquidate the settlement once and for all, and the baymen, as the English settlers called themselves, being forewarned, mustered their meagre forces for the defence. Reading of the remarkable disparity in the opposing forces one realises that here was the making of one of those occasions that are the very lifeblood of romantic history. The captain-general's fleet consisted of thirty-one vessels carrying 2,000 troops and 500 seamen. The defenders numbered one naval sloop, five small trading or fishing vessels, hastily converted for warlike purposes, plus seven rafts, each mounting one gun and manned by slaves – a total defensive force of 350 men. The resultant passage of arms has provoked a fair measure of armchair bloodthirst, flag-waving, and orotund speechifying on the annual public holiday which has commemorated it. In 1923 a Mr Rodney A. Pitts wrote a prize-winning poem called 'The Baymen', an ode in thirty-one verses, which, set to music, has become a kind of local national anthem. A sample stanza plunges us into an horrific scene of carnage:

> Ah, Baymen, Spaniards, on that day
> Engaging in that fierce mêlée –
> Ah, never such a sight before,
> They are all dyed in human gore –
> Exhausted, wounded, some are dead,
> They're sunken to their gory bed.

The cold facts of the case, supplied by contemporary records, paint a less murderous picture of the encounter. In an engagement which lasted two and a half hours, there were no casualties whatever on the British side, and the few bodies interred later by the Spanish on one of the cays were as likely to have been those of fever victims as of grapeshot casualties. One thinks of the dolorous quavering of generations of schoolchildren through such passages as:

> All died that this land which by blood they acquired
> Might give you that freedom their brave hearts inspired.

As usual, history turns out to be a fable agreed upon.

Modern times have brought with them a slackening in the idealised master-and-faithful-serving-man relationship of the past. A People's United Party (known as the PUP) has emerged, whose aim is total independence for British Honduras, and which, by way of a kind of psychological preparation for this end, urges the substitution of baseball for cricket, and the abolition of tea-drinking. The party's creator and leader is a Mr Richardson, a wealthy creole (a word that here describes an English-speaking person, mostly of mixed African and British ancestry). Mr Richardson's antipathy for Britannia and all her works supposedly originates in a grievance over some matter of social recognition – a familiar colonial complaint, and one that has cost Britain more territory than all her other imperial shortcomings combined. When recently the Government of Guatemala renewed its claim to British Honduras, the outside world speculated on the possibility of the PUP operating as a fifth column in support of the Guatemalan irredentists. The answer to this, I was told, is best expressed by the local proverb, 'Wen cakroche [cockroach] mek dance 'e no invite fowl.'

The party's official organ, the *Belize Billboard*, is a journalistic collector's item, combining the raciness of a scurrilous broadsheet with the charm of a last-century shipping gazette. It is particularly strong on crime-reporting, pokes out its tongue at the British whenever it can, and carefully commemorates the anniversaries of such setbacks in Britain's story as the sinking of the *Ark Royal*. It is regarded with sincere affection by the white members of the colony, some of whom keep scrapbooks bulging with choice examples of its Alice-in-Wonderland prose – full of such words as 'doxy' and 'paramour'. The trade winds blow right through the advertisement section of the *Billboard*, with its bald details of goods 'newly arrived', listed as if they were unloaded onto the quayside: clay pipes, lamp chimneys, medications for the bay sore and ground itch, beating spoons, cinnamon sticks, bridal satin, colonial blue-mottled soap and – in the month of March – Christmas cards. Dropped like a dash of curry into this assortment from the hold of a ghost ship are the announcements of the Hindu gentleman – with an accommodation address in Bombay – who promises with the aid of his white pills to add six inches to your height, 'provided you are not over eighty'.

In whatever direction the political destiny of British Honduras may lie, its economic future is dubious. In the past it has depended upon its forests; but ruinous over-exploitation in the half of the total land area of the colony that is privately owned has depleted this source of income and seriously mortgaged the future. The logical remedy would seem to lie in the switching over of the colony's economy to an agricultural basis. But it seems that the rhythm of seasonal, semi-nomadic work in the forest, sustained for centuries, has created what a government handbook politely describes as 'an ingrained restlessness'. In other words, the Hondurans tend to become bored with a job that looks like being too steady.

The eventual solution to this problem probably lies in the tourist industry, with a glamourised and air-conditioned nation emerging as another Caribbean playground of the industrial north – and anyone who has seen what has happened to the north coast of Jamaica in the last few years will know what to expect. All the ingredients for a colonial Cinderella story are present. Being just beyond the reach of the Cuban and Mexican fishing fleets, the Bay of Honduras is probably richer in fish – including all the spectacular and inedible ones pursued by sportsmen – than any other accessible area in the northern hemisphere. The average aficionado will lose all the tackle he can afford in a week's tussle with the enormous tarpon to be found in the river running through Belize Town itself. The forests, too, abound with strange and beautiful animals, with tapir, jaguars and pygmy deer, which await extermination by any smoothly organised hunting parties of the future. I was given to understand that even this year a tourist organisation calling itself The Conquistadors' Caravan was dickering with the possibility of including British Honduras in one of its 'Pioneer Conquistadors' itineraries, and was dissuaded only by the news that there was no nightclub, no air-conditioning, no Mayan ruins within comfortable reach, no sandy beaches, and that jaguars' tracks are seen most mornings on the golf course.

In the meanwhile, for the collector of geographical curiosities, there is still time – though probably not much time – to taste the pleasures of a Caribbean sojourn in the manner of the last century. As a matter of fact I can't think of any better place for someone seized with a weariness of the world to retire to in Gauguin fashion, than British Honduras. The intelligent recluse could even protect himself from the chagrins of the tourist era to come by renting an island, which can be had complete with bungalow and bedrock conveniences, for a few dollars a week. Here he would be able to knock down his own coconuts, ride on turtles, collect the eggs of boobies in season, put on a pair of diving-goggles and pick all the lobsters he could eat out of the shallow lagoon water, perhaps even note in his journal the visit of a transient alligator.

The reverse side of the coin is hardly worth mentioning. The drains *are* uncovered, but there are no mosquitoes, not much infectious disease, only an occasional plague of locusts; and for nine months of the year the heat keeps within bounds. Perhaps the hazard of the occasional hurricane should be touched upon. The last bad one blew up in 1931 on September 10th, the anniversary of a 1798 victory over the Spanish. A twenty-foot-high wall of water rolled over the town, and swept the houses off the cays, with a substantial death-toll of the merrymakers who were celebrating that famous victory. But taken over the years, hurricanes are a very minor risk. And while on the topic of winds, it might be considered reasonable, from an intending resident's viewpoint, to bear in mind that, however hard they may blow, they do so from a remarkably consistent direction, and that this direction, that of the Atlantic Ocean wastes, is one where a cloud of radioactive particles may be less likely to originate.

Slave-Labourers in the Vineyard

The Californian grape-pickers strike, which began in 1965, was in its fifth year when Lewis travelled to investigate it. The most prominent figure in the article is the Mexican, Cesar Chavez, who organised the strike, and fiercely campaigned against the dangerous use of pesticides. Even though exhausted, and with a bad back in bed, he agreed to meet Lewis.

Chavez had some success in improving the desperate lot of farmworkers. Five years after Lewis's visit, his efforts led to the California Agricultural Labor Relations Act of 1975, initiating the possibility of collective bargaining for the state's farmworkers. However, he grew increasingly autocratic and wayward, becoming linked to a dodgy cult, Synanon. On a journey to the Philippines, he praised the repressive President Marcos for introducing martial law.

He is nevertheless remembered for his bravery and formidable campaigning. In the year after his death, aged sixty-six, he was posthumously awarded the Presidential Medal of Freedom. His campaigning slogan 'Si Se Puede' ('Yes We Can') was taken up by President Obama for his first presidential campaign.

After this article was published, President Reagan wrote Lewis a rude three-page letter. Sadly, it has been lost.

First published in *The Sunday Times*, February 1, 1970

A Quiet Evening

THE TAXI-DRIVER who took me from El Centro, the southernmost airport in California, to Calexico on the Mexican frontier, provided in a thumbnail sketch of his life a hint of what was to come. He was a Chicano – a Mexican-American – now aged nineteen, with features a little blunted and blurred under the extinct volcanoes of his skin. This man had worked as a grape-picker since the age of ten and had three times been memorably and dangerously subjected to an accidental dousing with pesticides. The first of these, he said, was when he was about thirteen, and a plane had sprayed a vineyard with some primitive and ineffectual compound of ammonia. 'It didn't kill any bugs,' he said, 'but it half-killed the grape-pickers who got in the way.' On this occasion he was blinded for only a few hours, but a subsequent spraying with a more up-to-date formula had left him with an almost permanent headache. It was an allergy he developed to stings that made him give up work. 'They gave me injections, but they didn't do any good. The doctor said I'd croak if I went on with it. You can't imagine what grape-picking is like until you've worked at it. It's the crummiest job of the lot.'

It was 8 p.m. when we got into the border town, and Calexico lay brooding in absolute silence under the flares of neon. Most American cities are empty, and even a little sinister, at night, but here the feeling of human abandonment was so complete that it might have been a town undamaged, though emptied of its population by bacteriological warfare. The fact was that Calexico was already asleep. It is a place that lives fitfully on the logistics of shuttling an army of Mexican labourers daily from the border, deep into California and back, and, conducting its business in nocturnal outbursts, it thereafter sleeps. Up to five hundred ramshackle buses are used in these daily convoys. Some of the buses were already back in the car-parks, ready to leave again in the small hours. A number carried bumper-stickers making it clear that the labour-contractors who operated them made demands on their passengers that were sentimental as well as physical. 'The American Flag', said the notice under the Stars and Stripes. 'Love it, or leave.'

I checked into my hotel, picked up the key, along with the religious text that went with it. Because the restaurant had long since closed, and there was no-one about, and absolutely nothing to do, I surrendered to the custom of the town, and went to bed. At 3 a.m, however, I was awakened by pandemonium. The buses of the evening before, now crammed with Mexicans, and welded together in the street below in a thunderous and gear-grinding procession, were crawling northwards out of town towards the distant crops of California. Mixed in with the buses – like the parasitic pilot fish that swim with sharks – were Mexican nomad families, in their shacks on wheels, who would follow wherever the buses went, to snap up what scraps of labour they yield. Children, who were selling oranges and snacks, rushed for the traffic lights to hold up the food at the windows. There were few buyers, for the men in the buses were trying to snatch what sleep they could in preparation for the tremendous journeyings, the heat and the final exhaustion of the long day that lay ahead.

'The Hole' is to Calexico what the Piazza della Signoria is to Florence, or the Spanish Steps to Rome – as much institution as place – and one of the reasons for this journey to the frontier was to reconnoitre it as the marketplace of Mexican day-labour. It was five blocks from the hotel, and 4 a.m. before I reached it. 'The Hole' turned out to be an enormous sunken carpark, softly fogged with exhaust smoke, smelling of urine, and surely one of the last places in the world – outside the stinking, Mafia-ridden villages of western Sicily – where you can see one man feel another man's muscles before employing him. The hiring time was already past its peak, but a couple of hundred or so buses were still loading up with Mexicans who had just walked across the border, a few hundred yards away. The Mexicans had congregated reverently round the buses, like penitents at a shrine. They were small men in limp cottons, and full of submission; the huge Californian labour contractors who strode among them, picking this man and rejecting that one, were like visitors from a more advanced planet. When a Mexican received a nod of approval he shuffled forward, hat held respectfully in his hands, and climbed into the bus. Sometimes he put his hat on to cross himself before mounting the step, and occasionally a man ducked his head quickly in an attempt to kiss a contractor's sleeve. These peasants were in the big man's power and they knew it. Whatever the theoretical wage, he could deduct what he pleased for the journey. Most of the peasants were illiterate, too, with only a hazy idea of the laws of the land into which they would be taken. Few of them knew where they were going, or what the work would be. Whatever they were given in cash at the end of the day, they would take it without question. A common way of dealing with complaints is to throw the complainer off the bus, letting him find his way back as best he can.

It soon became clear by the size of the crowds they attracted that some buses were popular and others were not. When a man was rejected by a contractor he moved to a bus where there was less competition and sometimes resigned himself in the end to boarding one with seats going begging. These workers, who were obviously less in demand, were the older ones – men who in their early forties were reaching the end of their useful lives. They were obliged to take the jobs that were avoided as far as possible by the younger and more valuable workers. As it was, there would be a surplus each morning, and out of the 15,000 to 18,000 men who came here to offer their labour, a thousand or two wouldn't be taken. At 5.30 a.m., by which time the last of the buses had left, they would be forced to abandon hope and trudge back again across the frontier into Mexico.

Every Mexican labourer in Calexico hoped for the shortest possible journey to his day's work. These buses, most of them advanced in mechanical senility and travelling over slow, secondary roads, could average less than 30 mph. At this moment grape-picking was coming to an end in South California, but lettuces, cantaloupes and sugar-beet were being harvested, and the journeys involved could be as little as 70 miles or as much as 120 miles in each direction. This meant, at worst, that a man who left The Hole in Calexico at 3 a.m. reached his place of work at 8 a.m. He then worked an eight-hour day, to get back to Calexico by nine at night, and could expect to be reunited with his family in a shack just over the Mexican frontier by about ten, to prepare himself with four hours' sleep before the calvary of the next day's work. None of the waiting men I spoke to ever hoped to get more than six hours' sleep a night. They reported far harder cases than their own – of peasants who instead of living conveniently by the frontier, lived further away, and thus were obliged to travel more terrible miles in Mexican buses. Surely this must be the grimmest of labour markets since the abolition of slavery.

What sort of misery and poverty is it that can drive so many Mexicans to deliver themselves into bondage of this kind? To try to find the answer, I went to see Manuel Chavez at his office in Calexico.

Manuel is the cousin of Cesar Chavez, the chief organiser of the great Californian grape-pickers' strike, now it its fifth year. Every tentative attempt in the history of the United States to form an effective union of agricultural workers has been demolished by the simple expedient of importing foreign strike-breakers – Chinese, Japanese, Filipinos, and now Mexicans, and Manuel was in Calexico trying to persuade Mexicans not to take jobs in the vineyards. His task seemed to me hopeless. The immigrant workers he dealt with were too desperately poor to sacrifice a day's wage to such an abstraction as the struggle for someone else's betterment, and however much a man might prefer not to be used as a blackleg, in three cases out of four he was not told in advance whether he would be working in a vineyard.

I found Manuel in a tiny office encircled by worried men from across the border who had brought their problems to him. He turned out to be genial and extroverted; a Mexican who could have passed for a white American, unlike Cesar himself who is more of an Indian, full of reflection and delicate Indian reticence.

'I've just seen Cesar,' I told him.

He was delighted. 'That's great,' he said. 'How is my brother?' The bond of affection between the two men is such that Manuel always calls Cesar his brother.

'He's fine, except for his bad back. I saw him in Kansas City. He'd just got in from St Louis and his back was giving him trouble. He was in bed when I saw him.'

'How did he say the trip was going?' (Cesar was on a tour of the Eastern cities, drumming up support for his union's boycott of Californian grapes.)

'He seemed fairly cheerful about things in general.'

'So am I. I'm cheerful too. We're all cheerful here, because we know we're winning. Even if we have to carry on this strike for ten years, we'll get our union recognition in the end.' He got up. 'Come on, let's go and see how our people live across the border.'

The city of Mexicali began a hundred yards up the road from where we had been sitting in the United States. A long line of cars were waiting at the US frontier post to be searched for drugs. About a million people lived in this city on the Mexican side of the frontier, perhaps a third of them in spectacular poverty.

The area of the town taken over by the families of migrant workers displays its sores on the banks of the Rio Nuevo. At first glance this does not appear to be a river at all, but rather one of those empty watercourses one sees in the south of Spain, which carry a brief spate of water in the rainy season and dry up as soon as the rain ceases to fall. A closer inspection from an eroded, sun-scorched bank showed that the bed of the Rio Nuevo was not dry – it held rivulets of black fluid that might have been oil but in fact was sewage. The river is the town's sewer, and all through the long dry summer months a dreadful process of concentration of its contents takes place, from which only the short-lived deluges of winter bring relief.

No property-developer or landlord lays claim to these apocalyptic banks, nor to the temporary islands emerging in summer from the ooze in the riverbed, and here the migrants build their huts. These people couldn't even afford the usual bidonville materials, such as flattened-out oil cans and corrugated iron. Instead, Mexicali's slum was built entirely of cardboard – from boxes salvaged from the supermarkets. Children were pottering about, barefoot, among the hideous jetsam in the mud, while ragged, demoralised women scuttled away to hide as we came into sight. We walked with our handkerchiefs pressed to our noses. This was a fine coolish day at the beginning of the winter with the thermometer in the eighties, but it was hard to imagine what it would be like to live here in summer with the nearest source of fresh water half a mile away, and temperatures reaching 51 centigrade in the shade. Without any doubt this was the worst slum I had ever seen.

Mexicans came from all the depressed parts of Mexico, but chiefly from the slums of the cities as far south as Yucatan, drawn northwards to establish themselves in places like this along the US frontier. About ninety thousand of them pass over into the US each day in search of work, lured by the money to be made by those who are prepared to drive themselves to the limits of human endurance. In the vineyards of California a Mexican labourer can earn nearly two dollars an hour before the contractor takes his cut. Mexico, a land possessed by the few, where a middle-class has hardly emerged, pays a minimum of 3 dollars 20 cents for a ten-hour working day. But even the provision of this minimum rate is easily and constantly avoided under a regulation permitting an employer to pay his workers what he thinks fit during a sixty-day training period. Some factories simply sack a man at the end of this normal training and so run their plant with a permanent force of trainees. Nothing is simpler than to do this in the case of plants engaged in a simple assembly routine of components manufactured in the US. There are more than a hundred American-owned factories in Mexicali, most of them of this kind, taking advantage of limitless supplies of cheap labour.

The choice for the destitute Mexican lies between the sweatshops of Mexicali and the hard and often dangerous, but better paid, labour in the fields of the southern states of the US – and this is what most of them choose.

The Californian grape-pickers' strike that began nearly four and a half years ago under the leadership of Cesar Chavez was not, in the first instance, concerned with the succour and protection of these hungry multitudes of migrant workers, although its success would be certain to have a dramatic effect on their fortunes. What Chavez was primarily fighting for was the living and working conditions – to use his own word, the *dignity* – of more than two million Mexican-Americans living in California, and ultimately, no doubt, the five million in the whole of the United States.

US citizens of Mexican origin are an acutely depressed minority, 'the niggers of ten years ago', as one of their spokesmen puts it. Eighty percent of them are housed in slums, and one-third are classed as living below the poverty level. Their children average two years less at school than the children of the black population, and four years less than the children of whites, and they are still punished for speaking Spanish within earshot of their teachers. Few of them go to high schools, and only a handful to universities. Later they are likely to find that, whatever their scholastic achievements, only menial employment will be offered them. The picture of social and economic discrimination practised against them closely resembles that suffered by the black population of the eastern and south-eastern States, but the Mexican protest is only beginning to be heard.

Of all the jobs that almost automatically fall to the Mexican, whether migrant or US citizen, the most unpleasant is that of the grape-picker. Grape harvesting is done under a raging sun from which the low vines themselves offer little shade. The surface of a harvester's skin is soon coated with a sludge of grape-juice mixed with sweat and dust, which attracts so many winged insects that at worst the pickers appear to be working in a snowstorm. Chavez, from a bitter experience of this work, described it to me. 'It's degrading, de-humanising. After an hour or two everybody gets in such a mess you can't tell a man from a woman.' It is rare in the fields for proper drinking water to be available, or any form of latrine to be provided. More seriously, in recent years growers have taken to the use of toxic sprays that have caused innumerable cases of severe illness and some deaths. Workers are housed in compounds on company property which the members of Chavez's organisation are frequently prevented from inspecting, and are said, at worst, to resemble concentration camps.

To find working conditions in Great Britain paralleling those in the vineyards of California one would probably have to go back to the middle of the 19th century, before the first agricultural labourers' union was formed in 1872. The picture of exploitation on the Victorian model is completed by the presence of child labour. When I was in Delano last November the local paper, the *Fresno Bee*, reported the case of Theresa Arellano, a girl of eight, who worked a seventy-hour week on a grape ranch. The newspaper regarded this as exceptional, but by my own experience the spectacle of young children at work in conditions which are arduous and even dangerous for an adult is commonplace. The laws of the United Sates relating to the employment of young children are held in contempt. When the Giumarra Vineyards – largest of the growers – was tried and convicted of nearly forty violations of child labour and health laws, it was fined a total of a thousand dollars by the Kern County Superior Court. The fine was suspended.

At the root of these evils is the single staggering fact that the Government of the United States has never recognised the right of its agricultural workers to form a union. To quote the introduction to Cesar Chavez's statement in the House of Representatives last October: 'Farm labourers are excluded from minimum wage legislation and from unemployment insurance, and are at a disadvantage where social security is concerned. They are denied the collective bargaining rights guaranteed to non-farm workers, and are effectively cut off from every benefit of a negotiated contract. So the vast majority of Californian farmworkers have no contract, get no overtime, and may not even know their rate of pay... They are often victims of deception and graft. They get no time off with pay, no health or pension plans, no regular rest periods. Speed-ups and abusive supervision are common. Workers may be laid off at any time and for any reason, as for objecting to being assaulted by an owner...'

It was this situation Cesar Chavez set out to remedy. He chose the town of Delano in the San Joaquin Valley, 120 miles north of Los Angeles, as his headquarters, because it was in the centre of an area of 400 square miles of grape ranches. He started his work of organisation in 1962, had built up a union membership of a thousand by 1964, and by the autumn of 1965 he was ready to go into action. Every move he made had to be planned with extreme caution, to avoid crushing and bloody defeat. Tulare County, of which Delano is a main town, had a reputation for teaching troublemakers a stern lesson. The last strike in this area had been organised by cotton pickers who were quite literally starving to death at the depths of the depression in 1933. They had called a meeting at Pixley, about 10 miles from Delano, to demand an increase in their wages which, by compelling men to bid against each other, had been forced down to 10 cents an hour. On this occasion the cotton planters had shot three of the organisers dead out of hand, and the other leaders, after they had been tarred, feathered and thoroughly flogged, were handed over to the police who threw them into jail. Eleven cotton planters subsequently stood trial for homicide, but all were acquitted.

It would be unrealistic to believe that Californian ranchers, pushed a little too far, would be any slower on the draw now than in those days, and Chavez must have realised that his movement's only real hope for survival lay in non-violence. When the strike was called, he ordered that this principle was to be followed whatever the provocations that had to be endured.

One has to realise that the complacent social climate of the San Joaquin Valley is not greatly different from that of England in the days of the Tolpuddle Martyrs. Unions are seen by many Californians as 'unfair', and the grape-pickers' strike as a conspiracy organised by a handful of wicked and ungrateful men – almost certainly the agents of international Communism – to overthrow the decencies of American democracy. Governor Ronald Reagan called the strike 'immoral' and an 'attempted blackmail of free society', and his attacks have been echoed by righteous fulminations from many Californian pulpits.

What followed as soon as strikers began to appear at the vineyards was therefore to be foreseen, and has been described in Chavez's statement to the House: 'Strikers have been physically attacked; they have been shot at and kidnapped. They have been rammed by trucks. Their offices have been looted and destroyed. More often than protecting the rights of strikers, local police have condoned and even aided those that would destroy them.' Strikers who had left a field and then gone back to their camp to pick up their belongings were arrested for trespassing, stripped naked and chained together. Wholesale arrests were made, and some of the charges seem extraordinary. It was made an offence to shout the word '*huelga*' (strike), and one man was even hauled away for reading aloud – from a work of Jack London – a description of a strike-breaker that was unflattering.

Time failed to ease this situation. In March of the next year one of Chavez's organisers was run down and crippled for life by a truck. He has since been cited for obstructing traffic, although a case against his assailant for common assault has yet to be heard. Last July a Chavez supporter risked giving a slow handclap at a meeting at which a politician was denouncing the strike. For this the man got six months in jail. He was a sufferer from emphysema and, being denied the medicine necessary for his condition, he died in prison.

In the first few months of the strike thousands of workers were ready to brave violence and intimidation, and left the fields but, as they could be immediately replaced, this presented the growers with no real problem. They were secure in the knowledge, too, that even if Mexican immigrant labour should ever be cut off, there were plenty of Filipinos, Puerto Ricans, and even Arabs that they could fall back on. A labour recruiter who brought in a dozen or so good strong men could ask for, and expect to receive, a foreman's job, and the right to exploit his underlings in all kinds of small ways. There was no trouble in getting a resident's permit for a Mexican who said he wanted to stay, but many growers preferred not to do this because a new racket in illegal immigrants had developed.

Apart from the 24-hour pass handed out to fieldworkers who commuted daily from the Mexican frontier, a 72-hour pass was easily obtained by Mexicans who wanted to enter the United States on occasional short visits. Once inside the States very large numbers of these were induced to remain as illegals. In the year 1968 alone 142,000 illegal immigrants were caught, most of whom had entered in this way. Many weren't discovered, remaining to form a vast, docile and completely tractable labour-force, happy to take what they were given and determined above all, to keep clear of difficulties of any kind. There is no onus upon an employer not to take on illegals, and there has been no single case of any action being taken against one for so doing. This life of an illegal immigrant is likely to be hard – and not only in his working hours. A few days before I visited Delano, a 14-year-old boy without papers had been found lying in a ditch, desperately ill from the effects of poison sprays. He hadn't dared to go to a doctor for fear of being handed over to the police.

Since strike-breaking was ubiquitous, Chavez began a nationwide campaign for public support of his cause, urging Americans not to eat grapes, and this soon began to make some headway. At Easter in 1966 the strikers marched 300 miles to Sacramento, and they were welcomed there by several thousand new supporters, and a few liberal politicians. Following the publicity given to this march, the firm of Schenley capitulated and agreed to begin negotiations with the union. It was the first victory of its kind for agricultural workers in US history, but it produced immediate and sinister repercussions. Chavez had now turned his attention to the firm of Di Giorgio, which produced what it probably believed to be a trump card in an offer to allow the formidable Teamsters' Union to represent their workers. The control of spurious labour unions has always been a Mafia speciality. The Teamsters have been charged in the American press with Mafia affiliations, and Chavez believed a 'sweetheart' contract of the kind that the Teamsters would probably negotiate would be of more benefit to the growers than the pickers. It seems evident that Di Giorgio thought the same. The fight with the Teamsters was the turning point in the development of Chavez's movement, and the fact that he could defeat these redoubtable champions of the status quo testifies to his generalship as much as it does to his courage.

A few growers followed the Schenley example at this juncture, and made their peace with the union, but the majority held out. Giumarra, largest of all – the firm which had the convictions for employing child-labour – tried to get round the boycott by marketing its grapes under the labels of union firms. Chavez's reaction was to extend the boycott to all table grapes, and the firms who had signed union contracts went over to wine production. Grapes are hardly ever displayed for sale these days throughout the United States, but many devices have been tried for getting round the boycott. The percentage of grapes in tinned fruit salad is now remarkably high, while the importation of raisins from Greece has slumped. For all that, the growers are beginning to weaken, and Chavez says that only the government's indirect intervention in purchasing grapes for the armed forces has saved them.

Cesar Chavez's boycott against the grape-growers of California keeps him on the move, prodding the American conscience awake, and seeing to it that the grape remains forbidden fruit. At the end of last year he spent two months in the eastern states addressing public meetings and appearing before congressional committees. As a climax of the tour he spoke from the pulpit of the National Cathedral in Washington. On this occasion he made the charge that the boycott was being sabotaged by the Defence Department in collusion with the growers. The army's increased purchases – fifty percent up on the previous year – had largely compensated the growers for their loss of sales to the public. Chavez said that if the Pentagon could be induced to stop buying grapes, the strike could be brought to an almost immediate end.

'We cannot win this on our home ground,' he said. 'We know that, and we ask you to help us extend the strike. It is my firm conviction that banded together we shall bring justice to the tortured valleys of our land.'

When I saw Chavez last November, he was on his way back to California by slow stages, and it was arranged that I should see him, en route, in Kansas City. He was expected to drive most of the day from St Louis, where he had been addressing a rally, but when I rang the United Farm Workers office no-one knew when he was going to arrive. Chavez lumbers erratically across the continent in his aged car, and by normal American standards his progress is slow and uncertain. At 8 p.m. his office phoned to say that he had arrived, but was exhausted and suffering from a bad back. Would I mind if he went to bed? I could talk to him just the same.

The office turned out to be in a private house in a rundown and remote part of the city, and arriving one could practically sniff the rich, human aroma of Mexico through the wound-up windows of the taxi. The door was closed, and children in pyjamas were romping about in a passageway, and one of them, aged about six, subjected me to a giggling interrogation through the letter box before I was allowed to enter. Inside I found an enthusiastic confusion of secretarial activity as well as more bread-and-jam-munching children. A momentary crisis flared up when it was decided that Cesar ought to be propped up in bed for the interview, and a young lady was sent out to borrow an extra pillow from a neighbour. After this I was conducted through a kitchen in which four women were cooking the evening meal into a bare and cell-like bedroom beyond. Chavez lay on a narrow iron bed, showing his single, comfortable gold tooth in a smile of welcome. Although by this time it was nine o'clock and he'd been driving with a sore back since early that morning he seemed so genuinely pleased to see me that for a moment I thought some old friend of his must have come into the room behind me.

Cesar Chavez, when you meet him in the flesh, could be a pure Indian. He possesses a geniality that isn't always apparent in his photographs, and although there is something about the melancholy of the mouth which recalls the sad Indian faces of Mayan sculpture, this is instantly remedied when he smiles. He is small and fragile-looking, and his health may have been taxed by overwork, his habit of personal austerity, and by his fasts, the most recent of which, undertaken 'as an act of penance to recall workers to the non-violent roots of their movement', lasted twenty-five days. Unlike some other public figures of our times, who have acquired a veneer of saintliness, one's mind immediately clears him of any suspicion of theatricality.

We talked about his trip, and about the prospects of persuading the Pentagon to reduce its grape buying, at least to pre-strike levels, but above all about the threat to fieldworkers by pesticide sprays, which he now saw as more important than their economic struggle. The phasing-out over a period of two years of DDT had just been announced, but this had done nothing to relieve Chavez's pessimism, and he said that all this meant was that the even more toxic organophosphate pesticides would come into heavier use.

In such arguments one is immediately struck by the gentleness of Chavez's manner, his lack of obvious anger, his apparent incapacity for bitterness. Cesar invites you, soft of eye and voice, to share his boundless amazement at the wrongs practised on defenceless fieldworkers – and you do.

'They carried out tests on 774 people,' he said, shaking his head incredulously, 'and only 121 showed no symptoms of pesticide poisoning. And why can't we stop this kind of thing? Simply because of the interests of the pesticide manufacturers.'

There is nothing of the leftwing demagogue about Chavez. When he made this claim one felt that he saw the grape-farmers and the pesticide manufacturers not as capitalist hyenas, but as mistaken men, trapped in their materialist environment.

'Our people aren't even allowed to know what kind of poisons they're being sprayed with,' he said. 'The courts have made orders forbidding the agricultural commissioner from revealing the composition of some of these pesticides. They're trade secrets.'

Chavez himself began work as a grape-picker 30 years ago, at the age of ten, and it occurred to me to ask him how the grape-farmers got along in those days before the invention of pesticides.

'We didn't have any trouble,' he said. 'The vineyards were kept in good health by natural predators. If the fruit-fly population started to get too dense they used to send us kids to collect ladybugs in the alfalfa fields and put them on the vines. They did the job. All that happens now is that the sprays kill off all the ladybugs, and there are more flies than ever. You'll see for yourself, when you go.'

I asked Cesar whether he could describe the final objectives of his movement; he thought about this question some time before replying. 'They'll take a long time to be reached,' he said. 'And I imagine it's a thing I won't see in my lifetime. What we're fighting for can't be settled on a simple material basis. Our fight is for people's dignity. At the moment we see wage increases as far less important than our people's health. For example, we've been asking for two dollars an hour, but we'd reconsider this if growers would be willing to sign a strong health and safety clause. The day must come when the producers of food are the equals of anybody else.'

The question of Cuba came up. I mentioned that I'd visited it both before and after the Castro revolution, and it seemed to me that there, if anywhere, the problem of the peasant's human status may have been solved. Chavez seemed doubtful. 'The idea of any revolution worries me,' he said. 'It probably worries any Mexican from my background. We went through one, and there was a lot of killing, but nothing was really achieved. You've seen for yourself how things are in Mexico. As far as the peasants are concerned it hasn't made any difference at all. A Mexican boss is just as hard as any other boss.'

While Chavez remains at the head of the Mexican grape-pickers' strike, there will be no violence on the strikers' side. He has carried his non-violent philosophy to the lengths of a declaration that if his movement should be responsible for the shedding of the blood of a single grower or striker, it would not be worth whatever gains it made. Here speaks not only the disciple of Gandhi, but the intensely practical man who has chosen a strategy for survival in his perilous environment.

After leaving Cesar Chavez, I flew to Bakersfield, California, hired a car, and drove northwards up the freeway through the San Joaquin

Valley to Delano. All this part of southern California lay under a grey haze hardly distinguishable from the light smog normally present over the city of Los Angeles, produced by its factory chimneys and the exhausts of the three million cars. The landscape was one of hypnotic monotony; a visual treadmill of cement-coloured fields, devoid of trees and littered with mirages and the rusty detritus of harvested crops.

The Jesuit, Miguel Venegas, who explored southern California after years spent in the deserts of Mexico, reported on it in 1739 to the Spanish king. 'My eyes were restored by its brilliance and freshness,' he said. In the lakeside pastures and the woodlands he saw innumerable animals: deer of many kinds, elk, bears, cougars, and a great variety of ibises and wading birds. But now the touch of Midas lay heavy on the land and all the lakes Venegas spoke of had long since been drained, and every free-flowing river dammed and transformed into cemented canals and irrigation ditches. Even the low hills were braced with more concrete against erosion. Few parts of the earth's surface can have suffered a more baleful transformation in the century and a quarter since California, taken from Mexico by enforced purchase, was invaded by settlers, who massacred most of its Indians in two years, slaughtered its wildlife, turning its landscape into ranches and farms.

Towns such as Delano started life as raw encampments to serve gold rushes or land grabs. The clapboard has been replaced by brick, and suburbs added, but they have never acquired a style. 'When you get there, there's no there, there,' as Gertrude Stein once complained of nearby Oakland, and this might as well describe Delano. Arriving, one receives the warning, routine these days, that there is a section of the town – in this case the west side – which it is not safe to visit after dark. But east or west, the population drains from the streets as soon as night falls, and for that matter there are few pedestrians to be seen about at any time. Apart from essential visits to filling stations and supermarkets, there is little point in being out of doors.

It seemed certain, too, that Rachel Carson's *Silent Spring* had already arrived. 250 miles away, across the border in poverty-stricken Mexico, there were butterflies and hummingbirds in the bedraggled gardens of Mexicali, but not a single sparrow twittered in the eaves of Delano. Nor were there lizards on the dry walls, nor snails in damp places, nor frogs in ditches. Even domestic animals were rare. I remember after several days being stopped in the streets by the sight of a wasp crawling out of a hole in a post, which by this time seemed extraordinary. A blind and complaisant local patriotism prevents the average citizens of Delano from believing that anything can be seriously wrong with their environment. But while easily swallowing this camel, they strain at the gnat of minor perils to their health. Knives and forks in the leading restaurants are enclosed in sanitised envelopes. Butter – often considered dangerous – is hard to come by, and a 'creamer', which includes eight chemicals, replaces milk for one's coffee. These precautions seem excessive when one considers a sky that appears to be burdened with the ash of distant volcanoes, and ornamental trees must be hosed down to prevent the dropping of their leaves.

This was an average small town of the richest state in the richest nation – a state which, considered as a country in its own right, would be the sixth wealthiest in the world, and in which the per capita income is the highest. According to a recent issue of *Time*, it offers a preview of the America to come. "'I have seen the future,'" says the newly returned visitor from California, "and it plays."' We are informed that Californian affluence is so extraordinary that many women can afford to spend eight hours a day, every day, in health clubs. We already knew that every suburban house was equipped with more than one television set, that a thousand pounds can be spent on the funeral of a dog, and that California invented the lotus-eating hippy, supporting tens of thousands of them in unproductive bliss. Crouching next to this cloth-of-gold are the wretched figures of field workers and grape-pickers, including the eight-year-old Mexican girl who works a seventy-hour week.

At this moment of suddenly awakened interest in the pollution of the world's atmosphere, and its possible long-term effects on the human race, the grape-pickers – who have spent the last ten years in the front line of the assault by chemical sprays – offer a unique human laboratory for scientific study of the problem. In his statement to Congress, Cesar Chavez said that the adverse effects of chemical poisoning are so commonplace and pervasive that they are considered by farm workers to be part of their way of life, and accepted as such. However, one of the interviewers engaged in a secret State investigation into agricultural poisoning, leaked some of the facts to Chavez's organisation. Of nearly 800 workers interviewed, practically all showed signs of poisoning, and 163 reported *five or more* of the following symptoms: vomiting, abnormal fatigue, abnormal perspiration, difficulty in breathing, loss of fingernails, loss of hair, itching in the ears, nosebleeds, swollen hands and feet, and diarrhoea.

The first field worker I spoke to in the Delano area was a cook in one of the immigrant camps. We were talking of the wretched surroundings in which he lived, and no mention of pesticides had been made, and suddenly I noticed the man's hands looked like a leper's at the stage just before the fingers begin to drop off. He said that a month earlier, at the height of the harvest, he'd decided to make a little extra money by grape-picking and had been caught when a plane had sprayed the vineyard with the fungicide, Captan. 'The spray wasn't properly vapourised,' he said, 'and it fell in drops.' Mexican women were running screaming in all directions in fear of losing their looks. 'The thing is to keep the stuff out of your eyes. I covered my face with my hands, and this is what I got.' He was surprised that I should be surprised.

But sometimes an overdose of pesticides can be lethal, as in the case of the three-year-old Mexican child who dipped her finger in a can of TEPP, sucked it, and died in minutes. Chavez reports the case of a pilot who attempted a forced landing while spraying with a dust formula of the same compound. 'The pilot was not injured but was covered with dust. He walked a distance of about 50 feet to a field worker, said he felt fine and asked for a drink of water. After drinking the water he began to vomit, and almost immediately became unconscious. By the time the ambulance arrived the pilot was dead, and then the ambulance driver, the pathologist and the mortician became ill from handling the body.'

But as a general rule the consequences of poison sprays are easier to overlook, and officially to ignore. Of some thirty varieties of pesticides used in the vineyards and farms only DDT has been in use for longer than ten years, and no-one has the slightest idea of what the long-term effects of some of the more recent – and far more lethal – sprays are likely to be. Grape-pickers who bother to go to a company doctor can expect to be told that they are suffering from sunstroke or an allergy. Recently a Dr Lee Mizrah, of Woodville, near Delano, did a one-man investigation on 58 farm children, chosen at random, and found 27 of them suffering from pesticide poisoning. This was the first such study ever done on children. What the doctor reported on were obvious and dramatic symptoms, as he had no way of knowing what latent ill-effects were present.

And now the crowning irony emerges. The effects of spraying on the wildlife of California are self-evident and only reluctantly admitted. The peregrines, the brown pelicans and the mockingbirds disappear from the scene, and the ardent trout-fishermen of Delano are driven by declining rivers to migrate as far as Oregon in pursuit of their sport. But the one form of life that has had no difficulty whatever in surviving the spraying is the insect pest against which the pesticides were designed. A slow reproduction rate dooms the birds and the small mammals, but where the reproduction cycle is completed – as in the case of many insects – in a matter of weeks, the resistant mutant quickly appears, and the new strain soon reproduces millions of its own kind. Thus the chemists are trapped in a vicious circle, forced continually to invent stronger poisons against increasingly resistant pests.

When David Montgomery, this article's photographer, took his pictures of the grape-pickers he was astonished to find himself working in a blizzard of flying insects – survivors of all the dusts and sprays. So many of them were there that his problem was to keep his lenses clear, and, when he unscrewed a lens to change it, insects got into the body of the camera. These immortal fruit-flies, trips and hoppers found their way into the clothing, the hair, the ears and nostrils, and they formed clouds round the heads of the tormented pickers as they worked with desperate speed along the vines.

In another vineyard the tempo of work was slower. They were picking a late variety of grapes called Emperors, and most of the pickers were dumpy Mexican women, their faces muffled to the eyes to keep out the poisoned grime and the gnats. The grapes were exceptionally large, and a pale pinkish-purple colour. Each bunch had a few malformed or spotted grapes which had to be snipped out there and then. In many cases the film left by spraying could be seen, superimposed on the natural bloom of the fruit.

The work was being supervised by a pleasant young Filipino foreman, attended by a huge and ferocious-looking dog. Such Filipinos are considered the élite of the labour force, whose devotion to Californian agriculture was finally rewarded by the repeal of a law which prevented them from marrying white women or owning their own houses. This man was almost, but not quite, an American. He was dressed stylishly as a cowboy, with stetson hat pushed well back from his forehead, and tooled Texan boots with high heels. He was paler than the Mexicans who worked under him, whose language he did not speak although his English was fluent enough. Relations between such foremen and the pickers are feudal. The foreman gets a cash bonus from the grower for any pickers he recruits; but much of his income derives from shacks he rents to them, and deductions he makes from the weekly pay-packet for the blankets and food he supplies. In return for these sources of profit conceded by the grower, the foreman becomes the trustee of the system, and defends it with religious zeal.

'Strikers?' the foreman said. 'They don't trouble us here. Why should they? Everything is OK. Everything is great. If any strikers showed up here these Mexes would see them off. Nobody around here is looking for any trouble.'

Two Mexican women encircled by glum-faced children squatted miserably out of action at the end of the row.

'What's the matter with those women?'

'I guess maybe they got some kinda trouble with the stomach.' There were patches of vomit here and there among the squashed discarded grapes on the ground. 'Spray sickness?' I asked.

'Hell no,' he said. 'We don't have no spray sickness here.' He laughed. 'These Mexes all eat the wrong kind of food. Put chillies in everything, and it sort of twists up your stomach. What can you expect?'

The foliage of the vines all round was crisp and grey, as though it had been scorched by fire. Beautiful, symmetrical bunches of grapes hung with artificial effect among the withered leaves, and the foreman picked one and gave it to me.

'How's that for grapes? They don't come bigger.'

I took them and thanked him. As I was leaving, he called after me: 'Give them a rinse under the tap, eh? Just give them a rinse under the tap.'

It is a number of years now since the last of the signs 'No Mexicans or Dogs Allowed' were removed from the restaurants and poolrooms of the San Joaquin Valley towns, but the sombre stream of racial antagonism flows as strongly as ever in the underground channels. Nearly two-and-a-half million Californians of Mexican blood are still confined to their *barrios* in the cities, competing with the black population for largely menial employment.

I never met a white in Delano, outside of Chavez's own movement, who considered Chavez to be anything but a troublemaker, a Red, and the leader of a tiny group of outsiders who were not wanted even by the Mexicans they claimed to represent.

'We're Southerners,' was the claim continually heard in these parts, offered as a valid exemption from the restraints that applied elsewhere. 'We're Southerners – and we don't let anyone kick us around. We've put up with quite a bit from this Chavez outfit so far ... but if they give us too much trouble...' The speaker in this instance reached for his hip pocket, while making a crisp sound with his lips, in imitation of the impact of a bullet.

The Road to Hoa-Binh

This episode took place in 1950 when Lewis was travelling in Vietnam while researching material for A Dragon Apparent. *The book became one of his most successful, and much loved by later foreign correspondents in Southeast Asia.*

At the time of his arrival, Vietnam was still under the authority of the French colonial power, which was fighting to retain full control. In this article he travels with the French Union forces. After Lewis's departure Hoa-Binh continued to be the scene of fierce fighting until the French finally withdrew from the town in February 1952. Two years later, after the crucial battle of Dien Bien Phu, the relentless Viet Minh forced the French to abandon the entire colony.

First published in *The Changing Sky* (Jonathan Cape, 1959)

A BOUT TWENTY MILES FROM HANOI the security police were waiting in a cutting. They stopped the car and pointed to a line of tanks and lorries drawn up at the roadside. 'We are now at the front,' said the captain. Somewhere far away, bombs were thumping down with the sound of prize potatoes being emptied out of a sack onto an earthen floor. A few legionaries, rosy-cheeked and bearded like fierce Santa Clauses, awaited in boredom their turns to move off. When the security officer went fussily away, the captain told his driver to pull out of the line and go up to the head of the queue. Waiting until the military vehicle in front had a lead of a hundred and fifty yards, we started to follow it. This method of moving vehicles singly through the danger-zone had recently replaced the convoy system. Before reaching Hoa-Binh there were a hundred natural death-traps. Whenever the Viet-Minh felt like doing so, they could pick off an isolated car, but the new arrangement had reduced the regular massacres that took place when a solid jam of vehicles, immobilised, say, by the blowing of a couple of bridges, was annihilated at the pleasure of the attackers.

We climbed through a landscape of Hebridean harshness. Fishermen with muscular thighs protruding from their palm-leaf coats dipped nets the size of large handkerchiefs into muddy streams. The water's surface shivered with the explosion of heavy guns from nearby fortresses. The roar of our engine in a low gear muffled the sound of their discharge; but we felt the blast, a gentle, recurrent concussion as if rubber-insulated chassis-members had worked loose and were beating together under the car. Topping a low hill we saw the jungle flowing towards us through shallow valleys, and on the southern horizon arose a jagged denticulation of hills. These were the mountains of Chinese landscape painting, called *calcaires* by the French, who are as impressed by them as are the Chinese. The captain groped for his camera, then changed his mind as we passed an abandoned car with a sprinkling of clean holes in the bodywork.

The opportunity for photography soon came. A group of men holding their guns like tired deerstalkers barred the road. We pulled up in a spattering of distant firecrackers. A sergeant-major, grinningly invulnerable, came up saluting. 'A bit of a scrap up in the woods, sir.' 'What's that down there?' the captain asked, his eye caught by the tortoise-like shifting of something under a blanket. Following his gaze, the sergeant-major seemed a little surprised. 'A couple of chaps caught a packet.' A strange primness suddenly muted the professional insensibility of his voice. 'That one lost both legs.' Thin rain and plum blossom blew in our faces as the captain composed his picture according to the perceived photographic rules: the bodies at the dramatic intersection of the thirds, the middle-distance with its roofless pagoda, and the background of fabulous mountains. A click of the shutter, and we moved off.

Under some unseen compulsion peasants were working with feverish energy to clear the jungle back from the road. 'The verb *défricher* is not exactly correct,' the captain said in gentle correction. 'It means, to clear for cultivation – which is hardly the case. At all events, it is a waste of time. It would take years to clear the jungle back to beyond machine-gun range of the road.' He waved his hand towards a low hill-top, shaved of its vegetation and scarred with defence-works. 'Can you imagine what it is like to be up there waiting to be attacked? In this war you may sit in such a post for two, even three years without ever seeing the enemy. Then one night your first action happens, and often it is your last. I don't mind confessing that I get an eerie sensation. Is it cowardice? I hope not. It is the feeling that I am at grips with something ant-like rather than human. These unemotional people in the grip of some blind instinct. I feel that my intelligence and my endurance are not enough. Take, for instance, those fellows they send up to dig holes close to the wire, before an attack. You'd expect them to show some human reaction when our supporting guns start dropping shells amongst them; but they don't. They go on digging until they're killed, and then some other kind of specialist fellows come crawling up and drag the bits and pieces away. Sometime later that night you know the shock-troops are going to come up and get into those holes, and then you're in for it. Losses simply don't bother them. All they're concerned about is not leaving anything behind. Do you know, they actually tie a piece of cord to every machine gun, so that as soon as the chap who's using it gets knocked out they can haul it back to safety.'

The captain turned to the rearing shapes of the *calcaires*. On all sides these massive limestone ruins soared up from the matted jungle, their surfaces seamed and pitted like carious teeth. Whole armies could have played hide-and-seek about their bases, protected from the air in innumerable caverns, and on the ground behind an impenetrable palisade of tree-trunks. Even tank-crews might have felt themselves nakedly exposed when their road wound slowly through these sinister labyrinths. Beyond a natural gateway, in a miniature Khyber Pass where the *calcaires* closed in on the road, lay Ao-Trac, a principal defence-post and supply base on what remained of Route Coloniale Number Six. Hoah-Binh was ten miles farther on. Strong-points had been built at fifty-yard intervals in the bottleneck outside Ao-Trac, and here hundreds of Vietnamese suspects were purging their offence by forced labour, while Senegalese overseers, ebony-masked colossi, strode amongst them, armed with switches.

Sun broke through the clouds lying along the rim of surrounding hills, and shone on the steel, the canvas, and the earthworks of Ao-Trac. The officers' mess was in a dugout and every time the heavy guns fired, earth slid down like loose snow from the sloping roof. 'I echo,' said the colonel, 'General De Lattre's words, namely, that we are here to stay for ever. To these I add, in the humility befitting my lesser rank, a single stipulation that, if God wills it, they continue to send us a sufficiency of shells, and half a litre of wine per man per day.' He laughed suddenly; a full-blooded man, happily acclimatised to the proximity of death. The junior officers produced obedient guffaws.

'We try to look after ourselves here,' the colonel said, 'with particular emphasis on the rations. The men get the same food as the officers. Might as well be comfortable as long as –' The diabolical crash of 155mm howitzers drowned the rest of his words, and set the unfilled Burgundy glasses chiming thinly. Shells plunged with harsh sighs into the sky and exploded six seconds later in staccato thunder. 'By the way,' the colonel said, 'I'm afraid Hoa-Binh's quite out unless you feel like being parachuted in. The Viets are shelling the Black River ferry now. Sank the ferry-boat yesterday with the second round at two thousand metres. They use those recoilless mountain guns they make up themselves. Very easy to manhandle. Means they can keep shifting them all the time, and all we can do is to plaster the whole area and hope for the best.'

Suddenly a dull, grumbling undertone of heavy machine-gun fire had filled in the silences in the cannonading, and crashing echoes chased each other across the valley below. 'It's unusual for the tanks to be in action this time of day,' the colonel said. He pushed back his chair and got up. 'I'm afraid I must leave you. Hope you'll be staying the night. You'll find it a bit primitive – and, of course, noisy. We've hardly settled in yet, but I've great plans for the future. Come back and see us in a year's time, and I promise you won't recognise the place.'

Five days later, in Hanoi, the general called a press conference. Its point came towards the end of a discourse lasting forty minutes, but the correspondents, knowing what was coming, could take only a mild connoisseurs' interest in the peroration. 'People put it to me this way,' said the general, his fine brooding eyes fixed rather reproachfully on his audience. 'They ask: "Having achieved your purpose in forcing the enemy to give battle – having destroyed in that battle his two best-equipped divisions – why do you retain so many men in a position where, from lack of opposition, they can no longer be effectively employed?" In deference to this logic, which is unanswerable, I have decided to displace the centre of gravity of our forces, which will henceforward be concentrated in the delta. Hoa-Binh, which is now without value to us, has been evacuated. I do not wish to disclose the number of casualties we sustained in this totally successful operation (which caught the enemy completely off his guard), but I will inform you that less than ten were killed.'

Indulgent Burma

A Dragon Apparent had been such a success that Lewis's publishers encouraged his desire to explore Burma, then rarely visited by foreigners. He arrived in Rangoon eighteen months after returning from Southeast Asia, in 1951.

The merit-earning, which features so large in this article, is a cultural phenomenon that – at least in some respects – seems to have survived. In the Charities Aid Foundation's World Giving Index, Myanmar (previously Burma) came out top for three years running between 2014 and 2016, despite being one of the world's poorest nations. In 2023 it was still in sixth place out of the 142 ranked nations.

U Thant, the official who in this article is so helpful and friendly to Lewis, became one of the most famous people in the world ten years later when he was appointed Secretary-General of the United Nations. He was also widely admired for his role in de-escalating the Cuban Missile Crisis.

First published in *The Happy Ant Heap*
(Jonathan Cape, 1998)

I AM PROBABLY ONE OF THE FEW PERSONS to have been tipped by a taxi driver, instead of the normal reverse of this transaction. It was a small matter yet provided an unforgettable moment of illumination of a cultural and spiritual divide, which sometimes exists between West and East, here represented by Burma.

The driver, affectionately known locally as Oh-oh, charged reasonable sums for ferrying passengers in his canary-coloured taxi about the southern town of Moulmein, but offered his services free to foreigners deposited there for a day or two when the ship from Rangoon put into port. Most of these fares, Oh-oh had heard, were enjoying a temporary escape from the capital, where visits into the surrounding countryside were not permitted. Like so many of his countrymen he was constantly on the alert for an opportunity to acquire merit, and being kind to foreigners came under the heading of meritorious actions. When the *Menam* tied up, the yellow jeep would be seen waiting on the quay, with Oh-oh offering a free ride to the new arrivals to any part of the town, plus a visit to the pagoda at Mudon, a few miles away, so long as the road happened to be clear of insurgents.

At the end of such trips passengers received a small present in the form of an ornament cut from mother-of-pearl. In my case the gift was a superior-quality bird's nest. We had visited the caves where the earliest of the season's nests were being collected, and this was the first 'number one' nest of that day. It had probably been finished only the day before and was therefore spotlessly clean – a tiny amber saucer constructed from secretions in glands located in the bird's head. The collector gained merit, too, by giving it away. When Oh-oh passed it over, the collector shook my hands, congratulating me with a wide smile.

Oh-oh now proposed that we should take breakfast – it was by this time midday – by joining a party given by a local family to celebrate the entry of their son into the Buddhist novitiate. We found ourselves in a large hall in which we joined about two hundred people seated upon mats on a polished floor. Oh-oh assured me that our host had collected many of the guests at random off the streets. Girls dressed in old-style finery were going round distributing snacks of pickled tea-leaves, salted ginger and shredded prawns. Once again merit-gain was what mattered, and it was an occasion for the family to give a substantial portion of their possessions away. It might take them two years, Oh-oh thought, to settle the debts incurred by this entertainment.

It was Oh-oh who warned me, when I told him of my hope to travel in the interior of the country, that I should do something to modify the extreme pallor of my skin. 'They will not stare because they are polite,' he said, 'but the young people in the villages have never seen an Englishman before and they will believe you are Japanese. We are entertaining bad memories of these people.'

'What can I do about it?'

'You can make your face darker by keeping it as much as you can in the sun.'

I took this warning seriously, and after three days' exposure on the deck of the *Menam*, my skin was the colour of freshly cut mahogany, except for white circles left by the sunglasses round the eyes. This caused some amusement among the European passengers, but evoked the sympathetic concern of the Burmese, including the assistant purser, who confided in me his belief that I was the victim of witchcraft.

There was no outright prohibition on foreigners travelling in the interior of Burma at this time, six years after the conclusion of the Second World War, but those who arrived in Rangoon found that, such were the obstacles encountered, they soon gave up. When I presented my letter of introduction to U Thant, head of the Ministry of Information, he saw no reason why I should not go where I wished. Later he admitted that, this being his first experience of a request to travel in the country, he wasn't sure of the official procedure to be followed. Later still I was to be informed that the US Military Attaché had fared no better, and that a team sent by *Life* magazine for a photographic report had left after two uninteresting weeks spent in the Strand Hotel, Rangoon.

The days slipped away while I passed from office to office, handled always with wonderful courtesy, encouraged in my hopes and commiserated with upon my many frustrations. Escape was by the greatest of flukes. Someone told me that a certain powerful general was the only person who could do anything for me. I was admitted to his office to be received by a man overflowing with charm. My face was by this time covered in blisters, but whatever surprise he may have felt at this spectacle, nothing of it showed. The fluke consisted in his occupation at the moment of my arrival with the translation of a recently issued British military manual into Burmese, and the difficulties he had run into, for although he had been to Sandhurst, certain of the terms employed had since then been changed. 'Happen to know anything about this kind of thing?' he asked, and amazingly enough I did. One hour later I left his presence with the pass in my pocket that was open-sesame to any part of Burma. 'Damn interesting trip, I should imagine,' he said. 'Won't find it too comfortable, I think, but have a great time.'

The question was where and how to travel at a time when the Burmese army was at grips with no less than five different bands of insurgents in the provinces, and even the small town of Syriam, just across the river from Rangoon, was under attack by dacoits. The disruptions of war had apparently left a gap of a dozen miles in the main line connecting Rangoon with the old capital, Mandalay, and steamers which used the Irrawaddy to carry goods and passengers up-country were sometimes cannonaded. Travelling rough could still be undertaken on the lorries of traders generally supposed to have come to an arrangement with insurgent bands, but there was nowhere in the interior to stay, not even a single hotel, and the official *dak* bungalows that provided rough accommodation in the past were closed or had been destroyed.

Happily, Mandalay could still be reached by plane, and two days later I had landed there, to be met by Mr Tok Gale of the British Information Service, who told me that he had arranged for me to sleep in the projection room of the town's only cinema, and would do his best to find a seat for me on a lorry going north. I was astounded to hear that he lived in what was officially described as the town's dacoit zone, two miles away. Tok Gale instructed me in the protocol of travel by Burmese lorry. Drivers, he said, did not accept money, but it was suitable to present them with small gifts, and he suggested that I should carry such items as keyrings and plastic combs. Postcards of the coronation of George VI were also eagerly collected and he had brought along a selection of these. 'You will be seated next to the driver,' he said. 'Please take the trouble to compliment him on his driving skills whenever occasion arises.' There was a word of warning. 'Beware in conversation of disparaging dacoits. These persons may be respectably dressed and mingling unobserved with lawful passengers.'

The night of my arrival in Mandalay, while walking in the deserted main street, I was attacked by a pariah dog, which bit me calmly and quietly in the calf before strolling away. Fortunately the remaining place still open was a bar, where I bought a bottle of Fire Tank Brand Mandalay Whisky to disinfect the wound. I increased my popularity on the next leg of the trip by sharing the remainder of this with such of my companions who were not subject to a religious fast. From this experience I learned the usefulness of mentioning a religious fast when rejecting unappetising food, such as the lizards in black sauce sold by roadside stalls.

The first stretch of the journey was to Myitkyina, where the road came to an end in the north, followed by a route virtually encircling the north-east, through Bhamó, Wanting – almost within sight of China – and Lashio, then weeks later back to Mandalay. At Bhamó, in jade country, you could pick up beautiful pieces of jade for next to nothing, and to my huge delight a circuit house for travelling officials (although there were none) was actually open, run by a butler straight out of the Victorian epoch, who addressed me as 'Honoured Sir', and instantly provided tea with eggs lightly boiled, and later a bed with sheets.

A final adventure was protective custody, into which I was promptly taken in the small town of Mu-Sé. Once again I slept contentedly, this time in a police station, and by day was accompanied on pleasant country walks by a heavily armed policeman, who was as keenly interested as myself in wildlife and natural history.

Thereafter all was plain sailing. Children no longer were alarmed by my ravaged features, and no more pariah dogs were perturbed by my alien smell. At Bhamó again, I took the river steamer down the Irrawaddy to Mandalay – 'a pleasure-making excursion' as the man who sold the tickets described it, and he was absolutely right. For three days we chugged softly through delectable riverine scenes. We were entertained by a professional storyteller, musicians strummed on archaic instruments, and once in a while the girls put on old-fashioned costumes to perform a spirited dance. There was a single moment of drama that was more theatrical than alarming. Insurgents hidden in the dense underbrush at the water's edge fired a few shots. Those on deck took momentary refuge behind the tall bales of malodorous fish. No one was hurt and by the time I surfaced on the scene, our military escort, who had blasted away at nothing in particular, had put down their guns and gone back to their gambling.

Next day Tok Gale welcomed me back in Mandalay.

'No complications with journey, I am hoping? No bad effects from meeting with that dog?'

'None at all. Everything went off perfectly. Couldn't have been better.'

'I am relieved. Well at least something will be done now about all those dogs on our streets.'

'So, you're actually getting rid of them then?'

'For a while, yes. Abbot U Thein San is taking all these animals into his pagoda compound for feeding and smarten-up. They will be released in a better frame of mind. It is belief that they will give no more trouble. In Mandalay we are used to seeing them. We should be regretful to miss their presence.'

'That's understandable,' I said.

'So how are you planning return to Rangoon?' he asked.

'I'm taking the train.'

Tok Gale seemed doubtful about this. 'For train travel they are saying that things are worse than they were. Rangoon train never arrives at destination.'

'I've been hearing that, so I took the precaution of having a horoscope done at the stupa of the legendary King Pyusawhti.'

'Ah yes. This is famous King hatched from an egg. And was the result satisfactory?'

'Entirely so. The *ponggi* told me I was good for another thirty years.'

'Well, that is splendid omen,' Tok Gale said. 'So 6.15 to Rangoon is holding few terrors for you?'

'How can it after a horoscope like that?'

Tok Gale laughed and shook his head in mock reproach. 'Now I must tell you something, Mr Lewis. You are falling into our ways.'

Rangoon Express

These experiences are also from Lewis's 1951 journey, the same journey recounted in 'Indulgent Burma'. Burma had been granted independence three years before Lewis's arrival, but much of the nation was in turmoil. Although Lewis was an unfussy traveller, he later admitted that this journey had been exceedingly uncomfortable. His publisher Cape had given him adequate expenses but, before leaving, Lewis purchased a gold watch, for possible sale if necessary.

The resulting book didn't achieve the same success as A Dragon Apparent, but most reviews were enthusiastic. The Guardian said that it was 'hard to remember any better travel book published during the last twenty years ... the author combines delicate description with unobtrusive scholarship and robust but subtle humour'.

First published in *The Changing Sky* (Jonathan Cape, 1959)

PUNCTUALLY AT 6.15 A.M., to the solemn ringing of hand-bells, the train steamed out of Mandalay station and headed for the south. Its title, the Rangoon Express, was hardly more than a rhetorical flourish, since among the trains of the world it is probably unique in never reaching its destination. It pushes on, carrying out minor repairs to the track as necessary, unless finally halted by the dynamiting of a bridge. Usually it covers in this way a distance of about 150 miles to reach Yamethin, before turning back. Thereafter follows a sixty-mile stretch along which rarely fewer than three major bridges are down at any given time.

Here at Yamethin, then, passengers bound for Rangoon are normally dumped and left to their own ingenuity and fortitude to find their way across the sixty-mile gap to the railhead, at Pyinmana, of the southern section of the line. The last train but one had even ventured past Yamethin, only to be heavily mortared before coming to a final halt at Tatkon; but our immediate predecessor had not done nearly as well, suffering derailment, three days before, at Yeni – about ninety miles south of Mandalay.

Against this background of catastrophe, the Rangoon Express seemed invested with a certain sombre majesty, as it rattled out into the hostile immensity of the plain. Burma was littered with the vestiges of things past: the ten thousand pagodas of vanished kingdoms, and the debris of modern times; smashed stone houses with straw huts built within their walls, and shattered rolling-stock, some already overgrown and some still smelling of charred wood, as we clattered slowly past. In this area, the main towns were held by Government troops, but the country districts were fought over by various insurgent groups – White Flag Communists of the Party line, and their Red Flag deviationist rivals; the PVOs under their *condottieri*, the Karen Nationalists, and many dacoits. All these groups battle vigorously with one another, and enter into bewildering series of temporary alliances to fight the Government troops. The result is chaos.

Our train was made up of converted cattle-trucks. Benches which could be slept on at night had been fixed up along the length of each compartment. Passengers were recommended to pull the chain in case of emergency, and in the lavatory a notice invited them to depress the handle. But there was no chain and no handle. The electric light came on by twisting two wires together. Protected by the religious scruples of the passengers, giant cockroaches mooched about the floor, while clouds of mosquitoes issued from the dark places under the benches. According to the hour, either one side or the other of the compartment was scorching hot from the impact of the sun's rays on the outside. This gave passengers sitting on the cooler side the opportunity to demonstrate their good breeding and acquire merit, by insisting on changing places with their fellow travellers sitting opposite.

With the exception of an elderly Buddhist monk, the other occupants of my carriage were railway repairs officials. The monk had recently completed a year of the rigorous penance known as 'tapas', and had just been released from hospital, where he had spent six weeks recovering from the effects. Before taking the yellow robe he too had been a railwayman and could therefore enter with vivacity into the technicalities of the others' shoptalk. He had with him a biscuit tin commemorating the coronation of King Edward VII, on which had been screwed a plaque with the inscription in English: 'God is Life, Light and Infinite Magnet'. From this box he extracted for our entertainment several pre-war copies of *News Digest*, and a collection of snapshots, some depicting railway disasters and other objects of veneration such as the Buddha-tooth of Kandy.

Delighted to display their inside information of the dangers to which we were exposed, the railway officials kept up a running commentary on the state of the bridges we passed over, all of which had been blown up several times. It was clear that from their familiarity with these hidden structural weaknesses a kind of affection for them had been bred. With relish they disclosed the fact that the supply of new girders had run out, so that the bridges were patched up with doubtfully repaired ones. Similar shortages now compelled them to use two bolts for securing rails to sleepers instead of the regulation four. Smilingly, they sometimes claimed to feel a bridge sway under the train's weight. To illustrate his contention that a driver could easily overlook a small break in the line, a permanent-way inspector mentioned that his 'petrol special' had once successfully jumped a gap of twenty inches that no one had noticed. That reminded his friend to tell us that the other day *his* 'petrol special' had refused to start after he had been out to inspect a sabotaged bridge, and while he was cleaning the carburettor, a couple of White Flag Communists had come along and taken him to their HQ. After questioning him about the defences of the local town, they expected him to walk home seven miles through the jungle, although it was after dark. Naturally he wasn't having any of this. He insisted on staying the night, and saw to it that they gave him breakfast in the morning. The inspector, who spoke a brand of Asiatic-English current among minor officials, said that they were safe enough going about their work unless accompanied by soldiers. 'They observe us at our labours without hindrance. Sometimes a warning shot rings out and we run like hell. That, my dear colleagues, is the set-up. From running continuously, I am rejuvenated. All appetites and sleeping much improved.'

These pleasant discussions were interrupted in the early afternoon, when a small mine was exploded in front of the engine. A rail had been torn by the explosion, and after allowing the passengers time to marvel at the nearness of their escape, the train began to back towards the station through which we had just passed. Almost immediately, a second mine exploded to the rear of the train, thus immobilising us. The railwaymen seemed surprised at this unusual development. Retiring to the lavatory, the senior inspector reappeared dressed in his best silk *longyi*, determined, it seemed, to confront with proper dignity any emergency that might arise. The passengers accepted the situation with the infinite good humour and resignation of the Burmese.

We were stranded in a dead-flat sun-wasted landscape. The paddies held a few yellow pools through which black-necked storks waded with premeditation, while buffaloes emerged from their hidden wallows, as if seen at the moment of creation. About a mile from the line an untidy village broke into the pattern of the fields. You could just make out the flash of red where a flag hung from the mogul turret of a house that had once belonged to an Indian landlord. With irrepressible satisfaction the senior inspector said that he knew for certain that there were three hundred Communists in the village. Going by past experience, he did not expect that they would attack the train, but a squad might be sent to look over the passengers. When I asked whether they would be likely to take away any European they found, the old monk said that they would not dare to do so in the face of his prohibition. He added that Buddhist monks preached and collected their rice in Communist villages without interference from Party officials. This, he believed, was because the Buddhist priesthood had never sided with oppressors. Their complete neutrality being recognised by all sides, they were also often asked by both Government and by the various insurgent groups to act as intermediaries.

And in fact there was no sign of life from the village. Time passed slowly and the monk entertained the company, discoursing with priestly erudition on such topics as the history of the great King Mindon's previous incarnation as a female demon. A deputy inspector of wagons, who was also a photographic enthusiast, described a camera he had seen with which subjects, when photographed in normal attire, came out in the nude. The misfortunes of the Government were discussed with much speculation as to their cause, and there was some support for a rumour, widespread in Burma, that these were ascribable to the incompetence of the astrologer who had calculated the propitious hour and day for the declaration of Independence.

With admirable foresight, spare rails were carried on the train, and some hours later a 'petrol special' arrived with the breakdown gang. It also brought vendors of samosas (mincemeat and onion patties in puff pastry), fried chicken, and Vim-Tonic – a non-alcoholic beverage in great local demand. Piously, the Buddhist monk restricted himself to rice, baked in the hollow of a yard-long cane of bamboo, subsequently sucking a couple of antimalarial tablets, under the impression that they contained vitamins valuable to his weakened state.

Quite soon the damaged rail ahead had been replaced, and we were on our way again, reaching, soon after nightfall, the town of Yamethin, which is known as the hottest town in Burma. It was almost without water, but you could buy a slab of ice-cream on a stick, and the Chinese proprietor of the teashop made no charge for plain tea if you bought a cake. With traditional magnificence a burgher of the town had chosen to celebrate some windfall by offering his fellow-citizens a free theatrical show, which was being performed in the station yard. It was a well-loved piece dealing with a profligate queen of old, who had remarkably chosen to cuckold the king, replacing him with a legless dwarf. The show was to last all night, and at one moment, between the squealing and the banging of the orchestra, there could be heard the thump of bombs falling in a nearby village.

It was only here and now that the real problem of the day arose. Since we were to sleep in the train, who was to occupy the upper berths, now fixed invitingly in position? Whoever did so would thus be compelled to show disrespect to those sleeping beneath them; a situation intolerably aggravated in this case by the presence of the venerable monk who was in no state to climb to the higher position. Of such things were composed, for a Burman, the true hardships of travel in troubling times. The perils and discomforts attendant upon the collapse of law and order were of no ultimate consequence. The real importance was the unswerving correctness of one's deportment in facing them.

Siam and the Modern World

Kukrit Pramoj, who accompanied Lewis to the lunch in this article, was the son of a Thai prince and educated at Oxford. During the decades when military juntas ruled Thailand, he worked as a journalist and a banker. After a measure of democracy returned to Thailand, he became active in politics. Ten years after Lewis's visit in 1953, he starred in the film The Ugly American *as the Prime Minister of a fictional Asian nation. A decade later he briefly became a genuine Prime Minister, sandwiched between the two premierships of his brother. He is considered one of Thailand's great statesmen, and his house is now a heritage museum (google 'thaiser.com kukrit heritage'). He died in 1995.*

King Chalalongkorn, the grandfather of Norman's luncheon host in Thailand, was the first Thai monarch to reduce slavery. Despite the aspects of modernity that Lewis describes, the King retained five consorts, and more than a hundred concubines, who delivered him thirty-two sons and forty-four daughters.

First published in *Granta*, No. 26 (Spring, 1989)

FIELD MARSHAL PLAEK PHIBUNSONGKHRAM, a keen fan of Mussolini, became Prime Minister of what was then Siam in 1938. In the next year, he changed the name to Thailand as part of an intense programme of modernisation, including encouraging Thais to eat with Western-style utensils, such as forks and spoons, rather than with their hands, as was then customary.

Following his support for the Japanese in the Second World War, he was ejected from the premiership, but by the time of my arrival in 1953, the military had restored him to power, and he was keen to keep modernising the nation.

The order went out that the nation was to cease looking to the past and to take the future in a firm embrace. Hat Yai, a provincial town in the south within a few miles of the Malaysian frontier, was chosen for an experiment in instant modernisation, and I went there to see what was happening. There was a tendency in Siam for the words 'modern' and 'American' to be used interchangeably, so, when the decree was published for Hat Yai to be brought up to date, most Thais accepted that it was to be Americanised. Little surprise was aroused when the model chosen for the new Hat Yai was Dodge City of the 1860s as revealed by the movies.

In due course the experts arrived with photographs of the capital of the wild frontier in its heyday, and within weeks the comfortable muddle of Hat Yai was no more. Its shacks reeling on their stilts were pulled down, the ducks and buffaloes chased out of the ditches, and the spirit-houses (after proper apologies to the spirits) shoved out of sight. It became illegal to fly kites within the limits of the town, or to stage contests between fighting-fish.

Where the bustling chaos of the East had once been, arose the replica of the main-street made famous by so many Westerns, complete with swing-door saloons, wall-eyed hotels and rickety verandas on which law-abiding citizens were marshalled by the sheriff to go on a posse, and men of evil intention planned their attack on the mail-train or the bank. Hat Yai possessed no horses, so the hard men of those days rode into town in jeeps – nevertheless, hitching-posts were provided. For all the masquerade, Hat Yai in the fifties bore some slight accidental resemblance to what Dodge City had been a century before, and there were gun-fighters in plenty in the vicinity. It was at that time an official rest-area for Malaysian Communist guerrillas from across the frontier, tolerated simply because the Thais lacked the strength to keep them out. The communist intruders were mostly armed to the teeth, and

Thai law-enforcement agents – part of whose uniform included Davy Crockett fur caps from which racoon tails dangled – were few in number. Reaching for one's gun was a matter of frequent occurrence in the main-street saloons. Although it was largely a histrionic gesture and few people were shot, newcomers like me were proudly taken to see the holes in the ceilings.

The arrival of the cinema played its part in the vision of the new Thailand. In a single year, 1950, hundreds, perhaps thousands of movie theatres opened all over Southeast Asia, the first film on general release being *Arsenic and Old Lace*. With this the shadow-plays that had entertained so many generations of Thais were wiped out overnight. A multitude of mothers throughout the land had worked tirelessly at pressing back their daughters' fingers from the age of five to enable them to take a stylish part in the dance dramas such as the *Ramayana*. From this point on it had all been to no purpose, and the customers who befuddled themselves in the saloons with *mekong* whisky, drunk hot by the half pint, were waited upon with sublime grace by girls whose performing days were coming to an end. Real-life theatre demanded the imaginative effort of suspending disbelief; gangster movies did not.

Investigating the threatened disappearance of the puppet-show, a Bangkok newspaper reported that it had only been able to discover a single company, surviving somewhere in the north of the country, working in Thai style with life-size puppets manipulated not by strings but by sticks from below stage. It took forty years to train a puppeteer to the required pitch of perfection in this art, and it seemed worthwhile to the newspaper to bring this company down to Bangkok to film what was likely to be one of its last performances.

This was given in the garden of the paper's editor, Kukrit Pramoj, and attracted a fashionable crowd of upper-crust Thais, plus a few foreign diplomats, many of whom would see a puppet-show for the first and last time. So unearthly was the skill of the puppeteers, so naturalistic and convincing the movements of the puppets, that, but for the fact that their vivacity surpassed that of flesh and blood, it

397

would have been tempting to suspect we were watching actors in puppet disguise.

After the show most of the guests went off to a smart restaurant, filling it with the bright clatter of enthusiasm that would soon fade. Such places provided 'continental' food – the mode of the day. In this land offering so many often-extraordinary regional delicacies, found nowhere else in the world, successful efforts were now made to suppress flavour to a point where only a soporific vacuum remained. Kukrit, then, as ever since, a champion of Thai culture, made the astonishing admission that he knew little of the cuisine, now only to be savoured at night-markets and roadside stalls. In a flare-up of nationalist enthusiasm, he announced his determination to put this right. He made enquiries among his friends and a few days later I received an invitation to lunch with him at the house of a relation, a prince who was a grandson of King Chulalongkorn. The prince, said Kukrit, employed a chef trained to cook nothing but European food, and he could not remember when – if ever – he had tasted a local dish. Entering into the spirit of adventure he had tracked down a Thai cook with a popular following in the half-world of the markets, to be hired for this occasion. And so the meal offered, for him too, the promise of novelty and adventure.

The prince lived on the outskirts of Bangkok in a large villa dating from about 1900. It was strikingly English in appearance, with a garden full of sweet peas – grown by the prince himself – which in this climate produced lax, greyish blooms, singularly devoid of scent. He awaited us at the garden gate. Kukrit leaped down from the car, scrambled towards him and, despite a government injunction to refrain from salutations of a servile kind, made a token grab at his right ankle. This the prince good-naturedly avoided. 'Do get up, Kukrit, dear boy,' he said. Both men had been at school in England and, apart from their easy, accent-free mastery of the language, there was something that proclaimed this in their faces and manner.

My previous experience of Thai houses had been limited to the claustrophobic homes in which the moneyed classes took refuge, shuttered away in a gloom that sheltered the clutter of dark furniture from the menacing light of day. The villa came as a surprise, for in the past year an avant-garde French interior designer had flown in to effect a revolution. He had brought the sun back, filtering it through lattices and the dappled shade of house plants with great, lustrous leaves, opening the house to light, diffusing an ambience of spring. We lunched under a photo-mural of Paris – *quand fleurit le printemps* – and a device invented by the designer breathed a faint fragrance of narcissi through the conditioned air. The meal was both delicious and enigmatic, based we were assured on the choice of the correct basic materials (none was identified), and auspicious colours according to the phase of the moon. Kukrit took many notes.

The entertainment that followed was in some ways more singular, for the prince told us that he had inherited most of his grandfather's photographic equipment, including his stereoscopic slides, and he proposed that we should view them together – 'to give you some idea of how royalty lived in those days'.

King Chulalongkorn, who reigned from 1868-1910, was a man of protean achievement. On the world stage he showed himself to be more than a match for the French colonial power which entertained barely concealed hopes of gobbling up his kingdom. At home he pursued many hobbies with unquenchable zest; organising fancy-dress parties and cooking for his friends, but, above all, immersed in his photography. He collected cameras by the hundred, did his own developing, and drew upon an immense family pool of consorts and children for his portraiture. We inspected photographs taken at frequent intervals of his sons lined up, ten at a time in order of height. In the King's loving record of their advance from childhood to adolescence, all of them, including the six-year-old at the bottom of the line, were wearing a top hat. Toppers had only been put aside in one photograph, showing four senior sons crammed into the basket of an imitation balloon.

The queens and consorts were even more interesting, and here they were seen posed in the standard environment of Victorian studio photography, lounging against plaster Greek columns, taking a pretended swipe with a tennis racquet, or clutching the handlebars of a weird old bicycle. Fancy-dress shots, of which there were many, bore labels in French – the language of culture of the day – L'Amazone (Queen Somdej with a feather in her hair grasping a bow); Une dame Turque de qualité (the Princess of Chiang Mai, with a hookah); La Cavalerie Légère (an unidentified consort in a hussar's shako); La Jolie Cochère (another ditto, in white breeches and straw hat, carrying a whip). The impression given by this collection was that the Victorian epoch had produced a face of its own, and that this could triumph even over barriers of race. Thus Phra Rataya, Princess of Chiang Mai, bore a resemblance to Georges Sand, Queen Somdej had something about her of La Duse, while a lesser consort, well into middle-age, reminded me of one of my old Welsh aunts.

The prince put away the slides. Like his grandfather King Chulalongkorn, and his great-grandfather, King Mongkut – who was an astronomer, and invented a quick-firing cannon based on the Colt revolver – he had a taste for intellectual pleasures. He showed us his Leica camera with its battery of lenses. Candid photography was in vogue at the time. By use of such gadgets as angle-viewfinders it was possible to catch subjects for portraiture off-guard, sometimes in ludicrous postures. There was no camera to equal it for this purpose, said the prince. As for his grandfather's gear, it took up rather a lot of space, and he would be quite happy to donate it to any museum that felt like giving it houseroom.

We strolled together across the polished entrance-hall towards the door, where my attention was suddenly taken by what appeared to be a large, old-fashioned and over-ornate birdcage, suspended in an environment in which everything else was modern. I stopped to examine the cage, and the prince said, 'Uncle lives there.'

Although slightly surprised, I thought I understood. 'You mean the house-spirit?'

'Exactly. In this life he was our head servant. He played an important part in bringing up us children, and was much loved by us all. Uncle was quite ready to sacrifice himself for the good of the family.'

The prince had no hesitation in explaining how this had come about. When the building of a new royal house was finished, a bargain might be struck with a man of low caste. The deal was that he would agree to surrender the remaining few years of the present existence in return for acceptance into the royal family in the next. He would be entitled to receive ritual offerings on a par with the family ancestors. Almost without exception such an arrangement was readily agreed to.

'How did Uncle die?'

'He was interred under the threshold. Being still a child, I was excluded from the ceremony, which was largely a religious one. Everyone was happy. Certainly, Uncle was.'

I took the risk. 'Would a Western education have any effect at all on such beliefs?' I asked.

'That is a hard question,' the prince said, 'but I am inclined to the opinion that it would be slight. It appears to be more a matter of feeling, rather than conscious belief. Education is an imperfect shield against custom and tradition.' We stood together in the doorway and the cage swayed a little in a gust of warm breeze. 'In some ways,' the prince said, 'you may judge us still to be a little backward.' His laugh seemed apologetic. 'In others I hope you will agree that we move with the times.'

Namek's Smoked Ancestor

The incidents described in this article took place during a journey that Lewis made to Irian Jaya (West Papua) in 1991, when he was eighty-three years old. It was the first of three arduous journeys into the most troubled parts of Indonesia, later described in An Empire of the East *(Cape, 1993).*

The custom of smoking the corpses of revered ancestors suffered its greatest decline after the arrival of missionaries in the 1950s. Some of the mummies, however, still survive. Normally corpses would have rotted in the humid heat, but smoking for thirty days preserved them. A writer for BBC Travel *magazine (December 2015) was taken to a cluster of fourteen smoked mummies, some of them arranged on bamboo scaffolding in life-like poses (google 'bbc travel lloyd neubauer smoked corpses of Aseki'). For some striking photographs, and the story about the restoration of a chieftain's mummy, google 'Tia Ghose live science mummies'.*

First published in GQ, March 1993

The one thing that impressed me about Wamena's airport building was an enormous artificial flower placed in the path of arriving passengers. This, a four-foot-across polystyrene Rafflesia, had been so painstakingly created that for a moment I thought I detected a sickly floral fragrance in its vicinity, whereas the airport as a whole smelt of nothing but a powerful anti-mosquito spray. Behind the flower came the information desk, where I enquired for a taxi driver who, according to a Jayapura agent, could usually be found at the airport; but the information desk informed me that Namek was the only Dani in Wamena who spoke English reasonably well. I was taken to the back of the building, where he was pointed out to me, occupied with some tourists who were photographing him in national garb. He was short for a Dani, with glittering eyes and a black beard, and as he hurried forward to introduce himself, his limp translated itself into a skip. His flat fur hat, of the kind once worn by Henry Tudor, enhanced a dignity by no means impaired by his nakedness. Apart from this head covering, he wore nothing but a two-foot yellow penis gourd held in the upright position by a string round the waist. The scrotum had been tucked away at the base of the gourd, exposing the testicles in a neat, blueish sac. This did not surprise me, for as the plane taxied to its rest I had noticed half a dozen naked men unloading a cargo plane.

We shook hands. Namek repeated the Dani greeting '*weh, weh*' (welcome) a number of times, excused himself, went off and came back wearing ill-fitting ex-army jungle fatigues. It now turned out that the taxi – in which he had a quarter share – was the magnificent ruin of an ancient Panhard-Levasseur, formerly owned by a Javanese raja. At the airport gate it now awaited us, refulgent with polished brass. In this we travelled in some state to a small hotel he recommended, where I took the austere room offered for one night. After a quick tidy-up I joined him in a species of porch, opening onto the street, where we discussed the possibilities of an investigatory trip into the interior.

By chance we had arrived in the midst of a minor crisis. The town had been showered overnight by large flying insects which, although harmless, were of menacing appearance. Many of them had found their way into the hotel, where they hurtled noisily across rooms and down passages, colliding with staff and guests, and then, their energy exhausted, were added to the piles into which they had been swept, until time could be found to clear them away. Namek took a gloomy view of this phenomenon, promising, he assured me, a change in the weather, which was likely to be for the worse.

'How do you come to speak English so well?' I asked him.

'My mother was killed in an accident, then a Catholic father adopted me,' he said. 'From him I am learning English and Dutch.' He spoke in a soft singsong, eyes lowered, as if soothing a child, then looking up suddenly at the end of each sentence as if for assent.

'Now I am a registered guide,' he said, 'no other taxi has assurance. Also I work in my garden. Tomorrow I will bring you sweet potatoes.'

'Are you married?'

'I have two wives,' he said. 'My father had two handfuls. That is the way we say for ten. We are always counting on fingers.' He raised his eyes to mine with a quick, furtive smile. 'You see we are going downhill.'

'Catholic, are you?'

'In Wamena all Catholics.'

'Doesn't your priest object to the wives?'

'For Danis they are making special rules. It's okay for them to have many wives. I cannot catch up with my father. Times now changed. Maybe one day I will have one wife more. That is enough.'

There was a moment of distraction while the hotel's cat raced through the furniture in chase of the remaining fearsome insects. 'My friend says you are wanting to see of our country. May I know of your plans?'

'I haven't any,' I said. 'This is just a quick trip to get the feeling of the place. What ought I to see? Merauke-Sorong? The Asmat, would you say?'

'You may show me your *surat jalan* (travel permit). Did you put down these places?'

'I only put down the Baliem. Can the others be added here?'

'No. For that you must go back to Jayapura for permission to go to these places.'

'In Jakarta they said it could be done here.'

'They are wrong. Go to the police office and they will tell you.'

'It seems a waste of time. Let us suppose I go back – am I sure of getting the permissions?'

'Here nothing is sure. One day they are telling you yes, the next day they say no. They will not agree to tell you on the telephone. Now also the telephone is not working.'

'So what do you suggest?'

He was reading the *surat jalan*, going over the words, letter by letter, with the tip of his forefinger, each word spoken softly, identified, and its meaning confirmed.

'With this *surat jalan* you may go to Karubaga,' he said.

'And what has Karubaga to offer?'

'Scenery very good. Also you are seeing different things. There are women in Karubaga turning themselves into bats.'

'That's promising,' I said. 'How do we get there?'

'By Merpati plane,' he said. 'To come back we are walking five days. In Karubaga you may find one porter. Maybe two. Also one bodyguard.'

'Why the bodyguard? Cannibals?'

The thick beard drew away from his lips as he humoured me with a smile. 'No cannibals. Sometimes unfriendly people.'

The many frustrations of travel-on-impulse had left their brandmark of caution on me. 'What are the snags?' I asked. 'Tell me the worst.'

'Very much climbing,' he said. 'Heart must be strong. *Surat* to be stamped by police in five villages. At Bakondini no river-bridge. Porters may bring you on their backs across, or rattan bridge to be built one day, two days – no more. Every day now it is raining a little.'

As he spoke a shadow fell across us. Part of the porch was of glass, and through it I saw that, where a patch of blue sky had shown only a few minutes before, black, muscled cloud masses had formed and were writhing and twisting like trapped animals. A single clap of thunder set off a cannonade of reverberations through the echoing clapboard of the town, morning became twilight, and then we heard the rain clattering towards us over the thousand tin roofs of Wamena. Pigs and dogs were sprinting down the street, chased by a frothing current, then disappearing behind a fence of water.

Eventually, the rain stopped, the sun broke through, and the steam rose in ghostly, tattered shapes from all the walls and pavements of the town. Mountain shapes, sharp-edged and glittering, surfaced in the clear sky above the fog. 'In one hour all dry again,' Namek said.

We came back to the question of travel. 'I'll think about Karubaga,' I said. 'Any suggestions about using up the afternoon?'

'We may go to Dalima to visit my smoked ancestor,' Namek said. 'For this we may bring with us American cigarettes.'

In Wamena they smoked clove cigarettes, and there was a long search in the market for the prized American kind that were rarely offered for sale. By the time we found a few packets, the shallow floods had already dried away, and we set off. We chugged away on three cylinders into the mountains to the north, left the car sizzling and blowing steam at Uwosilimo, and trudged five miles up a path to Dalima. In these off-the-beaten-track places the Dani had held on to their customs long after they had been stamped out elsewhere – the exaggerated expressions of bereavement following the deaths of close relatives, such as the cropping of ears and the amputation of fingers. Persons of great power and influence, known as *kain koks*, weren't cremated in the usual way but smoked over a slow fire for several months and thereafter hung from the eaves of their houses, where they continued to keep a benevolent eye on the community for decades, even centuries, until the newly arrived Indonesians launched their drive against 'barbarous practices', took down the offending cadavers and burned them or threw them into the river.

Namek's ancestor had been one of the few successfully hidden away, and now, in a slightly relaxed atmosphere, he could be discreetly produced for the admiration of visitors with access to cigarettes from the USA, which he let it be known through a shaman, was the offering he most appreciated.

The entire village turned out for us in holiday mood, the women topless and in their best grass skirts, and the men in the local style of penis gourds, with feathers dangling from their tips. We distributed cigarettes, and the current *kain kok* tottered into view, overwhelmingly impressive with his bird-of-paradise plumes, his valuable old shells, and the boar's tusks curving from the hole in his septum. Beaming seraphically, he punched a small hole in the middle of the cigarette and began to smoke it at both ends. He was the possessor of four handfuls of wives, and about this Namek said in a sibilant aside, 'Now he is old, and his women play their games while they are working in the fields.'

With this the smoked ancestor was carried out, having been crammed for this public appearance into a Victorian armchair. One arm was flung high into the air with a malacca cane grasped in the hand. The other hand, reaching surreptitiously down behind his back, held the polished skull of a bird. The Tudor-style hat affected by all the clan's leading males was tilted jauntily over an eye socket, and the ancestor's skin was quite black and frayed and split like the leather of an ancient sofa. His jaws had been wrenched wide apart by the fumigant, and now the old *kain kok* lit a Chesterfield, puffed on it, and wedged it between the two molar teeth that remained. Behind him descendants of lesser importance awaited their turn to make similar offerings to the ancestor.

The scene was in part grotesque but abounding in good cheer. The women rushed at us giggling and happy to show off their mutilated hands, and the men seemed proud of the tatters of skin which were all that remained of their ears. The village was a handsome one, scrupulously clean and well kept, and I was fascinated to see that the villagers had uprooted trees in the jungle and replanted them in such a way that they drooped trusses of fragrant yellow blossoms over the thatches of their houses. These attracted butterflies of sombre magnificence, which fed on the nectar until they became intoxicated and then toppled about the place like planes out of control, and were chased ineffectively both by the children and the village dog. In such Dani communities it is more or less share and share alike, and it seemed that in the allocation every child over the age of seven had been given a half-cigarette. These they were puffing at vigorously, and the village was full of the sound of their jubilation.

The Cossacks Go Home

The horrifying voyage that Lewis describes here began on 23rd October 1944, when, as an Intelligence Officer in the British Army, he was charged with accompanying a group of Soviet prisoners of war on the Reina del Pacifico, *once a most luxurious vessel. After switching ships, they arrived on 15th November at Khorramshahr, where the prisoners were handed over to the Red Army.*

The eye-popping tales that the prisoners related about their previous treatment in prison camps were no exaggeration. Since this article more documents have been published, confirming that the Germans had a policy of starving to death those Soviet prisoners who hadn't been selected for forced labour. More than three million out of 5.7 million Soviet prisoners died, most from starvation. Food was minimal or non-existent and, on pain of death, civilians were forbidden to give the prisoners any food. In several camps starvation resulted in cannibalism, even though in all German-run prisoner-of-war camps the penalty for cannibalism was execution.

First published in the *Observer*, November 22, 1983

When I returned on the pellucid evening of October 2nd 1944 from a day in the apple-scented criminal villages circling Naples, the bad news was that I would be leaving next morning to take some three thousand Soviet prisoners back to the Soviet Union. Although they were Soviet prisoners, almost none of them were ethnic Russians but instead from the various Asian communities of Soviet Central Asia.

Our senior sergeant and I stood together under the vast, dusty chandeliers of the Palace of the Dukes of Satriano, where 312 FS Section had its headquarters, while the doleful cries of Naples reached us through its open windows. The sergeant said that this was something that couldn't be handled by an NCO. Arrangements would have to be made for an emergency commission, but the Field Security Officer was away, therefore it was too late to do anything that day. 'We'll try and fix it up in Taranto for you as soon as you get there,' he said.

Next morning I took the train to Taranto, a pleasant trip in the almost icy perfection of southern Italy's most splendid month, after the air is washed and cooled by the first autumnal rains. At Taranto there was no talk of emergency commissions. A major was waiting for me at Movement Control. He wore no Intelligence 'green flash', but the faint aroma of lunacy, and the fierce but vague eyes identified him almost certainly as a member of the Intelligence Corps.

'You're taking over three thousand shits,' the major said. 'My orders are these. If any man so much as attempts to escape, you personally will shoot him. Is that clear?'

At this point the correct reply should have been, 'With all due respect sir, this is an improper order,' but I said nothing, assuming the man to be mad.

'Show me your gun,' the major said.

I took the .38 Webley out of its holster and handed it to him.

'Is this the only weapon you have?'

'Yes, sir.'

'I see. Well, I suppose there are plenty of guns to be had. Are you a good shot?'

'No, sir. I'm a bad one.'

'Oh. Well, you must do your best. Where do you come from? Naples, is it? Do they have any foxes up there?'

'Not as far as I know, sir.'

'Pity. They do in Rome. That may surprise you. In the woods.'

Mention had been made a few weeks before by a Taranto section member on a visit to Naples of a concentration camp for Soviet prisoners having been put up outside the town. The news had surprised us because it was the first intimation we had received of the presence of any quantity of Soviets in Italy. I was too late to visit the camp, for just before my arrival they had been moved out – with some difficulty, as it appeared – and put aboard the *Reina del Pacifico*, a regular troopship providing stark accommodation for troops and '3rd class families'. Here they were confined in remarkably cramped conditions below deck, guarded by the infantry company that would travel with them, via Khorramshahr in Iran, whence they would entrain for the last stage of their journey to the Soviet Union.

I went aboard, and went below immediately to inspect the prisoners, finding a dispirited rabble of men in rumpled German uniforms, lying about wherever they could, and covering every inch of deck space. An army that has suffered defeat and captivity is like a man overtaken by a sudden illness. The change is dramatic and instantaneous. Men seemed to have shrunk in the uniforms which now no longer fit; movements had slowed down, and discussions become spiritless and desultory. To this familiar climate of demoralisation, the prisoners had added a depressing ingredient of their own, for a number were singing an endless and mournful song. This was explained, by a young Jewish interpreter, supplied by the army, as a tribal death chant from Soviet Central Asia.

According to Benjamin, the interpreter, the 'Russians' had not been told that they were to be repatriated to the Soviet Union, until Red Army officers had arrived in the camp to inspect them, like looking over cattle that would soon go to the slaughterhouse. The result had been panic and some attempts at suicide. These men's grievance lay in the fact that, although compelled – as they claimed – to serve in the German army in Italy, they had deserted as soon as possible to the Italian partisans. When rounded up by advancing British forces they had been promised Allied status, which was to include re-fitting them out in British uniforms.

A young Russian – the only obvious European in sight – presented himself. His name was Ivan Golik, a Muscovite with the rank of senior lieutenant, who had assumed command. Golik was miraculously spruce and untouched by the demoralisation which surrounded him, but the message he had for us contained no comfort. Golik said violence had been used, and a number of men injured before they could be driven aboard, and that if our promise of Allied status wasn't kept, we could expect mass suicides. I passed this information to the Officer Commanding Troops who promised to contact GHQ for further instructions. On the strength of this, I told Golik that I believed that any promise made to the prisoners would be honoured, but that as the ship was leaving forthwith, nothing could be done about relieving them of the detested German uniforms until we reached Port Said. The assurance seemed to cool the atmosphere, and the Kalmuck death chant ceased; but it was still thought prudent to search the prisoners for combustible materials that might be used to set fire to the ship, and to keep them below hatches until Egypt was reached.

In the meanwhile, there were urgent problems to be dealt with, one being the production of a nominal roll required by the Officer Commanding Troops – all original documentation relating to the prisoners having been lost. This task was complicated by the fact that many men possessed the same name, and there were accusations among the Russians that some men had given false names in an attempt to conceal their identities. The inference was that certain of the prisoners had committed crimes while serving in the Wehrmacht.

In the course of these routine enquiries the facts soon proved to be less simple than we had supposed. Practically all these Asiatics had been members of the 162 Turkoman Infantry Division, composed of Uzbeks, Khirgiz, Kazakhis and other Muslim racial groups, who had fought in Northern Italy under the command of Lt General Von Heygendorff. The mere fact that they had served in the Turkoman Division was significant, because that meant that they were 'volunteers', and not mere *freiwillige*, who normally served in non-combatant labour battalions.

The division had begun life in 1942 in German-occupied Poland under the command of an eccentric Wehrmacht officer General von Niedermeyer, who thought of himself as the German Lawrence of Arabia, speaking Turkic even in his home, embracing Islam, and dressing as a native. In 1943 the division operated in the Ukraine, and after the defeat at Stalingrad it returned to Germany for re-organisation, before being sent to Northern Italy.

Here it fought well under the German officers, but committed to battle in July 1944 against American armour it began to disintegrate. It took a bad mauling near Orbetello-Grosetto and Massa Maritima, and this provided the opportunity for many of the Asiatics to change sides. They were terrified, they said, of falling into the hands of the Americans who, as they thought, believed them to be Japanese auxiliaries under German command, and who – as the rumour went – ran over such prisoners with their tanks. For this reason they took care to surrender only to Partisans, and it was to the Partisans that they gave themselves up in large numbers on September 13th during heavy defensive fighting near Rimini.

Starvation and the most atrocious treatment in German POW camps, and the realisation that they were faced with the alternative of certain death, had induced these men to serve in the German armed forces. These ultimate survivors spoke in the most matter-of-fact way of their experiences, some of which had been macabre indeed, and I soon came to know that for every Soviet who had come through the fiery furnace of the POW camps, a hundred had found a miserable death.

It was a fire of a magnitude that Russia and central Asia could never have known in all history, for there were vast human surpluses to be cleared as rapidly and as economically as possible. Every prisoner was ready with his own personal recital of horror, but a typical account providing a hundred variations on the same theme was provided by a young Tadjik herdsman. This boy who had hardly ever seen a real Russian, and never heard of Germany, had been snatched suddenly out of the steppes, put into uniform and given the first train ride of his life, in the great westward scramble of an unprepared and ill-equipped army to face the Nazi tanks. The train stopped and the soldiers began to march towards a distant cannonading, but a few hours later were ordered to turn back. Marching thereafter in the direction from which they had come they were halted by soldiers in unfamiliar uniforms who disarmed them, put them in lorries and drove them a short distance to an enormous barbed-wire enclosure. It was only at this point that they realised that they were prisoners of the Germans. They then remained here for three days without food and water, before a body of Germans arrived accompanied by someone who addressed them in Russian through a loud hailer. The Tadjik remembered him as short, bespectacled and mild in his manner. 'There are far more of you than we expected,' he explained. 'We have food for a thousand, and there are ten thousand here, so you must draw your own conclusions.'

The prisoners were then lined up and the order was given for Officers, Communists and Jews to step out of rank, but no one moved. All the prisoners had by now torn off their badges of rank. The bespectacled German then invited any prisoner who wished to do so to denounce any of his comrades belonging to these categories. He promised that those who co-operated in this way would receive favoured treatment, including all the food they could eat, and after some urging and more promises and threats on the German's part, a number of men stepped out of the ranks and the betrayal began. Those selected in this way were marched off to a separate enclosure, and at this point the bespectacled German said that a further problem had arisen through the shortage of ammunition. The men who betrayed their comrades were given cudgels and ordered on pain of instant death to use these to carry out the execution.

The Germans, on the whole, contrived to have prisoners kill prisoners. There were not enough SS 'special squads' to go round, and it was found that regular army soldiers were reluctant to engage in mass murder. It was true, too, that ammunition for such secondary uses was running low. Later in the war an SS squad leader provided me with the official statistic that unless a man was shot in such a way that the muzzle of the weapon virtually touched a vital area of the body, it took on average three shots to kill him, and inexperienced squads armed with automatic weapons and firing at a range of 12–15 feet, used up double that number of rounds. At that time there were countless thousands to be destroyed.

In the disorder of those early days of the German push to the East, I learned from my SS informants that the method of selecting Jews for elimination was both rapid and unscientific. Prisoners, as soon as taken, were ordered to drop their trousers, and those found to be circumcised, shot on the spot. As all the Muslims composing the Asian units were also circumcised, these, too were butchered en masse.

Later, when the Germans came to realise that many of the Asiatics were fiercely anti-Russian, and therefore employable as required in the German armed forces, the selection became more careful, but mistakes were still made. At one camp, in which several of the men I questioned had been held, the shibboleth of old was still in use. Every captive in turn was made to repeat in the presence of a Muscovite collaborator the sentence 'Na garye araratye rostut krupniye vinogradi' (On the Mountain of Ararat grow great vineyards), and those who had difficulty with the Ps were assumed to be Jews, although in fact many Asian tribesmen found this as difficult to cope with as the Jews.

More than three million soldiers died in these camps, most of them of starvation, but for those men of iron resistance determined to survive, come what might, the first hurdle to be cleared was an aversion to cannibalism; and I was convinced that all the men on this ship had eaten human flesh. The majority admitted to this without hesitation, often – as if the confession provided psychological release – with a kind of eagerness. Squatting in the fetid twilight below deck they would describe, as if relating some grim old Asian fable, the screaming, clawing scrambles that sometimes happened when a man died, when the prisoners fought like ravenous dogs to gorge themselves on the corpse before the Germans could drag it away. It was commonplace for a man too weak from starvation to defend himself, to be smuggled away to a quiet corner, knocked on the head and then eaten. One of the Asiatic Russians I interviewed displayed the cavity in the back of his leg where half his calf had been gnawed away while in a coma. In these episodes there were certain privileges that fell to the strong, who like lions over a kill, were left to take their share of the meat before the hyenas were permitted to approach. Cruellest of the camps from which my informants had sought any way of escape, seemed to have been at Salsk in the Rostov province, on the railroad between Stalingrad and Krasnodar. Here, seven days of total starvation prepared the prisoners for what was to come. When bread finally arrived, they were forced to crawl on their hands and knees to reach it under the fire of German soldiers, who were being trained as marksmen. Jews were buried alive by their non-Jewish comrades, and very commonly drowned in the latrines. Naked prisoners were compelled to fight each other to death with their bare hands, while their captors stood by urging them on, and taking photographs with their Leicas. An innocent-faced Uzbek hardly out of his teens described these combats. A killer could earn a little favour, gather a little following among the Germans by developing his own murderous speciality. One man used to kill a defeated opponent by biting his throat out. Another would bring a man down by twisting his testicles, before breaking his neck by a kick to the head. The Uzbek claimed to have despatched one of the guards' favourites by braining him with a femur he had wrenched from a corpse and kept hidden until the moment came. For this he was much applauded, given a crust of bread and the chance to volunteer for service in one of the Muslim auxiliary units being formed at that time.

Deep divisions and animosities had developed among the prisoners as a result of their sufferings in the camps, where it had been every man for himself. I soon came to the conclusion that only Ivan Golik could keep them in order, and that our dependence upon him was total. Two days out from Taranto the interpreter, Benjamin, told me that Golik had asked to see me alone and I had him brought to my cabin.

A discussion was conducted with some difficulty in a mixture of Russian and German, but Golik managed to tell me that a mutiny was being planned by a mullah, who was held in great awe by the Asiatics. This mullah's official name on the nominal roll was Sultanov, but he was known to the prisoners as Haj el Haq (the Pilgrim of Truth). Although enlisted as a mere private in the Red Army, he had been a member of the royal family of the Emirs of Bokhara, who had been such a thorn in the side of the Kremlin, until the Russians had entered the city in the early twenties and had the mullah's murderous old great-uncle thrown from the tower of his own palace.

The mullah, a detester of all Russians, and enthusiastically pro-German, managed to convince most of the Muslim soldiers that Adolf Hitler was working secretly for their cause and had made the pilgrimage to Mecca. Golik explained that a leadership struggle had developed between him and the holy man, with both pulling in opposite directions. 'I am determined,' Golik said, 'to take these men home. I am their guardian angel. Haj el Haq believes in paradise. He wants us all to die.' Golik's view was that the mullah would be shot as soon as he set foot on Russian soil, and therefore the mullah had nothing to lose by instigating a revolt. Golik believed that as soon as he gained enough following, the mullah would order the men to force their way past the guards up onto the deck, and there commit suicide by throwing themselves into the sea.

As part of his strategy, he said, the mullah had ordered the men to resist the reintroduction of army discipline and to reject all orders given them either by Golik himself, or by the two junior lieutenants, or by men who had previously held non-commissioned rank in the Red Army. He believed that the mullah's influence could only be combated by the transformation of this rabble into something like the semblance of a fighting force. The argument came back to the British uniforms. If the prisoners got them, all would be well, because the mere wearing of them would proclaim to the Soviets when they arrived at the end of their journey, that the British had recognised them as their allies. The new uniforms, worn with Red Army badges of rank, would transform sluggards into soldiers and banish daydreaming and despair. Golik wanted to be able to hold inspections, arrange parades, award punishments, do a little ceremonial drilling on deck if that could be sanctioned, as soon as the uniforms came. If they did not, the mullah would have won, and the men would fall in behind him.

I asked for the mullah to be pointed out to me, and we went below together. Despite the late season, the weather remained hot. The ventilation had failed and the prisoners, crammed into the holds, and stripped to their underpants, lay in rows, as African slaves must have done, their limbs shining with sweat. The wooden partitions dividing up the holds released an ingrained sourness adding to the sharp odour of so many bodies in close confinement. There was a great shortage of water, because all the Muslims were obliged to wash ritually six times daily. One or two spaces had been cleared for the men to squat in circles to listen to their storytellers, and Golik called my attention to the mullah seated cross-legged in one of these circles, a small man with a polished ivory head and a face full of scepticism and malice. It was the mullah who led the audience's formal outcry of astonishment or alarm whenever the storyteller reached a dramatic crisis in his narrative; and whenever a man had to pass behind him he went over to kiss the prayer beads the mullah dangled from his hand. We noted men at prayer, taking care, Golik said, to make their prostrations well within the mullah's view. This in itself was a bad sign, he said, for public prayer was discouraged in the Red Army, and could cost a man promotion. If no uniforms came, they would all fall back in prayer.

The journey from Taranto was slow and tedious with the ever-present threat of trouble brewing in the holds. A hot wind from Africa breathed on the ship night and day. The Mongol Buryat tribesmen chanted interminably about death and paradise, and the water dripped evermore slowly from the latrine pipes. Golik felt his authority draining away. The two junior lieutenants, Pashaev and Genghis Khan (there was also a private with this distinguished name) pretended no longer to hear his orders, while the mullah terrified the men by his trances during which he prophesied doom for all of them.

On 28th October we reached Port Said, where we were told that there would be a delay of some days during which we would trans-ship to the *Devonshire*. Here we were joined by two more interpreters, Private Shor from Aleppo, and a Bulgarian Jew, Sergeant Menahem who had led the twelve-man demolition team in Colonel Keyes' unsuccessful commando raid on Rommel's headquarters.

With the arrival of these fluent Russian speakers, I saw my presence on the *Devonshire* as unnecessary. I had never been given any indication, except by the mad major at Taranto, as to what I was expected to do, and I had had virtually no contact with the Officer Commanding Troops, who was in all probability himself completely mystified as to what I was doing there, and had at no time sought to make use of my services.

I therefore visited Movement Control at Port Said to request permission to return to Naples, hoping to be favoured by the technicality that the movement order issued in Taranto instructed me to accompany the *Reina del Pacifico* to its destination, and made no mention of a further voyage on the *Devonshire*. My reception by the Movement Control Officer was a bleak one.

Seeing that my arguments were without effect, I produced for the first and last time an extraordinary identity document issued to members of my previous North African section when sent on the more absurd kind of missions. This authorised the bearer to wear any uniform and called upon all persons subject to military law to assist him in any way, etc. The effect on the officer was less than electrifying. He took the paper, glanced at it, and threw it down. 'This may have worked for you in North Africa,' he said, 'but it won't here, and it won't in PAI-Force, where you're going. Get back to the ship.'

Aboard the *Devonshire* again, I found that in my absence the British uniforms had arrived. Bound to the wheels of the military machine which, once set in motion could not be stopped, the quartermaster's department had spewed forth a wild assortment of stores, including not only the so-long-desired uniforms, but all the complex and in this case useless impedimenta supplied to troops, including anti-gas equipment, numerous entrenching tools, camouflage netting, long-johns, to say nothing of razors and shaving brushes, the uses of which were mysterious to these hairless men.

The prisoners swarmed like bees, buzzing with excitement over the piles of equipment dumped in their midst. Suddenly the fog of inertia and depression had been dispersed. Golik, in an evilly fitting battledress but full of martial zest, had become the hero of the day. Morale was ebullient, and even the heat and stench of the holds seemed to have subsided. When the men could find space to walk among these crowded bodies, they did so more briskly, and had straightened themselves up. The mullah had retired to the latrines, 'to await a great vision' and there he remained for the rest of the day.

Within a few hours the last of the Russians had been kitted out as British soldiers, and the tailors among them were given shears. They quickly set to work adapting garments made for the big-boned well-fed men of the West to the smaller bodies of Asiatics bred in the main from generations of mare-riding nomads. With their upgrading, the prisoners were to be given full army rations too, and although these men had eaten human flesh, on religious grounds they refused the liver – which was all the meat we ever received.

Our fully-fledged Russian Allies, as they now were, seized with the greatest delight on the remaining three-fourths of this gear, which one would have supposed to be quite useless, and began to convert these to their special purposes. Working with extreme ingenuity and skill, they dismantled such objects as zinc water-bottles, mess-cans, and above all toothbrushes, nail brushes and combs; and pierced, spliced and amalgamated them to produce a variety of miniature musical instruments: strange antique-looking fiddles, lutes, pipes and rebecs. Soon the bowels of the ship quivered with the wild skirl of oriental music.

We sailed from Port Said on 2nd November with hope fizzing like an electric current through the ship. Golik, transfigured with optimism, had one more request to make. Included in the kit issued to each Russian was a truly superb Canadian blanket of the finest and fleeciest grey wool, and Golik now asked if he could be permitted to have a pair of these transformed by the tailors into a Red-Army-style officer's great-coat, in which he would like to make his appearance at the celebrating concert to be given by the ex-prisoners next evening. This, he assured me, would set the final stamp upon his authority.

It was hard to refuse Golik anything, especially as in any case our interests interlocked. All that mattered was to come to the end of a trouble-free journey. The coat was made in a day: a garment fastening high in the neck, and falling to within six inches of the wearer's toes. It would have conferred dignity upon a trader in the old camel market at Ismailia. He came on deck to show it off when it was ready, standing at the rail against the hot glitter of the sea and the incandescent Arabian coastline, and a couple of off-duty members of the escort, sunbathing nearby, got up awkwardly as if undecided whether to stand to attention. When we went below most of the prisoners saluted him.

The concert given by the Asiatics was unlike anything I had ever seen before, or have seen since. It was an entertainment to fill the steppes' great emptiness, and hollow in time, transplanted perfectly here in the faceless surroundings in which we crouched. The art of the nomads had grown up without the aid of stage props, and depended on mime and masquerade, plus a dash of shamanistic witchery; it lifted the mind clear away from unacceptable reality to glowing new worlds of the imagination. Costumes were produced by magical adaptations of camouflage netting and gas capes. Supreme theatrical art had transformed a man who had tasted human flesh into a tender princess, stripping the petals from a lily while a suitor quavered a love song; we heard the neighing of the horses and the thundering hooves of a Mongol horde on their way to sack the town. Whatever these men had suffered in the camps, nothing had been able to take their art away. It was to be understood that this spectacle devised for the entertainment of the princes of Central Asia would have little appeal for the soldiers of the British escort, for not a man attended. What was less easy to understand was the boredom of a European Russian like Golik, who, sweltering in his coat, fell almost instantly asleep, snoring heartily to the accompaniment of arcadian pipes.

Next day the process of rehabilitation went ahead according to Golik's plans. The Russians were allowed up on deck in batches, and a little space was set aside for Golik to conduct token inspections, check haircuts, and lecture his NCOs on military tactics. The Officer Commanding Troops making his rounds of the Russians' quarters in the holds, noted that at last these had been scrubbed out to his complete satisfaction, and Golik was complimented, and some further relaxations decreed. The mullah had been forcibly put into a British battledress and for the moment little more was heard of him. We all began to breathe more easily. This interlude of calm was disrupted by a most singular happening.

The three interpreters were profoundly oriental in their backgrounds, an influence which especially showed in their attitude to gold. This they appeared to regard as a magical substance, quite apart from any value it possessed for its purchasing power. Sergeant Menahem wore a signet ring made from gold wrenched from the jaw of a dead Italian on the battlefields. This had become like an African ju-ju for him – something invested with its own spirit. He did not like the ring to be touched, and complained of feeling a slight headache whenever he removed it from his finger to wash his hands. Shor, from Aleppo, had been given his first bath as a baby in a bath into which one hundred gold coins had been showered; and his parents, holding his arms and legs, had made him go through the motions of swimming 'so that he should swim in gold for the rest of his life'. Benjamin had spent his boyhood in a religious community in which only the Rabbi handled money, and it was an unfortunate chance for all of us that this young man, for whom gold until now had been a legend, should have been the one to have smelt out its presence on the ship.

Benjamin was cheerful in appearance and sympathetic in manner, and the prisoners confided in him more freely than they did with us. It was this special intimacy that had sprung up that clearly induced one of them to show him a gold coin he possessed, and Benjamin borrowed it from the man and brought it to me, agog with excitement, for a ruling as to whether it was genuine. Of this there was no doubt. The coin was an Edward VII sovereign, but the mystery was where it had come from, and I asked Benjamin to do his best to find out. Questions were met with a smokescreen of conflicting stories, designed it was to be supposed, to cover up guilty facts. Piecing the evidence together, we concluded that the sovereigns had been taken from a British agent parachuted into Northern Italy, who thereafter, in all probability, soon vanished forever. We knew that agents sent behind the lines were normally supplied with gold, either in the form of sovereigns, or five-dollar pieces which had an accepted value wherever they might be offered.

What proved to be of fundamental importance in these events was that Benjamin, by his probings, discovered the existence, and eventually the whereabouts, of many more coins – about fifty in all – and immediately set about devising a method of persuading the prisoners to part with their treasure.

The ship possessed its own NAAFI, open for an hour daily, and selling a limited supply of such things as chocolates, sweets, cigarettes, stationery and depressing souvenirs stocked up in its call at Port Said. Despite their new status the prisoners were not allowed to visit this, perhaps because it was assumed that they had no money to spend.

Benjamin got his hands on a NAAFI pricelist, bought a sample of each article in stock and went in search of those with hidden gold. When he found one he pushed a square of chocolate into his mouth, and let him hold a toy camel, or work a lighter shaped like a sphinx. His offer was to supply one pound's worth of NAAFI goods for every gold sovereign. This was sharp practice, for everyone but the intended victims of the swindle knew that sovereigns changed hands in the bazaar of any Middle Eastern town at five pounds, five shillings apiece. Many of the prisoners were reluctant to pay up, and when one hung back, Benjamin brought into play a particularly disastrous form of salesmanship. His argument – as we learned too late – went, 'You'll be off this ship in a few days. After that what good will money be to you? Surely you know what's going to happen?' Sometimes at this moment, he went so far as to point a forefinger to his temple in a significant way.

In the end Benjamin succeeded in convincing most of the prisoners to discard hope in exchange for the pleasure of the moment, and they handed over the gold and went off chewing a Cadbury's bar, and often clutching a ridiculous toy. In this way the seeds of despondency were effectively sown, and soon the men began to go down with it, one after another, like victims of an epidemic disease.

The storm broke when the Straits of Hormuz were sinking below the lip of the sea behind us, and Khorramshahr was waiting, like a frown on the face of destiny, only two days away. The prisoners had been allowed on deck and a slow swing of the pendulum of authority back to Golik had left the mullah isolated, as one by one his adherents again placed their neck under the yoke of military discipline. The Pilgrim of Truth had got rid of his uniform once more, and now wore a kaftan with voluminous sleeves and a large turban, both made from British army underwear. He still received the unctuous attention of a hardcore of followers, most of them, it was said, having some special reason to fear Soviet retribution.

The mullah had professed all along not to understand Russian, so, when the final confrontation took place, and Golik ordered him to go below and put on his uniform, the Battalion Commander took care to be seconded by Junior Lieutenant Genghis Khan, still sullen, but finally subdued, who repeated the order in the Uzbek language.

Golik had prepared himself for what followed. The mullah, an agile man, jumped to his feet, shrieking to his supporters to follow him, slipped through the ring of Golik's guards, and jumped into the sea. Golik, close at his heels, went in after him. A number of men intent on suicide had been inspired to climb the rails, but their resolution was demolished by the general outcry of *akoola!* (shark). In fact, the twisting grey shapes of large fish were to be seen everywhere, swimming close beneath the surface. The mullah's kaftan billowed in the water, he spread his arms feebly as if trying to fly; his eyes were closed and the sea washed the memory of fury from his face. Golik had reached him in a vigorous dogpaddle and kept him afloat, while the ship hove-to, and a boat was lowered.

For the mullah, when he was lifted aboard, this was the end of the road. The Uzbeks had gone dashing along the rail for a last gaze into eyes full of the rapture of paradise, but all they saw was a man fighting to fill his lungs with air and wincing and puking like a drowning kitten. He had not been permitted to die, and his survival was a matter for humiliation and sorrow. They watched the artificial respiration being given on the deck, saw the mullah's limbs move and his eyes open; then they turned their backs, and went away.

The last day on board was spent in preparation for the handover, which was to be elevated to a military occasion; the men of the escort fussed endlessly with their equipment and practised the arid drill movements with which they hoped to dazzle their Soviet opposite numbers.

I saw Golik as he readied himself for the fateful confrontation.

'What do you feel about things now?' I asked.

'Optimism. As long as you people stand by us. At worst I'll do ten years in a camp. I'm twenty-five now. I've still plenty of life left.'

We squeezed through the narrow waterway of the Shatt El Arab and tied up under a cold drizzle in Khorramshahr. In this threadbare city the Russians and the West were in daily mistrustful contact. It was the military show-window of nominal allies who hid their aversion between unbending correctness and skin-deep affability.

We looked down over a glum prospect of marshalling yards under the soft rain. All was greyness, befitting the occasion. In the middle-distance the strangest of trains came into sight, an endless succession of pygmy trucks, like those used in the West to transport cattle, but a quarter their size. It was drawn by three engines, the leader of which gave a sad and derisive whistle as it drew level with us. It stopped, and this was the signal for a grey cohort of Soviet infantry to come on stage and change formation before deploying to form a line between us and the train.

The escort party and the returning prisoners now disembarked, and there was more ceremonial shuffling of men, slapping of rifle stocks and stamping of boots. The Officer Commanding Troops and the Soviet Commander then strutted towards each other, saluted, shook hands, exchanged documents formalising the completion of the handover, and the thing was at an end.

With the three interpreters, I had been quite left out of this. Our presence had always been an anomaly, a suffix to the Officer Commanding Troops' authority for which the Army had provided no rules. Excluded from the ceremony, and ignored by both sides, we went our own way. Sergeant Menahem had actually passed through the line of Soviet machine-gunners, cast like identical tin soldiers, to inspect the trucks they were guarding, in which our Russians were to be transported back to their Fatherland. He came back to say that they had been used to transport pigs, and from the smell of them, he believed that they had recently served this purpose.

The British had about-turned and marched away back to their ship, but no objection was raised when we stayed on to watch the Soviet Commander and a following of goose-stepping subordinates inspect the front rank of the Russians, who were now prisoners once again. They came to Golik, standing, immensely stylish in ultimate defeat, at the head of his battalion. The Soviet commander circled him slowly in absolute silence. Both men were of the same height and build, and their greatcoats were identical in cut and length, but Golik's was the better of the two. The Commander then turned in my direction and signalled to me, and I went over to him. He spoke good English, and his manner was pleasant. 'Comrade liaison officer,' he said. 'Please do me a favour. I prefer to avoid speaking to these pigs. I ask you to give them the order to board the train.'

I refused to do this, but told him that one of the interpreters might oblige him, and in the end, Benjamin did.

There was a bar in the port just out of sight of what was happening, and I sat there and listened to the sound of the train shunting, the clash of bumpers, the pig-trucks rattling over the points, and the train's whistle as it pulled out.

The three interpreters came in out of the rain.

'Any trouble?' I asked.

'Not a peep out of anybody, not even the mullah,' Benjamin said. 'They're going to be shot. Most of them anyway.'

'How do you know that?'

'I had a chat with the Major. He turned out to be quite a character. Full of jokes. Took a great fancy to Golik's coat. "Whatever happens," he said, "I'll see to it they don't spoil *that*." Russians have a funny sense of humour. It may have been one of his jokes, but I don't think it was.'

Heroes and Villains: Lord Kitchener

In the mid-1980s the Independent's *magazine started a column called 'Heroes', in which writers were invited to describe the hero of their choice. When in September 1989 Lewis was invited to contribute, he declined, replying that instead he was prepared to write a column about his chosen villain. The suggestion was accepted, and henceforth the column was called 'Heroes and Villains'.*

First published in the *Independent*, September 23, 1989

STROLLING IN THE COOL OF THE EVENING in the desert outside Khartoum, a friend from the city's university drew my attention to points of red light skipping among the sand dunes. It was an atmospheric phenomenon not uncommon at the time of the year, though understood by the villagers as denoting the presence of *afreets* – in this case the ghosts of ancestors fallen on the battlefield of Omdurman, over which we now wandered. It had been the scene of a victory famed both for its carnage and its economy, and one that established Kitchener as a national hero of a magnitude unequalled since Wellington.

Horatio Herbert Kitchener, selected from many candidates as my personal villain-in-chief, was born in 1850 into a military family of the kind given to character-building austerities. His father and mother slept in all weathers under a blanket of newspapers sewn together. Their four children were encouraged to devise a system of punishments to inflict upon one another, resulting on one occasion with Herbert being staked out in the sun – a predicament from which with difficulty he was rescued by his mother.

As a military cadet at Aldershot he drew attention to himself by his extreme and eccentric religiosity, and later in life became adept at that least appealing of the Victorian arts – the concealment of private iniquity by public devotion. He was a cruel man. In South Africa he invented the concentration camp, into which Boer civilians were herded after their farms had been burned down. Out of a population of 100,000 some 26,000 – mainly women and children – were to die of malnutrition and disease in the camps.

Announcing that the infant casualties were the result of maternal neglect, Kitchener threatened to prosecute the most neglectful of the mothers for manslaughter. He was addicted to retribution. In the course of building a railroad to carry troops into the Sudan, he condoned – some say ordered – the punishment by amputation and hanging of slackers among the Egyptian labour force.

Kitchener was also a thief on a remarkable scale, though his thievery was to be excused as part of the eccentricity of a great man. He was a collector of antiques, boldly filling his pockets with everything that took his eye. Antique dealers, hearing that he was in town, hastened to put up their shutters, while hostesses rushed to hide their treasures away. When, after the occupation of Khartoum, Gordon's palace was rebuilt for Kitchener's use, he charged his staff with its embellishment. 'Loot like the blazes,' he ordered. 'I want any quantity of marble stairs, marble pavings, looking-glasses, doors, windows and furniture of all sorts.'

Accumulation of booty crammed warehouses in every country to which he had been sent in the service of the nation. After he was made a freeman of the City of London and presented with a jewelled sword of honour, he warned aldermen elsewhere that such presentations should in future consist of antiques and gold plate.

The victory at Omdurman, which marked the end of a series of one-sided colonial wars, was described by one participant, Winston Churchill, as the triumph of equipment over valour. Churchill – whose *The River War* contained perhaps his finest writings –

described the forces of the Khalifa, advancing under their flags to the attack over a five-mile front, as reminding him of cavalry in the Bayeux Tapestry. Churchill was generous in his praise not only of the great courage of the Khalifa's forces, but also of their tactical skill. These, however, were suited only to medieval conflict. At three thousand yards the British artillery at Omdurman opened up with modern weaponry, resulting in the deaths of some two thousand dervishes within the first five minutes. Only 'one brave old man carrying a flag comes within 150 paces of the shelter trench'.

Determined to ensure their first battle honours, the 21st Lancers made an unnecessary charge against a shell-blasted cavalry group. 'All [the enemy] who had fallen,' he wrote, 'were cut at with swords until they stopped quivering, but no artistic mutilations were attempted.' Thus the butchery went on throughout the day. By sundown the Sudanese had lost as many as 12,000; the British had lost 48. In a letter to his mother, Churchill wrote: 'Omdurman was disgraced by the inhuman slaughter of the wounded, and Kitchener was responsible for this.'

Kitchener then ordered the body of the Sudanese religious leader, the Mahdi, to be dug up, dismembered and thrown into the Nile, and his skull handed over to the College of Surgeons to be turned into a drinking cup. Only Queen Victoria's protest prevented this from being done, and Kitchener was obliged to send a letter of apology to the Queen.

Massed bands playing 'See the Conquering Hero Comes' greeted Kitchener on his return to England. *Punch* caricatured him as Sir Galahad in shining armour, attended by a fawning Britannia. Only in Cambridge, when Kitchener went there to receive an honorary degree, did a handful of liberal students dare to protest against the massacres in the Sudan. They were quickly rounded up to be flung into the icy, mid-winter waters of the Cam.

High Adventure with the Chocos of Panama (six hours required)

In 1957 Lewis flew to Panama from Cuba to cover the presidential election.

First published in A *View of the World* (Eland, 1986)

PANAMA CITY ON ELECTION EVE was like a medieval town placed under an interdict. My plane touched down at about eleven-thirty, and at midday all the bars and cafés closed down while this torrid land went dry for thirty-six hours. A sullen, fevered preoccupation with politics had entered the Panamanian blood. People sat at home listening to interminable radio speeches, preparing themselves to go to the polls on the morrow – which in Panama can be an exciting and dangerous experience. Having decided that these two days might be well spent in seeing something of the country's interior, I was dismayed to learn that travel was restricted during this period. As a precaution against citizens attempting to register their vote in more than one district – a fairly common practice in the past – all movement between voting districts was under a ban. The prospect of seeing anything of Panama in the short time I had to spare seemed a gloomy one. The discomforts of the moment were increased because the rainy seasons had just started and therefore it could be expected to rain punctually every afternoon. In fact on that first afternoon in Panama I sat on the hotel's veranda watching a deluge that, according to the local papers, was unequalled in a single day since 1905. That evening in desperation, and for the first time in my life, I visited a tourist agent in the hope that he might be able to suggest some way of escape from this burden of other people's politics, and rain.

The travel agent, a genial New Zealander named Kemp, who seemed amazed that anybody should actually require his services, at first half-heartedly recommended a visit to the ruins of old Panama. 'That's if you like ruins,' he added. 'Some people seem to get a kick out of them. Why I can't imagine. The trouble is you've got a whole morning to use up, and the ruins only take a couple of hours at the most – and even that's too long.' He thought again. 'You could go to Colon. There's no restriction of movement in the Zone itself. Can't say I particularly recommend it though. It lay down and died when they closed the naval base. It's not a place that had much appeal for me at any time, but the way it is now, it's like a wet Sunday afternoon. Say now, how about a trip to see the Indians? You can at least use up six hours that way, and be back before the afternoon rain starts.'

He handed me a sheet of paper headed: 'Tour No 6. See the Choco Indians (six hours required).' I read on: 'This exciting excursion takes you right into the heart of primitive, untouched Panama. Here you will see primeval man, strange birds, extraordinary beasts, and rare animals, all in a breathtakingly exotic jungle setting. Transported by commodious motor launch, you will be the privileged spectator of the mysterious way of life of the Choco Indian – intrepid hunter, and most remarkable of all Panama's indigenous peoples. With your experienced native guide (English speaking) you will penetrate to the hidden places of the jungle rarely seen by the white visitor. The keynote of this tour is high adventure. Especially recommended to those prepared to tolerate some discomfort in exchange for a unique travel experience. Bush clothing, and rope-soled shoes for walking over shingles are suggested.'

'The bit about discomfort is to warn off elderly ladies,' Kemp explained. 'This is supposed to be a real expedition in miniature. I mean you could get your feet wet or have a wasp sting you. I can't take any chances about someone starting a court action over a case of sunburn.'

'Have you ever done this trip yourself?' I asked him.

'No,' he said. 'When you're in the business you don't. You can't be bothered. Haven't got the inclination. In any case I can't say I'm wild about Indians. If you go for that kind of thing you'll enjoy seeing the Chocos, though. There's nothing like them. You'll see them going around by the hundred without a stitch on. Paint their bodies all the colours of the rainbow. Mind you, they can be dangerous if you go too far into the jungle. Liable to shoot a poisoned arrow into you. You'll be all right with our guide though. Whatever you do, don't forget to take plenty of colour film, or you'll never forgive yourself.'

Next morning at six, the experienced native guide, Dominguin, called for me in a station wagon. I was surprised by the cut and quality of his clothes, his signet ring, his brown suede pointed shoes with white inserts, his bow tie, his fairly competent English and his well-bred reserve. I was unable to associate this man with jungles. To me he offered in his way an example of adjustment to an environment as delicate as that of the armadillo. His natural environment seemed to be the city, and I soon discovered my suspicions to be correct. Dominguín turned out to be an expert on what the New Zealander called his 'Slumming Tour' which took in Panama's nightspots (three hours required). He frankly admitted that he'd only seen a jungle from the outside. It turned out that the jungle tour expert was working that day as an officer at one of the polling stations. However, this presented no problem, because all Dominguín had to do, he said, was to drive us to a point where the road touched the shore of Lake Madden, where an Indian would be waiting with the commodious launch. Thereafter the Indian would take over.

'These Indians,' he explained, 'live on the water. They know their way round the jungle like we know our way round the town. If you and I went into one of those jungles by ourselves we'd never come out again. These Indians know every creek, every tree. It's their life.'

This seemed reasonable enough to me. Our road now, in fact, entered the jungle, and for the first time I saw the rainforest typical of South America, the authentic Green Hell, which begins here on the east bank of the Panama Canal, and extends as a trackless sub-continent of vegetation for more than a thousand miles to the edges of the Pampas of the Argentine. Suddenly the grass on the road's verge had become monstrous – a hedge of green bayonets. Beyond, the jungle pressed forward under its armour of varnished and sculpted leaves, held back only at the very edge of the tennis courts and playing fields of the Canal Zone. I stopped the car, got out and enjoyed this confrontation. Twenty paces and I had passed through the green wall into the gothic solemnity of the forest's interior, into the borderland of millions of acres of twilight and vegetable decomposition. The odours of sap and mould, of roots and rain and leafy decay hung as thick as a London fog. Nothing stirred, but the distances were full of the chuckling and jibing of hidden birds. I came out and struck my foot against a flower, a confection of white waxen tiers jutting leafless from the wet earth. It shattered like glass. Huge morpho butterflies, blue and iridescent, were using the road like traffic, flying along it, up and down, very deliberately, in straight lines, and so slowly that I was able to reach up and snatch one out of the air as it was passing by.

A few miles further along, a side road took us down to the shore of Lake Madden. Here a small handsome indifferent Indian waited with a dugout canoe, of the type known as a cayuco. The cayuco was fitted with a neglected outboard engine, and it held an inch of brown water, along with two or three small dead malodorous fish in its bottom.

The Indian, splendidly bronzed and muscular, a miniature Apollo in faded bathing trunks, appeared not to notice our arrival. He was standing on one leg with the other flexed, and the sole of his foot pressed against the knee in a posture commonly adopted by the Nilotic tribesmen of the Upper Sudan. Dominguín went up to him and started a conversation in Spanish. 'Are you Juan?' Dominguín asked him.

The Indian brought his leg down. 'No,' he said.

'I was told to look for Juan. Where's Juan, then?'

'Juan's away voting so they sent me.'

'They sent you. All right, well look here. This man wants to see Indians.'

'Why?' the Indian asked.

'Don't ask me. He just wants to see Indians. He's on a tour.'

'There aren't any Indians around here.'

'What are you talking about? The jungle's full of Indians. They're all over the place.'

'There aren't any Indians here. They don't come down here. There's nothing for them. There's only one boat comes down here. They sell cabbages. There's no business to be done round the lake.'

'We can go up the side creeks and look for them, can't we?' Dominguín said. 'This man is paying to see Indians.'

'You can't get this cayuco up the creeks,' the Indian said. 'There isn't enough water in them. It hasn't rained enough yet. Why don't we go fishing? There are plenty of fish in the lake.'

'What kind of fish?' Dominguín asked. 'Catfish – tarpon?'

'Mojarras,' the Indian said. He pointed to the small, shrivelled, sardine-like objects in the bottom of the boat.

Dominguín now supplied an English version of what had passed between them. 'He says there aren't any Indians around just now. He says you come back next month when it's rained some more and he'll show you all the Indians you want.'

At this point I decided to take a short cut in the conversation and tackled the Indian in Spanish.

'Where do you live?' I asked. I assumed that he lived in a village that might be worth a visit.

'I live on Mr Coronado's farm. I'm the one that does the odd jobs.'

'Coronado's the guy who hires the boats to my boss,' Dominguín explained.

'Yes,' I said. 'But where is your home? Where do you come from?'

'I'm a Cuna. I come from San Blas,' the Indian said.

'Then you don't know these parts very well?'

'I've been here a month. I've been up the river once, that's all.'

'That's the River Chagres,' Dominguín said. 'It's full of Indians.

I know people who've been up that river. The Chagres is where you see the Indians. They have a big village up there.'

'Did you see any villages up the Chagres?' I asked the Indian.

He thought about this. 'Yes,' he said. 'I saw villages. There were many villages.'

I detected a trace of eagerness in his tone. Indians tend to be over-anxious to be the bearers of pleasing information.

'What were the houses like?'

'They were houses.'

'Yes, but with walls, or with open platforms, and just a roof?'

'I cannot remember. Perhaps they were open.'

'And the Indians – did they wear clothing?'

'Some wore clothing. Others did not.'

'This man doesn't know anything,' Dominguín said. 'What's to stop us going and taking a look ourselves? I tell you, if you want to see Indians, the Chagres river is the place where you're going to see them.'

'Can you find the Chagres river?' he asked the Indian.

'I think so,' the Indian said.

'He thinks so!' Dominguín said. 'Can you imagine that? A guide who thinks he can find a river half a mile wide. Well, anyway, let's go.'

We lowered ourselves cautiously into the cayuco which responded to the slightest imbalance with a violent rocking. Dominguín manoeuvred a handkerchief into position between his posterior and the seat. He took off his shoes and placed them in his lap, put on a pair of darker glasses than the ones he had hitherto been wearing, and then grasped the gunwale with both hands. Perhaps fifteen minutes passed while the Indian tinkered with the engine, and then we were off.

The lake was majestic and unruffled, darkly mirroring the firm clouds of summer. We roared over its surface alternately spray-soaked and sun-scorched towards a curving horizon of forest, rising out of the water. Shortly, the rampart of trees ahead of us divided, and we entered the mouth of the Chagres River. Half an hour passed comfortably, a pleasantly monotonous passage through an ever-narrowing channel. Small trusses of lavender flowers relieved the unchanging green façades on either side. A few herons broke into flight ahead of us at each bend in the river. And then, with an unpleasant jolting, the propeller shaft struck the shingle of the riverbed. We were in shallow water. The Indian stopped the engine, fixed it into the tilted position with the propeller clear of the water, and got out a single paddle with a wide, spear-shaped blade.

Paddling against the stream our speed was reduced to less than walking pace. In the next half-hour, we covered only two or three hundred yards. Then the bottom of the cayuco began to drag on the shingle. First the Indian got out and began to wade, pulling the cayuco along. Then Dominguín and I found ourselves in the water too. I now remembered Kemp's warning suggestion about taking rope-soled shoes on this trip. It was agony to walk barefoot over the sharp pebbles, especially when hauling at and half lifting the cayuco. At practically the same moment Dominguín and I decided to sacrifice our shoes. Relieved of the pain of bruised cut feet we staggered on, sometimes plunging waist-deep into a pool, sometimes being forced almost to carry the cayuco through shallows where the water was only a few inches deep. Finally we came to a stop, exhausted, and Dominguín called to a mulatto, the first human being we had seen since the start of the boat trip; he was sitting in a hammock just above the water's edge.

The mulatto pulled himself to his feet, straightened his body joint by joint, and came wading out to meet us. He was shaking his head and clicking his tongue, roused into a kind of easy shallow anger at the spectacle of our foolish ineptitude. This man was a member of the half of humanity that lives for ever on the slippery edge of bare subsistence. A short life – perhaps thirty-five years – of utter want had left him toothless, lacklustre of eye, with sagging body; and his mahogany skin was blotched with an ugly yellow as if he had been splashed with acid. Despair hung about him like a ragged shapeless garment. A grey crone of a wife and a brood of sickly children crept out of a shack in the background to watch us.

'Greetings Uncle,' Dominguín said. 'How far do we have to go to get into deep water again?'

'Deep water?' the mulatto said. 'There isn't any. From this point on upstream it's nothing but pools and rapids. It beats me what makes people like you try to get up this river in whaling ships the size of yours, when there's not enough water. I'm always being dragged away from whatever I'm doing to give someone a hand. Where are you going anyway? There's nothing but a few million trees up there.'

'This man's looking for an Indian village,' Dominguín said.

'There's an Indian village, all right,' the mulatto said, 'but you'll never get to it in that boat. You'd better leave that deep sea vessel with me and wade upstream if you want to go. You'll probably find the Indians fishing on the other side of the rapids. Maybe they'll take you up in their canoes.'

This seemed the only solution to our problem. We dragged the cayuco across to the bank under the mulatto's shack and tied it up.

'Who are you voting for?' Dominguín asked the mulatto.

The mulatto's angry laugh turned into a cough. 'Who'm I voting for? Why, I'm voting for the only chap who's ever done anything for me. Myself.'

'You should vote,' Dominguín said coldly. 'This is your chance to show your sense of civic responsibility.'

'Listen to me,' the mulatto said. 'I don't throw my vote away. If they want to do the right thing by me, all well and good. Last time I only got a half a sack of rice out of it. They'll have to do better than that. They offered to send a boat to pick me up. But I told them, I said it's got to be something substantial this time. We never know where our next mouthful is coming from.'

We left enough of our food with this man to hold back the frontiers of hunger for another few hours. He was a colonist of a forgotten world; limited by his depleted store of energy to burning off a few square yards of jungle a year, thereafter depending on the meagre return of seeds sown in uncultivated soil, left to themselves. He and his like – and there were millions of them – were condemned by malaria, semi-starvation, and by civilisation's total neglect, to an endless servitude of years spent lying in a hammock by the waters of such rivers. His cash income, on average, would hardly be more than fifty dollars a year.

We began to splash on upstream again. 'They haven't much civic feeling,' Dominguín said. 'The Chocos are like that, too. You can't get hold of them to make them vote. The Cunas vote because it's easy to round them up on the islands, and then set up polling stations.' He began to reminisce happily. 'Last election, I was working as an agent for Don Fulano' (he mentioned the name of a presidential candidate). 'He was interested in buying the votes of the Cuna Indians in bulk, so we went down there to fix up a deal with their cacique. Well, we figured it was no use telling these people about the improved drains we were going to put in in Panama City, so we told the cacique we were fixing up a rain-making ceremony and we would like to have his people's help. The cacique was pretty flattered to have us come to him with a proposition like that. They're nice people, the Cunas. They like to help. We gave the cacique an outboard engine and he saw to it that we got every Cuna Indian vote.' Dominguín made a face. 'The way things went, it didn't do Don Fulano any good. The opposition got to hear about it. They hijacked the boat taking the ballot papers back to Colon, and threw them into the sea.'

Fortified by the opium of his memories, Dominguín struggled on philosophically. By now we would certainly be unrecognisable as the men who had left the urbane precincts of the Panama Hilton hotel only four hours earlier. Our clothes were wet, shapeless and bedraggled, and smeared in places with the blood from minor abrasions suffered when we had slipped and fallen among half-submerged boulders. All the exposed parts of my skin were brilliant with sunburn. At this stage of our journey, we experienced a single dramatic moment when we saw a long thin pink snake swimming vigorously towards us. Our Indian, alerted by Dominguín's yelp of horror, reached under the surface, found a large stone and with what seemed to me incredible marksmanship hit the small agitated target represented by the snake's head fairly and squarely at a distance of perhaps twenty-five feet. The snake, seemingly stunned, lay still on the surface for a few seconds, and then began to wriggle towards us again, whereupon the Indian coolly repeated his feat. This time the snake gave up and turned back. Dominguín said that he recognised it as the deadliest of all snakes, an aggressive monster known in Spanish as *cuatro narices* whose bite produces infallible and agonising death in ten minutes. The Indian, on the other hand, said that it was quite harmless.

This whiff of Kemp's high adventure reminded me of the other extraordinary animal life his leaflet had promised.

'Do you have any jaguars in these parts?' I asked the Indian.

'No. We have small cats. No jaguars.'

'Tapirs, then?'

'What are tapirs?'

I described one, modelling the form of its pygmy trunk with my hands.

'There is no such animal. They do not exist.'

I didn't want to argue about it. 'Deer?' I asked.

'There are no deer.'

'There must be some animals,' I said. 'What animals are there?'

He thought for a few seconds. 'Rats,' he said. 'There are many rats. And iguanas. We eat the iguanas.'

'I know,' I said. 'They taste like chicken.'

'No, no,' he said. 'Much better than chicken.'

This conversation was going on sporadically while we were bypassing the rapids, forcing our way through the vegetation along the riverbank. Many thorny bushes and saw-edged reeds blocked our advance. These left the Indian unscathed but Dominguín's hands and my own were soon bleeding freely, and rents began to appear in our clothing.

Fortunately, as the mulatto had promised, two Indians were fishing just above the rapids. Our Cuna spoke to the two Chocos in some Indian lingua franca, and they agreed to lend us their canoes. The Cuna was to take me in one canoe, and Dominguín was to go with one of the Chocos in the other. This Choco was taller and slimmer than our Cuna. He had the face of an Eskimo, a polished helmet of black hair, and wore cotton shorts. I noticed a pair of ordinary skin-diver's goggles in his canoe along with his spear, and his morning's catch of two tiny fish. Dominguín asked him why he did not paint his body, and the Choco said that he did not have the time.

Our progress, although faster now, was still laborious. The river had become a series of pebbly shallows, separated by fairly deep pools. We paddled through the shallows where the water was just deep enough to float the empty canoes, and when we came to a pool the Indians ferried us across. It must have been midday before we reached a steep path leading up to what was described as the Choco village. On arrival we found a single house surrounded by the usual meagre, weed-infested fields, and half-charred tree trunks. The house was in reality an open-walled platform, made of branches roped together and thatched with reeds. As we came up the steep path, the Choco Indian shouted, and the house was suddenly full of running people who were either completely or nearly naked. I soon realised that they were rushing for their clothes.

Skirts were hastily going on and blouses being pulled over heads. Only a row of children with enormously distended stomachs, who stood peering down at us, were not involved in this excitement.

Presently a young woman carrying a baby came down a ladder. She was dressed shapelessly in a kind of shift made from a sack that had contained fertiliser, and she wore gypsy earrings of city manufacture and several necklaces of bright, crude beads. Her face was completely devoid of expression. The Choco who had brought us conferred at length with her, and then turned to me and announced in pidgin Spanish that she and the other members of the household agreed to be photographed with their clothes off for four dollars.

Dominguín was able to throw some light on this proposition. He discovered that several months previously a camera-armed party of tourists had been here, and through these the Chocos had learned that their normal state of nudity had become a marketable asset.

And were their bodies in fact painted in the geometrical designs one had heard about? I asked.

No, they were not it seemed. These were civilised Chocos who had learned to despise barbarism of that kind. They lived by growing a little grain and fruit – principally bananas, for the market, and while they waited for the maize to come up and the bananas to ripen they had nothing much to do but keep alive. Recreation? The Choco's face was incapable of amazement, but he clearly didn't know what I was talking about. Music? – Surely the young men still piped tunes on their primitive flutes? He shook his head. They'd had an old gramophone once, but it was broken now. Dominguín became impatient of my naivety. 'What do they do? Why, they sleep, of course.'

A powerful, heavy-faced matriarch now appeared. She was the cacique's wife. The cacique was away voting. With some pride she admitted that his vote had earned a pair of trousers and a vest.

'But the village?' I asked. 'Where are the other houses?'

'They've fallen down,' the woman said. 'There used to be many houses but the people all died or went away. Now what's left of us live in the one house. The cacique forbids the young men to go and work in the town, but they go all the same, and leave the women and children to look after themselves.' This year alone a woman had died in childbirth and the fever had carried off two children. Doctors? Unemotionally, the cacique's wife continued her saga of neglect. 'Two years ago a doctor came here and stuck a needle into everybody's arm. Why, I couldn't tell you. We haven't seen one since. The uncivilised jungle people have a medicine for fever, but it takes two days to get there, and the shaman expects something like a pig in return. They're supposed to be Indians like us, but they're just as bad as the townspeople the way they take advantage of us.'

My eyes went back to the primitive shelter, only one degree removed as human habitation from the cave. Around us for a distance of twenty or thirty yards, the earth was scattered with stinking debris. Pygmy diseased-looking pigs squealed and rooted among the rubbish. 'Why do you live here?' I asked.

'This is our home,' the woman said. 'We're civilised people. Planters. We believe in God. We grow bananas, and sell them in the town. The only trouble is they see you coming when you're an Indian. If it's a load of bananas you've got to sell, they'll tell you they're the wrong kind, or they're past the season. Supposing you offer them a monkey you've caught in a trap, they'll tell you it's going to die. It's a question of half price, take it or leave it. We're Indians. We have to take whatever they like to offer us. They know we can't argue with them.'

A shadow fell across us. I looked up. The sun had just fallen out of a last enclave of blue sky into curdling clouds. Thunder came galloping to meet us over the treetops, and raindrops began to spatter all round. We said goodbye to the Chocos, scampered back down the path to the canoes, and began the journey back. Now I knew what the real high adventure of Kemp's tour was to be. It was to be an adventure of rain.

When the rain started in real earnest, it seemed to close in on us until we were enclosed in a prison cell of water. At first the trees were still visible, as if lightly sketched in by a Chinese artist behind the curtains of rain. Then the rain washed out all the landscape. Dominguín's canoe, only a few yards ahead, had vanished. We slid forward over a vapour of pulverised water. Lightning glared all round us in prismatic colours, but a soundproof wall of rain held back the roar of thunder. Huge severed leaves came flapping down and fell in my face and on my lap. I found it helped to hold a hand over my nostrils to avoid breathing water. Presently I felt the canoe's bottom scrape on the shingle. We got out and began to grope our way ahead, repeating the laborious procedure of the upstream journey, alternately hauling the canoes along and then dragging ourselves into it them to ride a few yards as soon as the water deepened.

Suddenly the sky had emptied itself. The clouds overhead were torn apart, and sunshine poured through. Dominguín and the Choco took shape in a brilliant mist, standing knee-deep beside their canoe. There was a faerie-like quality in this scene that transformed them into the creatures of some watery Celtic legend. It seemed that we had reached the head of the rapids, and the Choco had a suggestion. It would save time, he said, as well as being a perfectly safe and reasonable thing to do if we shot the rapids. The Chocos themselves, he said, did it as a matter of course, and without a second thought, every time they went down the river. He would go first to show us the channel to take, and all we had to do was to follow him.

We set off and headed for the waters prancing and leaping between the high banks ahead. Somehow the rapids had taken on a fiercer vitality than they had possessed when we had passed them going up stream. Our plan of action instantly collapsed when the Cuna, instead of remaining behind the Choco, shot into the lead. I was startled at the speed we were travelling at, and tried hard to comfort myself with the thought that the Cunas were an island people, miraculous watermen, who passed nine-tenths of their lives in boats. The canoe sat so low in the water under our weight that it only had about three inches of freeboard, and among the rocks, the races and the whirlpools of the rapids, small agitated wavelets broke continually over the side. With growing disillusionment and concern, I watched the water in the canoe's bottom deepen. It would have been impossible to bale because the slightest movement on my part would have upset us. Black rocks crested with flying water hurtled past on each side. The six inches of water in the bottom implacably deepened to a foot. The Cuna, who had been inclined to show signs of amusement since the beginning of this difficult passage, was now laughing outright. This chilling sight prepared me for the inevitable. Indian impassivity is rarely disturbed by anything short of catastrophe. I remembered agonisingly that the last Indians which I had seen laughing in this way had been the survivors of a bus crash in Guatemala.

Water poured over the gunwale now. Our weight slowed us and I saw the Choco with Dominguín bearing down. The Choco, too, was convulsed with laughter. He waved his paddle in what was perhaps a farewell gesture as our canoe sunk under me, and I found myself being carried away at such a speed that it was impossible even to influence my direction by trying to swim. A moment later

I was bumping on shingle. For another twenty yards I scrambled, slipped and struggled in the water trying to find my feet before I landed on a sandbank. The current tugged like wrestlers' hands at my ankles. Only when the water was less than knee-deep could I stand upright. Dominguín and the Choco had passed but were clearly in extremis. As I watched, Dominguín, still seated bolt upright, and strangely spruce and dignified in this moment of truth, appeared to be lowered gently below the surface. An instant later water, too, cancelled out the Choco's happy grin. Fortunately for Dominguín, who couldn't swim, he was carried straight on to a small island, where he squatted rather miserably, until he could be rescued. Both Indians were washed up a hundred yards or so away, and came clambering back over the rocks to recover their canoes. They were in high spirits.

Perhaps an hour later we reached the mulatto's shack. By this time our clothing had dried on us. We found the mulatto waiting for us. 'I've been thinking about doing something about voting after all,' he told Dominguín. 'Any chance of your finding anyone who can do better than a half sack of rice, if I come along with you?'

Dominguín with his connections in the city thought he might be able to do something.

We took our seats in Kemp's commodious launch and the mulatto excused himself. He came back carrying something like one of those straw capes which Japanese peasants are depicted as wearing in the old colour prints. This he fastened round his shoulders.

'Going to rain in a minute again,' he said. 'And it won't be a shower this time either, like the last one.'

I looked up. The sky was again turning into porridge.

'Another couple of hours,' Dominguín said, 'and we'll be home.'

The Burning of the Trees

Lewis's previous journeys to Latin America had mostly concentrated on the horrendous plight of the indigenous Indians. In December 1978, however, he read an article in the New Scientist *describing the devastating consequences of tropical deforestation.*

He contacted Peter Crookston, editor of the Observer *magazine, to suggest an article. Lewis told Crookston that he thought the deforestation could have worldwide implications, that burning the forests in the Amazon would be like the world losing a lung. Crookston had never heard anyone say this before and, bowled over by the idea, he 'busted the* Observer's *budget' to pay for the travelling.*

When the subsequent article was published, a photograph by Colin Jones ran on the magazine's cover emblazoned with the words 'How the rape of Amazonia could ruin life on earth'. Lewis was the first non-specialist writer to predict the possibility of this future environmental disaster. After his 'Genocide' article, he believed this to be the most important that he had written.

Nevertheless, the trees have continued to be burnt, especially after Jair Bolsonaro became President of Brazil in 2019. During his three years in power, the Amazon forests lost an area larger than Belgium.

First published in the *Observer*, April 22, 1979

THE AMAZON FOREST leaves its mark on the imagination of all who see it. It is one of nature's exaggerations, matching the great river which, viewed across white beaches, could be an ocean drifting eastwards.

One third of the world's trees grow in the forest's five million square kilometres – an area larger than Europe. It extends its umbrella of shade over half Brazil; the cool, damp, crepuscular corner of a continent. It is shown as blanks on the map between the veining of rivers, having little legend and no history.

Romantic explorers and holy madmen like the celebrated Colonel Fawcett paddled their canoes up creeks, hacking brief trails in the jungle in search of hidden cities, but there was never anything there but trees and painted, feathered Indians, almost as much a part of the jungle as the trees themselves. The first satellite photographs revealed more of the details of the forest than had three centuries of exploration. The trees provide the final refuge of about a hundred Indian tribes, numbering perhaps 40,000 people. It is believed that almost as many have disappeared since the turn of the century – many as the result of outright murder, more through the white man's diseases, against which the Indian has no immunity. Occasionally a new Indian group is found – flushed out of the trees by pioneers who cut the trails ahead of the road-building gangs – and when this happens about a quarter of those 'contacted' can be expected to die of one commonplace Western ailment or another within the year. Indians are entirely dependent upon the forest. They cannot survive properly outside it.

Theodore Roosevelt, spokesman and clairvoyant of the world of quick profits, pondered over the Amazonian vacuum and predicted what was to come. He had written a book about the pleasures of ranching, but had little use for trees. 'The country along this river,' he wrote in 1914, 'is fine natural cattle country, and one day it will see a great development.' His judgement as to the suitability of the land for cattle ranching was abysmally wrong; his prediction at least half correct. The ranchers arrived and the great attack on the trees began.

Sixty years later the Brazilian people were to learn that about a quarter of the forests of the Amazon Basin and Mato Grosso had already been destroyed. Eleven million hectares of trees had been cleared in the preceding decade alone, and it became a matter of simple arithmetic, if this were allowed to go on, to forecast a date when the forest would eventually cease to exist.

Most of the clearing was done by foreign enterprises such as Daniel Keith Ludwig's Jari Forestry and Ranching Company, the Italian firm Liquigas, Volkswagen do Brasil, and King Ranch of Texas. These and many more had been encouraged in their attack on the forest by financial incentives offered by the Brazilian Government. Great fires – some of them ignited by napalm bombing – raged all over Amazonia, consuming trees by the hundred million, and for months on end travellers on planes on their way from Belém to Manaus or Brasilia saw little of the landscape through the smoke.

There were few parts of the world left in this century where uninhibited commercial adventures of this kind were still possible, where land could be picked up for next to nothing, where wages were about a tenth of those paid in Europe or the US, and where a modest investment in stock and equipment offered the prospect of spectacular profits. The government's early enthusiasm for the giant ranches began to falter when it was found that, like the trees they replaced, they seemed to be self-sufficient, producing little surplus to help with the balance of payments. Nor did they relieve unemployment, because when a ranch became a going concern it took only one man to look after a thousand head of cattle.

With the growing suspicion that the multinationals were little concerned with the long-term problems of the nation, voices were raised to enquire whether this rape of the forest, so apparently devoid of sustainable economic reward, might not in the long run have some undesirable effect on the climate. The mild obsession, familiar in northern latitudes, over the possibility of the return of the Ice Age is replaced in the tropics with a conviction that the reverse is likely to happen. In Peru a loss of permanent snow has been recorded from the Andean ice-peaks. Bolivia has suffered from declining rainfall and searing winds, while in Brazil itself parts of the north-east in the area of Ceará have been reduced to near-desert.

These misgivings were given alarming substance by the publication of figures based on seventeen years' field studies in Amazonia. This research was done by Harald Sioli, Director of the Max Planck Institute of Limnology, in West Germany, and his calculations showed that the Amazon forest contributed through photosynthesis half of the world's annual production of oxygen. He argued that it could not be sacrificed without a dramatic, if not fully predictable, deterioration in world climate. He calculated that the forest contained about 300 tons of carbon per hectare, and that its total extension of 280 million hectares, if burnt down, would allow sufficient carbon dioxide to be released into the atmosphere to cause a 10 per cent increase of the gas. The threat was two-fold; the loss of the forest's important contribution of oxygen, and of its capacity to absorb carbon dioxide. Sioli noted that the burning of fossil-fuels had already caused a 15 percent increase in carbon dioxide over the past century, and that the forests were failing to contain the increase. He concluded that destroying the Amazon forest would be like getting rid of one of the world's major oceans – environmental suicide.

There has been some scientific bandying of arguments over these figures, which have been wholeheartedly endorsed by some experts and received with caution by others. One climatologist, for example, has argued that one third of the carbon dioxide released by burning fossil-fuels remains in the air, while another believes that the proportion is two-thirds. These are matters for discussion by the illuminati. On the danger represented by oxygen reduction, Dr Mary McNeil, an American specialist in laterite soils, says that were all tropical forests – of which the Amazon forest is a major component – to go, the earth's atmosphere would eventually be denuded of oxygen.

To turn to the other problem of the excess of carbon dioxide, Norman Myers, a consultant in environmental conservation, wrote in *New Scientist*, last December, 'Widespread deforestation in the tropics could lead to increased reflectivity of sunlight in the equatorial zone (the "albedo effect") and to a build-up of carbon dioxide in the earth's atmosphere. Both these processes could upset global climatic patterns....'

Eneas Salati, professor of physics and researcher in agriculture at the University of São Paulo, was quoted in *Critica* of Manaus as saying that the destruction by burning of the Amazon forest, and consequent increase in the carbon dioxide content of the atmosphere, could result in heightened world temperature, the melting of the polar ice-caps, and a sufficient rise in ocean levels to bring about the inundation of hundreds of coastal cities throughout the world.

The predictions of all these experts are deeply worrying in their various ways. It seems clear that the least we have to fear from the loss of the Amazon forest is undesirable meteorological changes, and the worst is the catastrophe promised by Sioli and Salati.

The threatened forest offers the paradox of an area into which is crammed the greatest abundance and diversity of living things to be found anywhere on earth, yet is potentially a desert. Only the thinnest skin of humus covers the laterite floor. Apart from what is derived through photosynthesis, the trees live almost by what can be described as self-cannibalisation, upon nutrients furnished by the litter they themselves provide, made rapidly available through the action of insects, worms and fungi. The forest recycles 51 percent of the rain that falls on it, and produces little more energy than it consumes. It lives then, almost independent of the soil, in a state of equilibrium.

Remove the trees and the average temperature of the area where they once stood increases by 30°F; rainfall declines sharply (by five per cent per year over the last 10 years in some areas recently deforested in neighbouring Bolivia), yet flooding becomes a recurrent hazard, because with the loss of the 'sponge effect' of the forest's root-mat, the soil can no longer contain the excess water. With the rains, such nutrients as the forest floor contained are instantly washed away, and the laterite, laid bare to the sun, oxidises and loses every trace of fertility. This is no equatorial replay of the slow process of the formation of dust-bowls in ruined, US prairies: this is instant desert unless immediate and costly counter-measures are taken.

From the time of its discovery until now, while it faces the threat of obliteration, the Amazon jungle has remained a scientific void. Those who have penetrated it – in the main rubber tappers and diamond prospectors – have been acquisitive rather than curious. Rough counts of its fauna have been made. Bates, the Victorian naturalist and contemporary of Darwin, collected seventeen thousand different insects before giving it up, and it has been estimated that there may be two thousand species of birds.

All the figures attempting to define this vanishing abundance are vague. Forest ecologists at INPA (the National Institute for Amazonian Research) had counted about four hundred species of trees growing in a single hectare. With almost every week, new plants and trees were being discovered. Even to the scientists it was beginning to appear pointless to continue with the labour of classifying and cataloguing all these living things, so soon to vanish.

Where the destruction of trees threatens a commercial resource, the government is sometimes moved to act. In the province of Acre, where the worst deforestation took place, ten thousand Brazil nut-gatherers had lost their livelihood and only a handful of them could find work on the new ranches. A law was therefore passed prohibiting the cutting down of the *castanheiras.* But forest fires are not selective, and as before, the nut trees went with the rest. Where it was possible to leave one standing in isolation it was soon found that such solitary survivors failed to produce nuts. It was discovered that without the presence of certain insects the pollination of their flowers couldn't take place, and the pollinators had gone with the rest of the forest. Even had nuts been produced, such trees couldn't have increased their numbers, because this called for the co-operation of a species of rodent, also defunct, which had been programmed by evolution to chew the hard coating off the nuts, then distributing them in places suited to their germination.

These recently discovered mechanisms responsible for the production of the Brazil nut provide a clue to the dimensions of our ignorance of the workings of the forest. Some trees will not fruit without the aid of a single specific bird, others are fertilised by bats, yet others by moths, and a number of seeds receive their germinative impulse in some selected animal's digestive tract. The macaw, an agent of this kind, is relentlessly hunted for supply to the pet trade, and its disappearance from any area eventually damages that area's ecology. There is a tree producing pseudo-fruit, with no biological function other than the reward of its private army of insects, kept to ward off the attack of such predators as leaf-cutter ants. Strangest of such arrangements is the case of the *Inga edulis*, a gigantic runner-bean, dangling from the tree that bears it in the waters of a creek; the bean can only germinate after a spell in the gut of a fish, which eventually defecates it into propitious mud.

Such are the marvels of these vegetable-animal alliances for survival: the tree providing shelter and food, the animal offering those complex biological services without which the tree could not reproduce. INPA's Department of Forest Ecology was involved in an experiment to establish plantations of certain valuable trees, such as the rosewood, source of an oil used in the manufacture of perfumes, and exploited to the verge of extinction in its forest habitat. The rosewood project promised success, but other seedlings wilted and flagged when transferred to INPA's facsimile jungle. One of the many defence mechanisms developed by the Amazon jungle is the dispersal of its species, which grow in isolation, sometimes only one to a hectare.

There may exist in these particular trees an inbuilt dislike for the proximity of others of their own kind, causing them to frustrate all attempts to grow them in the nurseryman's row. So extraordinary is their sensitivity that no more than four trees per acre can be cut down without causing environmental trauma, and when disturbed a tree may suspend its growth – no one knows how – for up to 20 years. 'The fact is that we don't really know how to plant forest trees,' said an INPA scientist. 'There just isn't enough money for research.'

INPA employs at present 43 such specialists, although according to Dr Warwick Kerr, its director, a minimum of 1,500 would be required to solve the problem of the forest's economic development without its destruction.

Above all there was an urgent need for research into the therapeutic utilisation of the many types of chemical defences developed by tropical trees against insect or virus attack. Very few species have as yet been studied, but they had provided quinine, cortisone, new types of oral contraceptives, and what was hoped would prove to be effective anti-cancer drugs. In all such discoveries, the drugs had been found to be in use by the Indians who, it might be supposed, were the possessors of other valuable therapeutic secrets yet to be communicated. To quote Dr Paulo de Almeida Machado, the previous director of INPA, 'Whatever science can learn before the forest is destroyed will mean the difference between short-term prosperity and sound economic development.' He added that although Indians might survive, their culture will not.

The approach of Doomsday for the Amazon forest was signalled by the launching of the great road-building programme of the early 1970s intended to slice it into easily accessible segments. This enterprise, undertaken in haste and with small forethought, followed a visit in 1970 by President Medici to the chronically distressed north-east, at that time proclaimed a disaster area after one of the worst droughts in its calamitous history. The President spoke movingly to the large crowds who had flocked into the town of Recife to hear him. By chance he was standing within a few miles of that once-enchanted spot in Pernambuco of which Darwin had written in 1832, 'Forests and flowers and birds I saw in great perfection, and the pleasure of beholding them is infinite.' Of these Arcadian delights nothing remained, replaced as they were by treeless wastelands, ruined smallholdings among the thorny scrub, and the shanty villages of peasants who had to live on an average of 50 American cents a day.

The President promised, in effect, to spirit these wretched people away from their dour surroundings and deposit them in sylvan glades of Amazonia, where they would receive one hundred hectares of land apiece. Along the new highways he proposed to build low-cost but cheerful housing, and supply credits and facilities of all kinds, and thus they would be encouraged to lay the foundations of new and fruitful lives.

As a safety-valve for the chronic poverty of the north-east the project was a failure. The intention had been to settle five million peasants in Amazonia by 1980, but after two years only a hundred thousand had arrived, and many of them were already beginning to slip away back to the badlands they had left.

Like poor city dwellers induced to leave the companionable slums for the antiseptic planning of a garden suburb, too often they soon yearned for the familiar squalor they had left. They lacked the energy and the improvising genius necessary to come to terms with soil that produced two crops – three at most – before giving up the ghost. Rain, the greatest of all blessings in the north-east, was now the enemy. Life had to be shared with a multitude of stinging insects; new sicknesses defied the familiar remedies, and there were snakes in the back garden. Thus, the planning floundered and collapsed, leaving Amazonia littered with destitute homesteaders.

The new highways – the desert-makers as they have been called – did nothing to improve the lot of Brazilian subsistence farmers, but they fulfilled the wildest hopes not only of the multinationals, but of a new breed of predators who knew the true facts of the expendability of Amazonian soil and made it clear that they were not there to stay. As one American rancher put it to Robin Hanbury-Tenison, Chairman of Survival International: 'You can buy an acre of land out there for the same price as a couple of bottles of beer. An acre! When you've got half a million acres and twenty thousand head of cattle, you can leave the lousy place, then go live in Paris, Hawaii, Switzerland, or anywhere you choose.'

Inevitably, these roads open the way to the destroyers not only of trees but of men, for they pass through a number of Indian reserves, promoting the contacts that are so often fatal. Where thought necessary the Indians have simply been picked up and put down elsewhere, despite the fact that the new environment may not provide a living.

Brazil abounds with vigorous and articulate conservation groups, but they are powerless in the presence of one crushing fact: the desperate need of the country's many poor. In January *Journal do Brasil* published figures showing that in Rio de Janeiro alone 918,000 people were living in 'absolute poverty', and in the city's total population of nine million, more than a quarter live in 'relative poverty'. These are the statistics that often overwhelm the scruples of conservationists. There is a constant pressure to develop more sources of food, joined to an irrepressible belief that sooner or later a way will be found to turn the relatively unproductive five million square kilometres of the Amazon Basin into a bottomless larder.

The Government seeks to put a brake on the excesses of 'developers' by measures that are too often evaded or ignored. Official approval must be obtained for large-scale forest clearances, but nobody seems to bother. Regulations exist prohibiting the burning-off of forest close to river banks where animals tend to congregate. These go unheeded. Slopes are not allowed to be cleared, because to do so is to guarantee immediate erosion, but landowners tend to give priority to clearing the slopes on their estates. They do so because it is easier to drag or roll the tree trunks down the slopes and leave them to rot at the bottom, than to go to the trouble of extracting them. A promising law forbade the clearing of more than 50 percent of any concession, but many methods exist by which it is dodged. A common one is to clear half one's land in compliance with the regulation, and then sell the forested remainder, a half of which will be cleared by the buyer in his turn – and so on.

Implementation of these laws is left to IBDF, the Brazilian Institute for Forestry Development, but this body is said to possess less than a hundred inspectors to police an area in which twenty thousand might be too few. It has been announced that its hand will be strengthened by the end of the year by Brazil's new independent space satellite, which will report back on violations of the forest code. Here it might be argued that by the time photographic proof of such violations is supplied, the damage will have been done. But how, it may asked, is the feeble David of the IBDF to stand up, if needs be, to such Goliaths as Daniel Keith Ludwig, absolute ruler of an area somewhat larger than Belgium (referred to sarcastically as 'The Kingdom' in the Brazilian press) from which, incidentally, observers are debarred?

IBDF's one semi-success was with Volkswagen do Brasil, a pigmy by comparison, with a ranch no bigger than Luxembourg. The US Skylab satellite photographed a huge unapproved fire that had been started there, and it was reported that a million adjacent hectares of the forest were alight. Reprimanded for carelessness or wilfulness by the head of SUDAM, the Superintendency for Amazonian Development, Volkswagen's director is reported to have replied that burning was the cheapest way of clearing land.

Of course it is the cheapest for those who aren't residents, but the short-term yields they derive are paid for by a huge and irreversible loss suffered by Brazilians who have to live with the results. Ninety percent of the soil of the Amazon Basin is so poor that it can't be converted into adequate pasture without the addition of costly fertilisers, and as soon as these fertilisers are no longer forthcoming, the coarse African grass it supports will vanish. For a multitude of small-scale ranchers three acres of land can hardly feed a single cow, and after two or three years, when the rains have washed the nutrients from the soil, over-grazing and over-trampling starts turning the land into a dustbowl.

Wild animals are of less than secondary importance in developing countries, and little is said in newspaper polemics on the fate of the animals in the path of the fires. Their lot is death on a hardly imaginable scale. Birds, monkeys, jaguar and deer may escape the racing flames, but most of the Amazon animals – armadillos, anteaters, porcupines, sloths and frogs – are slow moving and doomed to incineration. Innumerable bats, whose beneficent function it is to keep insect populations in check, are disorientated by smoke, and fall into the flames. A great variety of small mammals, rabbits, agoutis, pacas, and rodents of all kinds are moved by instinct to take refuge in burrows or in holes quickly scuffled out in the forest litter, and in these they are roasted. Animals that live by, and in the water, otters, water-opossums and fish are doomed, too, not always directly by the fire, but the loss of forest cover, causing streams to overheat, becoming clogged with algae and silt; devoid therefore of both oxygen and food. An INPA scientist said: 'We are threatened with possibly the greatest ecological disaster in world history.' He had calculated that between 5,000 and 20,000 vertebrate animals were killed per square kilometre when the forest burnt down. But the burning goes on.

When Colin Jones, the photographer and I arrived in Manaus, capital of Amazonia, the January weather provided evidence of secondary inconveniences to be expected as deforestation progressed. It had been raining torrentially off and on for some weeks, and could be expected to go on doing so for at least a couple of months. The Rio Negro, on the banks of which the city is built, was rising at the highest rate ever recorded: 7.5 centimetres a day. The Hydrographic Service was able to predict at this stage the strong possibility of a repetition, by the middle of February, of the floodings of the past three years, when the city's centre had been under water. The news from the Manaus-Cuiabá highway was that six hundred lorries were stuck in the mud without hope of immediate rescue, filling the air with the stench of their decaying cargoes of food. The passengers from a stranded bus had taken refuge in an Indian village. Such floods, occurring at rare intervals in the past as a result of freak weather, are now coming to be accepted as a normal feature of the rainy season, lending weight to the theory that they are caused by the forest's loss of capacity to absorb water.

We took advantage of a break in the weather to hire a plane and in this we flew northwards from Manaus, keeping in sight the new BR174 Highway.

It was a good time to see the forest from the air, because by January the dry season's smoke has cleared away. A little late burning-off and tidying-up goes on, as the weather allows, but all the essential details of the landscape below are visible in the clear, rain-freshened air.

We flew at about a thousand feet over the limitless spread of trees, having from this height the appearance of sparkling moss. Across this the highway was a red line, ruled to the horizon. Immediately beneath, the road was close enough for the erosion to be visible, biting into its margins, and there were swamps created in its making, bristling with dead trees and gaudy with stagnation. Fires appeared as blue smudges here and there, and there were never less than a half-dozen in sight; many charcoal scrawls and flourishes showed on the green pages of the jungle where others had burnt out. Such clearances were often the work of rich businessmen running plantations as a side-line, or hobby-farmers from the city. Land here costs too much to attract ranchers thinking in terms of 20,000 to 30,000 hectares, and their operations would show up on a satellite photograph. Close to Manaus it was a matter of 100 hectares here and 200 there, but it was sad to think just how many small fires must have been alight all over the Amazon Basin on a fine day like this.

What our bird's-eye view made so startlingly clear was that the process the scientists called 'desertification' was even more rapid than we had been led to expect. In many places where patches of forest had been left, strewn with ash to await replanting or cultivation, the arid ochre of the subsoil already showed through. There were old, abandoned fields, too, now totally eroded, and from them the new desert spread like a creeping tide in all directions.

Later, accompanied by an INPA scientist, we drove along the BR174 Highway with the object of studying the close-quarter effects of what we had seen from the air. An hour's drive brought us to the scene of a recognisable fire on the 100-hectare estate of a successful Manaus shopkeeper. We had passed several estates like it along the road, often bearing romantic names such as 'My Blue Heaven' and sometimes furnished with a swimming pool by which the *fazendeiro* and his friends can sit to enjoy the view across a wasteland resembling the aftermath of the Battle of the Somme.

The estate we visited would be largely for the owner's weekend use, but a manager had been put in charge to also make sure of a reasonable return on investment. When cleared of forest the estate would be planted with trees for an attempted early profit.

The manager had no objection to describing and even demonstrating his methods. Clearing the forest with machete, saw and axe, he said, was a slow and laborious business. Instead you could do it with a tenth of the effort by getting a big fire going; all that was needed was a good supply of combustibles. His own method was to buy up all the old lorry tyres he could find. Having cut down the big trees and left them to dry for a few weeks, the lorry tyres would be stacked at intervals through the dry foliage, doused in petrol, and then set alight. The heat generated by this kind of fire took care of all the small standing trees and the undergrowth, and all that anybody had to do after that was to go round and saw down a few stumps that might have been left.

A final hectare or two was about to be cleared, and we looked on while this was done. It was a sad but spectacular business, with small risks to be taken if workers were so inclined, and these risks were clearly enjoyed by the young fireraisers who went in to the attack, cans of petrol in hand. As fire roared up from the underbrush a whole panorama of foliage shrivelled, leaving the bone-structure of the trees sheeted in flames. We were startled by the speed of the fire's advance, dodging and leaping into the branches, clinging to lianas, turning into small bonfires the bundles of orchids and ferns lodged in the forks and axils. Everywhere the wood was exploding under pressure from the boiling sap. The noise was terrific.

If there were a real danger here, it lay in the chance of getting in the way of a frantic snake as it came twisting out of the flames. One or two of these were quickly despatched by the club-armed workers. This was the moment providing the manager's children with the opportunity to own an occasional baby boa-constrictor, much in demand as a pet. They had managed to capture one a week before, and it was already quite tame; an engaging and endlessly curious creature, which slithered away to inspect any new arrival at the farmhouse, and was overfed on live lizards by its admirers. But few animals were as lucky as this, and a short walk into the most recently burnt-off area revealed a number of small charred skeletons.

The estate was to be planted with fruit, rubber and avocado pear trees, and the soil here was so poor that an inch or two of humus and ash could be scuffed away with the boot to expose subsoil like yellow concrete. Several hundred seedling trees waited in containers. To give them as good a start as possible the manager had dug holes by the hundred and replaced the arid subsoil with good earth, and in this the young trees would be planted. Not a single forest tree was to be left. Asked if he had not heard of the Forest Code, and the fifty percent rule by which no more than one half of the forest cover could be removed on any property, the manager agreed that he had. In a case like this, he said, it did not apply. It was not as if he had cleared the land for pasture or growing crops. Trees had been cleared, but others would be planted in their stead. What was the difference?

Loke's Merc

Lewis of course realised that the guest in his spartan, toilet-free cottage was immensely rich; but the article was written in pre-Wikipedia days, so he probably hadn't fully grasped the astonishing scale of Loke's wealth or his flair for business.

Loke Wan Tho was the founder of Cathay Enterprises, which grew to own more than eighty cinemas. He was also a film producer, but from the end of the Second World War Loke was increasingly immersed in the business world. Besides his own companies, his chairmanships included Malayan Airways, the Singapore Telephone Board, and Malayan Banking. He was also on the Board of Directors of numerous other companies, including Sime Darby, Straits Steamship, and so on. He was the ultimate Far Eastern magnate.

Loke also became one of the world's finest and most intrepid bird photographers after inspirational, often arduous, tutelage by the great Indian naturalist, Salim Ali. In Malaya Loke once risked his life by spending hour after hour at the top of a swaying 130-foot-high bamboo tower, in an unsuccessful attempt to photograph the hatching eggs of the white-bellied sea-eagle.

One minor point about this article: could Lewis have been mistaken in his surprising claim that rats reached the cottage bedrooms by climbing a ladder? A search on YouTube offers convincing evidence that rats can indeed manage this feat.

First published in the *Independent*, December 24, 1988

B ACK IN ITALY IN 1937 someone sold me an Alfa Romeo car, a recent winner of the 24 Hours Le Mans race. Having collected it off the train in London I set off somewhat cautiously to find a suitable road to try out its paces. This turned out to be the A120 going northwards through Epping Forest and virtually empty. The car accelerated easily up a slight gradient to about 110 mph, with power clearly in hand, which seemed good enough at the time. The test at an end, I pulled in at the Wake Arms at Loughton, and almost immediately another car drew up at my side. This was an astonishing Mercedes of a kind I had never seen before. Out of it stepped a smiling and immaculate young Chinese who introduced himself as Loke Wan Tho. Loke wanted a close look at the battle-worn Alfa. Encircling it, excitement leaked from him like an electric current. 'Oh, absolutely!' he said. 'How absolutely!' It was a form of commendation lifted from P. G. Wodehouse, then reaching the end of a long vogue. For fifteen years young men of the upper classes, and especially foreigners in Britain, tried to talk like Bertie Wooster.

I invited him to try the Alfa; he was breathless with gratitude ('Jolly sporting of you') and returned entranced by its bleak functionalism, its lack of concession to driving comfort and the sheer noise generated by the combination of the regulation track-silencer, supercharger and straight-cut gears. Meanwhile I had floated up and down the road at the wheel of his Mercedes – in a silence and smoothness so unnatural as to foster a moment of illusion in which the landscape appeared to slide away while the car stood still. Loke asked me if I liked the Mercedes and, when I told him that I did, he suggested an exchange there and then and without further ado.

It was a proposition that astounded me. His lustrous and extraordinary machine with its voluptuous display of exhaust-pipes and its leopard-skin upholstery would have been worth, as I saw it, two or three times as much as the battered Alfa which had never recovered, and probably never would, from 2,000 miles covered at an average of 87 mph. I had never met a Chinese before, and for a moment I suspected a conventional oriental courtesy by which admiration was expressed, but which was not to be taken seriously.

I had no way of knowing that this was a very rich man, prepared at this moment to indulge what to him was no more than a trivial whim. Loke took my hesitation to mean that he had not offered enough, so hastened to add a cash inducement. I explained that it was not a matter of relative values but the fact that I was half committed to a project to be undertaken in partnership with a friend. This was to convert the Alfa for racing on Brooklands. The explanation satisfied him, and he gave me an address in Cambridge in case I changed my mind.

I was now presented to his companion, Miss Dovey, a neat and sparkling English girl, who had stood aside while these transactions were in progress. Miss Dovey thought we should have a drink. We went into the Wake Arms where we sipped orange juice and nibbled sociably at potato crisps, which Loke, trying them for the first time, responded to with what struck me as no more than simulated pleasure. I picked up a few scraps of information. Loke was at Cambridge, reading English. Of Miss Dovey I learned little except that she collected shoes, and had a hundred or so pairs. She made mention of Loke's interest in rubber plantations in Malaysia, and that he kept an apartment in a Park Lane hotel for his use when in London. This, she added, with a touch of proprietorial satisfaction, she had helped him refurbish according to his taste.

During a further half-hour's amiable exchange of ideas Loke said that they were both interested in birdwatching, and that the visit to Epping Forest was to facilitate their study of the European wren (*Troglodytus troglodytus*), found in its woodlands and glades in exceptional concentration. He opened up the boot of the Mercedes to display a collapsible hide imported from Switzerland into which, when the occasion presented itself, he and Miss Dovey would creep to take photographs and record the bird's song. This, he agreed, most laymen would regard as an uninspiring twitter. It was none the less of great scientific interest since the troglodytus was believed to possess extraordinary ability to vary it according to the environment.

We parted company. The Mercedes stole away down the road, turned off into a track and embedded itself in a likely thicket. I made for Weybridge where an expert on the staff of the firm undertaking conversions of the kind I had in mind, took one look at the Alfa and shook his head. It would be cheaper to buy a car designed for track use than to convert one that was not. They had an ERA, and a Maserati in reasonable shape in their used stock, but both were far and away outside my price.

Faced with this verdict I decided to go ahead with Loke's proposed exchange if by this time he hadn't changed his mind, and thereafter sell the Mercedes to raise the cash required. I wrote to him but there was some delay before the reply came. He had been away touring in Germany and a photograph enclosed with the letter showed the wreck of his car. It had been hit by a train at an unguarded railway level-crossing and, while neither he nor Miss Dovey had been hurt, the impact had sliced away the Mercedes' rear wheels. It would have to be rebuilt, he wrote, and this would take some months. It was clear that the matter of the exchange was not ruled out. Mentioning that a complete repaint would be required, he added, 'What colour do you prefer?'

Almost with that, it seemed, the Munich crisis was upon us, changing not only our plans but those of the world. Loke, obliged to drop everything, was called back to Singapore. Escaping subsequently from the Japanese invasion, he was on the *Nova Moller*, sunk in an air attack, and rescued from the water with severe burns and temporary loss of sight. In the meanwhile, I was in North Africa then Italy, and it was 1947 before we met again.

I was in Pembrokeshire, where I spent that summer rock-climbing, when a letter from Loke, forwarded from London, announced that he was back. When he heard where I was, he wanted to come down. The cliffs of Pembrokeshire – although I had no idea that this was the case – were a famous venue for birdwatchers. He arrived overbrimming with enthusiasm and, apart from an area of pink new skin surrounding the eyes, little changed in appearance. The conditions in which I was living must have been among the most primitive in Europe; certainly far beyond anything Loke had ever experienced. The three-room cottage I had rented in a fishing village possessed no running water, no sanitary arrangements of any description, and no electric light. It was a scene into which he plunged with relish, although unable to believe his eyes at his first sight of one of the mild and contemplative rats that were a feature of the place, even climbing up the ladder to ascend to the bedrooms. The villagers seemed not to notice their presence and Loke, always on the lookout for virtue lurking beneath everyday attitudes, saw this as evidence of a latent, intrinsic Buddhism in the Welsh character. The truth was that due to a superstitious local aversion to cats, rats were tolerated in their place for the efficiency with which they cleared up the mess left on the quayside after the fishermen had boxed up their catch.

Loke was in his element. Littlehaven was brilliant with wildlife, with seals in every cove, the morning fox on the beach in search of anything left over by the cats, a stream with an otter at the back of the hill and a bluster of wax-white gulls always in the sky. Ensconced in brambles and bracken Loke trained his 20-inch telephoto on a 1½-inch bird. Despite all the ravens and peregrines around him he was back to his first love, the common wren, hunting its microscopic prey just out of reach of the spume. He informed me that three island sub-species of *troglodytus* were to be found on St Kilda, the Hebrides and the Shetlands, and his hope was to identify a fourth variation, based on the large and deserted island of Skomer, a few miles away. This, despite many days of fieldwork, he never achieved. When I made some mention of the Mercedes, he seemed momentarily puzzled. Then he remembered. 'I had to leave it behind,' he said. 'I expect my people will have dealt with it.'

He was now in control of his family empire, of its cinemas, rubber, tin and real estate, and had become one of the world's rich elite, yet he admitted that when the time came for his return to Singapore he would do so with reluctance. It evidenced a personal schism never to be repaired. He was committed by custom to a pursuit of wealth for which he had little true inclination, and removed from the convention of his background, his tastes were frugal, even austere. In the introduction to his book, *A Company of*

Birds, published when he was seen as the best bird-photographer in Asia, he lays blame on destiny: 'I was destined to be a businessman.' Of his ornithology, he adds an explanation, 'Every man needs some invisible means of support.' How sad that the empty ritual of a man of affairs should have usurped so much of his life.

A year or two later he invited me to join him in an expedition to northern India, the proviso being that I must first learn to skin birds, but this it proved impossible to do in the time. We met again in 1954 in Singapore when I stopped off there to snatch a night's sleep on my way to North Vietnam. I phoned Loke from the hotel and shortly afterwards an extraordinary cortège of cars drew up outside. It was impossible not to be reminded of a top mobster's funeral, minus the flowers. Loke had clearly arrived, but I found it hard to associate him with this arrogant display. The explanation was simple, and his smile apologetic. 'We're going to my sister's birthday party, dear boy,' he said. 'Black tie – white Cadillac. Hop in.'

The main feature of the open-air restaurant taken over for this event was a rotunda set in a garden of flowering trees among which birdcages were artfully concealed, some furnished with real songsters, others with vociferous mechanical bulbuls used to encourage natural song. About two hundred Chinese guests were seated at long tables forming a hollow square. Within the rotunda a pianist in tails seated at a grand piano worked his way with some panache through a repertoire of Sankey and Moody hymns. Loke explained that most of the guests were Christadelphians, members of an American fundamentalist sect that attempts to inculcate severe morality upon its adherents, including – despite the example of the marriage at Cana – an absolute ban upon alcohol. For this reason all believers present were invited to wash down the exquisite food placed before them with Dr Pepper, Coca Cola, or other such blameless refreshment. Nevertheless, discreet arrangements had been made for the 'unsaved', such as myself, to be provided with whisky contained in antique bowls of great beauty and doubtless of great worth.

A Quiet Evening

It was a lively affair, the guests exchanged jokes, pulled funny faces, and punned in English – probably in Chinese too. They were easily amused; a feature of the ingenious and interminable banquet was a partridge served to each guest in which a simulated and edible bird's nest had been inserted. Someone stood up and said, 'Normally the bird is to be found in the nest. Now we are eating the nest that was discovered in the bird.' Everybody clapped.

The real and mechanical bulbuls warbled in their cages, the pianist charged for the third time into *Through The Night Of Doubt And Sorrow*, played as if it had been a wedding march, and a pentatonic tittering arose from the guests, encouraged at this stage to indulge in horseplay of a decorous, almost formalised kind. A Chinese lady in a white robe upbraided us all in a brief sermon which appeared to fall on deaf ears. Towards dawn the scent of frangipani strengthened, and at sunrise Loke drove me in one of the white Cadillacs to the airport. At our parting I promised to take a crash course in taxidermy, and Loke agreed to include me in his next expedition. We continued a regular correspondence but were never to meet again, for a year or two later he was killed in a plane crash in Taiwan.

Last August a national newspaper published a piece about a 1936 Mercedes Grand Tourer – described as rusted all over, with mouse-gnawed seats, and full of holes – being sold at Christie's vintage car auction for an unprecedented £1,595,000. A Mr Davies, a butcher from Walsall, had been left the car by an uncle in 1956, since when it had slowly mouldered away in the garage at the back of his shop. It had gone to a Swedish financier, and the extraordinary price depended, the newspaper had ascertained, on the crucial fact of this car's possession of a right-hand drive, being one of only two models thus built out of a total production of 356. Loke's 500K, too, had been a right-hand drive version. Was this the self-same car? Possibly, and, if not, it could only have been the twin – equally valuable to a financier one supposed – if it had managed to escape the scrap heap.

Two Generals

The Cuban War of Independence against Spain, in which these two heroes fought, was the last of three uprisings against the colonial power, lasting from 1895 to 1898.

*Lewis met the ancient heroes in May 1960. Because they were still alive when he wrote the article, he may have been circumspect about one revelation during his interview. He relates later in an autobiography (*The World, The World*) that General Garcia Velez showed him an album that had belonged to the remarkable revolutionary, Francisco de Miranda. The album contained snippings from the pubic hair of fifty-one women, all of them Miranda's conquests.*

One of the snippings was identified by a letter K, allegedly referring to Catherine the Great. This may seem a wild story, but respectable sources now confirm that Miranda conducted a love-life that could rival Casanova's. Indeed in 1787 the impecunious thirty-seven-year-old Francisco de Miranda did meet Catherine the Great, then fifty-eight years old; she made him a count, gave him a Russian passport, and beseeched him to stay in Russia. A worried Spanish plenipotentiary minister wrote to Madrid that Miranda was the Empress's favourite foreigner. It seems highly possible that she was yet another of Francisco de Miranda's many conquests.

First published in the *New Statesman*, July 2, 1960

A Quiet Evening

GENERAL ENRIQUE LOYNAZ took me to see General Garcia Velez, the other surviving hero of Cuba's War of Independence against Spain. General Velez sat in the cool, vaulted marble of his library, in his pyjamas surrounded by piles of magazines, mostly English. He had been ambassador to Great Britain for twelve years. A softly groaning symphony of distant car horns and loudspeakers came through the open window. There was a big military parade on in the city.

'I'm commonly stated in the press to be 94,' General Velez said. 'It's not true. I'm only 93.' General Loynaz was in his late eighties. Before leaving his house he had shown me a slightly bent Toledo sword. 'It's out of shape from whacking cowards on the backside,' he said. 'To keep their faces to the enemy.'

'The very opposite in fact, of my own methods,' General Velez said. '*He* used to bully his men. I believed in kicking them along. In my opinion he was guilty of faulty psychology. How none of them ever had the guts to shoot him in the back I shall never understand.'

These two old men had sat quietly in the shadows for sixty years, watching with sardonic eyes the comings and goings of the politicians and the big businessmen who had gathered like vultures over their victories. They had sat through the revolutions and the coups d'état, had seen tyrants rise to power and fall, seen poor, honest men become rich and corrupt, seen young idealists transformed into bloody dictators, seen the vulgar image of Miami stamped over the soft, grey, baroque elegance of the Havana of their youth.

'Above all, my boy, don't get old,' General Velez said. 'It imposes an excess of reflection. I do practically nothing these days but read and think. See that pile of *Edinburgh Journals*. I've every number since 1764, and I've read them all. Mostly I read history with the inner reservation that it's largely romance and lies. At least nearly everything that I can check on from my own personal experience is a lie. Did you ever see the film *A Message to Garcia*, for example? The Garcia in the film was my father.'

'Calixto Garcia – the liberator of our country,' General Loynaz explained.

'I didn't see it myself,' General Velez said. 'As a matter of principle I've never been to the cinema. Nor have I even seen the television. I've always believed in living my life, not watching how other people are supposed to live theirs. But from what they tell me about this film, and the book it was based on, it was pure rubbish.'

'The actress was Barbara Stanwyck,' General Loynaz said. 'A very pretty girl. I much regret never having met her.'

'You cannot awaken the interest of Americans without a big fraud,' General Velez said. 'It was supposed to be some secret mission to my father, shown as carried through in the face of all kinds of nonsensical adventures. My father was depicted as a sort of romantic bandit hiding in the mountains. How they managed to bring Barbara Stanwyck into it, don't ask me. The real truth is there was no adventure. The American agent met my father in a hotel in the town in Bayamo. I don't think the message was particularly important either, whatever it was. My father certainly never bothered to mention it to me.'

'Our old friend's a sceptic,' Loynaz said. 'He's lost the power of passionate conviction.'

'A circumstance in which I rejoice,' Garcia Velez said. 'Our war was terrible enough, but when I say that it was conducted with the utmost brutality, I say this of both sides. Thousands of our people died of starvation and disease in the concentration camps the Spanish set up. Their *guerrilleros* didn't spare our women and children. But let's at least admit we weren't much better.'

He knocked the ash from the end of his cigar into a tin which had held herrings.

'Mind you, it wasn't particularly comfortable to be a general in those days. If you got into a mix-up in a battle the enemy always went for the uniform. I can't remember how many times I was wounded. General Maceo collected twenty-seven wounds. We seemed to be indestructible. When they took my father prisoner, he shot himself in the head. The bullet came out of his mouth. He still lived for seventeen years. Tell us about that famous wound of yours, Enrique.'

General Loynaz said: 'Those were the days when generals died with their boots on. I was in 107 combats. The 107th was at Babinay in '98 – the last stages of the war, when our American deliverers had belatedly decided to come in. I was in command of an infantry brigade.'

'He was a real general, I might say,' Garcia Velez said. 'He was always at the head of his troops on a white horse. An admirable spectacle, but not for me. Not for my father, either.'

'On this occasion there was no white horse,' Loynaz said. 'I'd been on one earlier in the battle, but it had been shot under me. I can't remember the colour of the second horse, but it certainly wasn't white. Anyway, there I was on the horse, as usual, with a cavalry escort, and the Spanish *guerrilleros* were waiting for us behind a stockade. We were undergoing heavy rifle fire and a moderate artillery barrage. I gave the order to charge.'

'In true Cuban style,' Garcia Velez said. 'Light Brigade stuff. Into the Valley of Death. The kind of thing they love. I shudder at the thought of it.'

'When you run up against cannon fire at point blank range, it's the only way,' Loynaz said. 'Above all things, you want to get it over with. I was the first over the stockade. Unfortunately I was never much good at jumping, and this time I landed on the horse's neck. A Spanish *guerrillero* brought his machete down on top of my head.'

'You should have led your charges from the rear,' Garcia Velez said.

General Loynaz took my hand and placed it on his scalp. I felt a shallow trough in the skull, about six inches long.

'Three American Presidents have asked to touch that wound,' Loynaz said; 'Teddy Roosevelt, Hoover, and I can't remember the name of the third. I managed to scramble back into the saddle holding in the few brains I possess with one hand, and I sat there not able to contribute much to the course of the battle, until it was over.

'They took me to a hut where a honeymoon couple had installed themselves, and I commandeered their bed. The effects of this wound by the way, after the initial pain quietened down, were wholly beneficial. Up till that time I was a martyr to headaches, but

I've never had one since. It probably made more space for my brains. That was pretty well the end of the war so far as I was concerned. The Americans decided to come in after that. They were just about a year too late. We should have welcomed them in '97.'

'Friends are always welcome,' Garcia Velez said.

'We had won the war,' Loynaz said. 'The whole country was in our hands.'

'But not the towns,' Garcia Velez said. 'The Spanish still held the towns. You speak as a patriot, not a historian.'

'For six years the foreigners ran our country,' Loynaz said. 'They bought up the best land in the island. Do you know how much they paid? Ten cents a *caballería* of 33 acres. The price of two bottles of Coca-Cola.'

Garcia Velez shook his head at him. When Loynaz had left the room, he said: 'My old friend has always remained a Cuban, whereas the twelve years I spent in London has wrought a profound change in my character, I see things calmly now; almost I believe, through Anglo-Saxon eyes. Moreover, living abroad, I became wholly a pacifist. Had my twelve years in England come before our war, I don't believe I'd have fought in it. The things I was forced as a patriot to do, now seem to me to be hateful – against nature.'

Our conversation was only a few weeks after Fidel Castro's capture of the capital, and from where I sat I could see through the window a squad of feminine militia come marching down the road. Half of them were in uniform, the rest in pretty dresses. A sergeant marching beside them called out the marking time. With them came a blare of martial music from the speakers of an escorting van.

'Please close the window,' General Velez said. 'The noise oppresses me. What do their banners say?'

I read: 'Fatherland or Death. We will fight to the last drop of our blood against foreign aggressors.'

'And they will,' the general said. 'And they will if necessary. Alas, haven't I seen it all before.'

The Bay of Pigs

Lewis went to Cuba in September 1961, five months after the Bay of Pigs invasion. While there he interviewed Commandant Enrique Carreras who, with three barely airworthy planes, had played a crucial part in repelling the invasion.

Enrique Carreras was born in 1922. He became one of the very first Cuban pilots, but under the dictator Batista's regime he had refused to bomb some insurrectionists. This caused him to be imprisoned, and soon afterwards he became an ardent supporter of Fidel Castro. He died in 2014 at the age of ninety-one.

By googling 'Critical Past Enrique Carreras talks to Captain Bourzak' you can find a five-minute interview (in Spanish) with Carreras which was conducted during the invasion. The film includes some grainy, but rare footage of the fighting.

First published in *The Happy Ant-Heap*
(Jonathan Cape, 1998)

ON A VISIT TO CUBA IN 1961 I went to see Enrique Carreras, the new air force chief, and found him at his office desk at Havana airport. He was a man in his late thirties, although perhaps in appearance a little old for his years, and there was something in his studious, concentrated expression that reminded me – until one of the frequent smiles broke through – of a friend who studied postage stamps through a magnifying glass. His secretary warned me, 'He's very unassuming. Don't call him Captain. Just say "You".'

Chaos still had the upper hand here, since a recent attack by Cuban opponents of the regime who had flown here from Miami. A chair for me had to be dislodged from a pile of salvaged furniture. 'You see how we live these days,' Carreras said with a laugh. 'Still, things are on the upturn.' He spoke good English, from which I noticed that the Americanisms locally in common use had been expunged.

Carreras had agreed to talk about his personal contribution to the defeat of the invasion attempt at the Bay of Pigs earlier that year. This, as expected, he seemed determined to play down. 'It was largely a matter of luck,' he said. 'They weren't prepared for what little resistance we were able to put up, and it took them by surprise. The fact is that by the time I took over, all the best pilots had cleared off to Miami. I was sad to see them go because we were all friends together until the last moment. The worst of it was they took all the planes worth having, leaving us with the junk. I was particularly close to Rojas, who was chief here before me. He left a letter for me ending, "Goodbye sucker". There was nothing malicious about it. Rojas was fond of a joke. He had his future to think about, but I couldn't help seeing what was happening as the beginning of the end.'

A secretary came in with coffee. She was one of the new type with short hair, no make-up and flat shoes. This gave me the opportunity for a quick glance round at the surroundings, which revealed a shattered window still awaiting replacement and a door that hung askew. What might have been Carreras's personal possessions had been shoved into a corner, including a fuzzy photograph of a woman with a child, probably his wife, and a voodoo shrine-idol of the type that people now collected, with a fat cigar stuck between the idol's straw lips.

'They came over at six in the morning,' Carreras said. 'My first thought was that this was another earthquake. I'd taken to sleeping here to be on the safe side and I rushed out to see my personal T-33 jet-trainer burning like a hayrick, and the best of our Sea Furies blown all over the parking bay. The bombing was very accurate and whoever planned it knew just what to leave and what to take out. I couldn't help feeling that Rojas was in on this. Somehow or other he managed to knock out the generator and the phones. They're working on them now. You can still smell the fused wires.' He pushed open the nearest window just at the moment when some mechanics, believing themselves unobserved, had put aside their spanners and reached for their guitars. I listened to the thin, sweet music of Cuba as the flutes picked their way through the background noise of riveters at work. Carreras attempted an indulgent smile.

'Was it five planes you were left with?' I asked.

'Three,' he said. 'All of them ready for the scrap heap. We were down to two Sea Furies and an even older B-26. They were suffering from metal fatigue and their engines were worn out. The Sea Furies had defective brakes. The guns on the B-26 jammed and it was doubtful if it had the power to get off the ground. The next thing was that the news came through that the invasion fleet had been sighted in the Florida Straits heading south. That gave us exactly one day to get ready for them.'

'Goodness, what a problem!'

'We called up every spare mechanic in the city and had them working all day and most of the night. No one was allowed to stop to eat. We took rice and beans out to them and had the base doctor mix ground-up amphetamine pills with it to keep them awake. We cannibalised old passenger planes and managed to adapt a few parts stripped from lorries and tractors. The old B-26 was practically rebuilt. I put in a few hours myself to help all I could. At about four in the morning I went to sleep in a chair under one of the Sea Furies, and just before five they called me to say that Fidel was on the phone. "Enrique," he said, actually calling me by my first name, "the sons-of-bitches have arrived. They're landing at the Bay of Pigs. How many planes can you get into action?"

'"Three, Commandante," I told him. "Three, well, more or less. If they can be started up."

'"How soon can you get them down to Largatera?" Castro asked.

'"In twenty minutes, Commandante," I said.

"'Good," he said. For a moment he was cut off. Then he was speaking again. "And Enrique," he said, "don't let a single one of those sons-of-bitches get away. Don't let them get away. I'm relying on you."

"'At your orders, Commandante," I said. "They'll be lucky if they do.'"

Carreras laughed in a slightly apologetic way, as if on the verge of a damaging admission. 'The truth of the matter is that phone call did something to me. I supported the revolution, but up to this time the leader was a kind of remote figure. I'd seen him at a distance in the Plaza making his speeches, but I'd never spoken to him and he'd never set eyes on me. And here he was now, talking to me like a member of the family and calling me by my first name, as if he knew I was on his side. "Enrique," he said, "don't let them get away", and I had the feeling that I really mattered to him and the revolution. "They won't, Commandante," I told him. "Not if I can help it."

'With all the work done on the planes I still wasn't quite sure of them,' Carreras said. He explained his decision to lighten the B-26's weight and to limit the bomb-load to four 250-pound bombs, and the two Sea Furies, now in their sixteenth year of service, would carry two similar bombs apiece, plus eight five-inch rockets. No time had been left for final checks and Carreras's persistent fear was that undetected faults would be revealed at the moment of take-off when the veteran planes were exposed to maximum strain. He recounted that at one stage he had absented himself for a few minutes to write a note to his wife, with recommendations for the disposal of his possessions in case he was killed. He then climbed into the cockpit of one of the Sea Furies and noted with huge relief the smoke billowing from all three planes, which proved that they were about to take off.

Almost as soon as he had promised Castro, they were over the Bay of Pigs. The sun had just come up and the sea was patterned all over with ships. To the west the great swamp known as La Largatera began immediately at the edge of the beach, which curved all round the bay and stretched unbroken, except for a single narrow road, all the way to the horizon. Tiny white puffs showed here and there along the road, and Carreras concluded that these were shell-bursts, and that invaders already ashore were under fire from the defenders. In a matter of seconds Carrera was over the invasion fleet, followed by the second Sea Fury and the B-26. His orders were to avoid attacking any invaders already ashore and to concentrate on the incoming ships. Carreras picked for himself the target of a large troop transport escorted by two frigates. It was moving slowly towards the beach, its deck crowded with men. This he identified as an 8,000-ton Liberty ship probably carrying, in addition to the maximum number of fighting troops, the key personnel of the operation. It would have been held back until the landing had been secured by commandos already ashore. It was at this juncture that, in an attempt to check his height, Carreras discovered that his altimeter had stuck at 3,000 feet.

Even an experienced pilot finds difficulty in judging altitudes when over the sea, so guesswork now took a hand. Coming in for a bombing run at an estimated 800 feet, Carreras ran through the anti-aircraft barrage of a dozen ships. He let go his bomb, missing the Liberty ship by twenty yards, and, caught in a hailstorm of iron and fire, the Sea Fury juddered over cobblestones of air, ducked, wobbled and finally climbed out of range, punctured and ripped in a dozen places in the fuselage and wings. A stray thought about Rojas forced itself on Carreras. The force's top pilot would never have missed.

With a single bomb left and his confidence on the verge of collapse, Carreras decided that his only hope of putting the Liberty ship out of action was a rocket attack at close range under the anti-aircraft fire and at maximum speed, whatever the risk of ending in the sea.

He banked, turned and went into a dive known in the training school as an 'ultra', something he had never attempted before. For a moment he felt weightless, hanging in space; his joints cracked, blood vessels snapped in his temples, there was pressure on his eyes, and iron fingers had been thrust into his ears. At about 300 feet, as the ship rushed up to him and with the plane's nose held steady and pointed at the centre of the cattle-stampede of human bodies, he fired the rockets, blotting out the scene with smoke, before pulling out of the dive and into a sky that was buttoned all over with bursting shells. His friend Mateos passed below in the second Sea Fury, blasting with machine-guns at a nearby transport, and at this moment his accompanying B-26, seen for the first time, drifted backwards into view. An arc of gunfire from which Carreras presumed it was too slow to escape had followed it round the bay. Now it simply disappeared.

Carreras turned back for a final view of the Liberty ship. 'They must have stacked up ammunition on the deck,' he said, 'for it was on fire, still struggling for the shore, and dragging with it in the water a black encrustation of drowning men.' He described this scene with no trace of satisfaction. 'I'm squeamish by nature,' he said, 'I wanted to do something to save these poor men, not kill them.' A few, he said, had reached the beach, but there was no cover from the fire of Fidel's shock troops on the narrow road from Cienfuegos, and they were all mown down.

'The swamp here,' he said, 'comes down to within feet of the sand, and a lot of them took the chance of hiding in it.' He sighed. 'I'd been there hunting in the old days, and I knew only too well that there were crocodiles everywhere.'

Fidel's Artist

Herman Marks, the executioner in Havana's La Cabana fortress, was born in Milwaukee in 1921. Before he arrived in Cuba to fight alongside Castro, he had been arrested at least thirty-two times for numerous crimes, often drunkenness, but also robbing an elderly lady.

By all accounts he was a sadist, greatly enjoying his eventual role in La Cabana, where he oversaw more than two hundred executions. While in charge of these frequent events, he became a prominent and somewhat popular figure in Havana. Errol Flynn (who may be 'Shiralee Shepherd' in the article) watched several of the executions – which Herman Marks called his 'festivities'. Lewis would not have been surprised by material that has since surfaced in the Paris Review *(November 2016), claiming that Ernest Hemingway, along with friends, sometimes attended the executions, taking along chairs and a cocktail shaker.*

At the time of writing this article, Norman believed that Marks had in turn been shot by the regime which became fiercely anti-American. But we now know that Marks escaped Cuba in time, though after reaching the United States he had to fight several legal battles to prevent deportation. The last record of him is in 1966, when he was accused of child molestation. Soon afterwards he disappeared, and it is believed that he may have died impoverished in Mexico.

A fascinating article was written about Herman Marks by Tony Perrottet in 2021 (google 'Tony Perrottet butcher of Havana').

First published in the *New Statesman*, December 17, 1960

I N DECEMBER 1959, shortly after the Castro victory in Cuba, I attended several of the trials of war criminals conducted in the Cabaña fortress of Havana, in the course of which I was subjected to an extraordinary encounter with Herman Marks, the American who had become the Cuban executioner. Marks spent some time justifying his activities and expounding his personal philosophy, probably in the hope that I might help to rectify his image 'in the world's eyes'.

'Well all right, all right, we know all about the stretches I may have done. I was waiting for that one. You may say I was a no good son of a bitch when I was a kid, and I might agree with you. But I suppose you've heard of such a thing as moral regeneration? I guess you'd say that any guy has the right to do what he can to put himself in the clear with society. Maybe that's why I'm doing what I'm doing – in other words a necessary job that nobody else wants to take on.

'I guess I feel that in this way I'm doing something to clean the slate, and I figure that's the way the people here see it too. They accept me. I'm regarded as a useful citizen. People like to be seen going round with me. If I happen to feel like taking an evening off and going to some place like the Riviera, for example, I get the best table that's going. Some guy I don't know is always picking up the tab for my drinks. Even Fidel gives me the big hello when he sees me. I do my job conscientiously, and I'm respected for it. That's the way it is.

'Listen, the way I figure it is, you have a job to do? OK, do it well. Maybe you know the Cadillac and Limousine Service on Nott Street, Zenith? I was with that bunch as a senior servicing operator for five years and, believe me, I was always noted for the pride I took in my work. Anyone there will tell you that. And if you think that anyone could do my present job – boy, you just can't imagine how wrong you are! Believe me, it calls for everything you've got. You're up against the human element all the time. The kids they send me to work with: you'd break your heart if you saw them. As a technician – that's how I see myself – I hate a bungled job.

'Listen, I'm only supposed to put the finishing touch – that and give the word of command. Not to have to check up on every detail with the deadbeats they send along for these parties. What I mean is they're supposed to be volunteers, but most of them turn out to be strictly chicken when it comes to the point. If I didn't watch them like a cat, you'd get half these characters only pretending to fire and then quietly unloading as soon as I turned my head. That kind of thing puts extra work on me. Believe me, I drive myself, I really do. Way back last year when we had our busy spell when I've been on special missions half the night, I've worked some nights from midnight until five or six in the morning. You can't rush this kind of thing. It takes time. And I might add, I don't touch a drop of liquor when I'm carrying out a mission. The most I have is a cup of coffee sent down every hour or so. With milk. Sleep well? Oh, sure I do.

'Another thing might surprise you, and that's the trouble I've put myself to – so that I make the whole thing go as smoothly as is possible in the circumstances. For example, whose idea do you think it was to fix up for these jobs to be done in the old moat under that big statue of Jesus Christ they light up at night? Why mine of course. I can't claim to be a religious man, but at least I understand the way other people feel about these things.

'When I put up that idea to the revolutionary committee at the Cabaña they said it was a masterpiece. You know the statue, don't you? It stands up right over the wall. It must be sixty feet high. You can see it ten miles away at nights. It struck me as a kind of nice idea that that would be just about the last thing these poor guys would see. Now you see what I mean about giving all I've got on the job?

'The fact is, I suppose I feel somehow like a doctor does with a patient. I go easy with them. Put it this way, I don't go in for rough talk, or wisecracking, and I don't let any of the kids either. I'm ready to spend half an hour with a man kidding him along, just to see it goes smooth – you follow me? He wants to make a speech? OK, he makes a speech. He wants to give the orders himself? That's OK too. Anything within reason goes by me.

'This business about giving the orders themselves seems to be a sort of craze these days. They nearly all want to do it. I figure it's a kind of last minute show off. Search me why anybody should want to show off at a time like that when there's only me and a bunch of stupid kids to see it. To tell you the truth, I wish they wouldn't do it. I warn them to space out the orders properly; to count up to six slowly between the take-aim and the fire. But they always make it too fast and what happens is the kids loose off before they get a chance to take proper aim. That way you get a really crummy result and it puts it all on me. Anyway, what I'm trying to get round to is this. I go out of my way to show consideration. These guys are in a highly nervous state.

'As I said, they can have half an hour to shoot the breeze. More if they want. Well of course, some of them try to drag things out. They're liable to beef on about their innocence. "Sure you're innocent," I say. "I know you're innocent. All right, fellow, all right. Now how about standing over here where we can get a look at you?" That's the way I kid them along. You have to be ready for anything in the way of propositioning. You get rich guys who want to give you a million dollars to fix it so that they go out of the Cabaña some other way but in that box. Some guys never give up hope. I mean that literally. I've known the time when a fellow's gone on trying to talk his way out right until I put the finishing touch – and that, by the way, throws a light on the quality of the workmanship I have to put up with. You'll get another customer who wants to shake hands.

'"I forgive you," he says.

'"Thank you, thank you," I tell him. "That sure makes me feel better." While I'm holding his hand I'm sort of strolling along with him, manoeuvring him into position, in a way like he doesn't realise what's going on. About one in three of them wants to pass you something they like to hang onto until the last moment, maybe it's a rabbit's paw or a locket with a picture of their mother, or something like that. Personally, I make a strict rule not to touch anything of value. "Give it to one of the boys," I say. I don't object to the kids taking a locket or a ring or something like that if it's offered to them, but what I won't stand for is the racket that those kids used to go in for – of selling spent shells to those niggers who use them for some sort of voodoo stuff. The regular price used to be five bucks a shell till I stamped on it.

'I know what you're going to say now. You're going to bring up that story that I have cufflinks made out of them myself, and hand them out to my friends. Sure I do, and why shouldn't I? It's not a racket. I don't take payment for them, and nor do I see anything morbid about it.

'Listen, if you want to talk about people being morbid, maybe I should tell you about some of the characters who come and ask me about letting them come along to one of the performances, and I don't mean two-bit journalists either. I mean guys whose names you read every time you pick up a newspaper. If I could mention some of those names you'd certainly be surprised. Maybe you'd changed your mind about who's morbid, or put it this way – who'd like to have the chance to be morbid.

'I had a case the other day. Two fellers came up to the Officers' Club and asked for me. I knew one of them quite a bit. He was a big wheel at one of the embassies. I don't want to say which one. He always wants to buy me a drink, whenever he sees me. "Good evening, Captain," he says, "I want you to meet a very distinguished friend of mine, and a very great creative artist. This is Mr Shiralee Shepherd."

'"Not *the* Shiralee Shepherd," I said. To tell you the truth, although I'd seen this guy on the films he looked somehow different. "I saw your last film," I said. "It sure was a gas."

'"Thank you. Thank you indeed," Mr Shiralee Shepherd says. "As from one artist to another I take that as a great compliment."

'I got a bang out of the artist stuff. "An artist," I said. "Well, I guess maybe you're right. I wouldn't say I was a creative one, though."

'We all had a laugh, and the diplomat fellow says, "Shiralee's been hearing a great deal about you, Captain, and I was wondering if we couldn't get together. I guess you understand that a man engaged in his kind of imaginative work requires a diversity of experience out of which to fashion his material – experiences that others might wish to go out of their way to avoid."

'I knew what was coming and I particularly liked that bit about experiences that others might wish to go out of their way to avoid. I could have given him the names of a hundred guys who had their name down on my waiting list. "You mean Mr Shepherd wants to come to a gala evening," I said.

'"If it can be arranged," Shepherd says. "Discreetly, of course."

'I looked at the guy, and I wasn't too crazy about him. He looked kind of fat-lipped off the screen. I didn't go for him in a great way, but his pal from the embassy was a good enough guy, and I wanted to do what I could for him. "It might be arranged," I said.

'"When?" says Shepherd, in a very anxious manner.

'"Ah, that I can't say," I told him. "Business has been pretty slow lately. It's only just beginning to look up again. It looks like we'll be getting a few candidates again before long, but even then certain formalities have to be observed. As for example the guys are supposed to be tried."

'"Yes, of course," Shepherd says, "but tell me, these trials, and so forth – are they likely to take long? I mean do you think it would be any use if I arranged to stay on another week?"

'For Christ's sake, what was I to say to the guy? Did he think I could have someone knocked off specially for his benefit? When I told him it might take a month you should have seen the look on his face. I've never seen a man look so disappointed. I found out later he came all the way from New York on the chance of seeing me slip somebody the pill.

'Now please do me a favour, will you? After that, don't talk to me about a guy being morbid.'

A Mission to Havana

Ian Fleming, then Foreign Manager of The Sunday Times,
*sent Lewis to Cuba in December 1957. Fleming had recently
met Lewis, and became a fan of his writing. Having been
sent a proof-copy of* The Volcanoes Above Us, *he wrote to
Lewis's editor, 'Volcanoes is a wonderful book…showing a
fascinating mind and really startling powers of description
and simile. I haven't enjoyed a book so much since I can
remember.' Separately he wrote a letter with helpful criticism
to Lewis, adding, 'I am every time arrested by your genius for
intellectual photography in prose.'*

*Fleming had another motive for sending a writer to
Cuba. He was working for MI6, with a special interest in
the Caribbean, where he had a house in Jamaica. Earlier
than most, he realised that Castro would probably triumph,
and considered the Secret Service's intelligence-gathering
inadequate. He reckoned that the adventurous, Spanish-
speaking Lewis would do better.*

*Ernest Hemingway committed suicide three years after
the meeting described in this article.*

First published in the *Observer*, November 15, 1981

I MET IAN FLEMING IN 1957 at a party given by our mutual publisher,
Jonathan Cape, which Fleming had attended with ill-grace.
A shortage of space at the Cape headquarters in Bedford Square
made it necessary to spread the occasion over successive days. We
found ourselves immersed in this rump of the party, reserved, Ian

suspected – though certainly without justification in his case – for Cape's less prestigious authors, and he retired, disgruntled, to a corner, where I shortly joined him. He asked if I wrote poetry, and when I said I didn't, he seemed disappointed.

Although already famous as the creator of James Bond, Fleming seemed to extract less pleasure than one would have expected from the writing of successful thrillers. He craved the society of what he considered 'serious' writers, above all poets, like William Plomer, who had introduced him to the firm of Jonathan Cape, and through whom all his business with Cape was done. Jonathan Cape himself much disliked Ian Fleming's writing, and refused to meet him, and could only be persuaded to publish his books by a united front established in Fleming's favour by the firm's other directors, and by William Plomer, their reader. Michael Howard, the junior director, told me that the decision to publish *Casino Royale* gave Cape sleepless nights, and a guilty conscience.

The acquaintance made at the party developed into friendship, and Fleming and I saw something of each other over several years. I found him genial and expansive, although many people did not. His habitual expression was one of contained fury, relieved occasionally by a stark smile. He seemed to wish to inspire fear in others, and on several occasions said of some person under discussion, 'He is afraid of me,' a conclusion seeming to give him satisfaction. Another habit, which didn't endear him to women, was frequently to explain in their presence that he had only taken up writing 'to make me forget the horrors of marriage'.

For some reason I couldn't at first understand, Fleming showed much interest in the fact that I had travelled in Central America, more particularly in Cuba, which I had visited quite often. At that time he was Foreign Manager of the *Sunday Times*, and one day he asked me to come to his office to discuss a potential article for the paper.

He wanted me to visit Cuba for him, to see as many people as I could, including some to be named by him, and investigate the possibility of the success of the Fidel Castro revolt, of which little at that time had been heard in this country. It seemed that Fleming's desire for information was not only on behalf of the *Sunday Times*. It was generally known that he had been assistant to the Director of Naval Intelligence during the war, so I assumed that he was still involved in one or other of the intelligence organisations, probably in a department concerned with Latin American affairs. He said that he was unhappy with information about the progress of the revolt received through the Foreign Office, and also with the reports from his personal contact, Edward Scott, who lived in Havana. He showed me Scott's most recent letter. The revolt, said Scott, was contained in a small mountainous area, the Sierra Maestra near the far-Eastern tip of the island, and should give no cause for concern. Scott predicted that with the United States solidly behind the dictator Fulgencio Batista, the revolutionaries would shortly be rounded up, and massacred to a man in local style, while the world turned its back. Fleming said, 'I simply don't believe it.'

For some reason he was convinced that Ernest Hemingway, who had been living outside Havana for several years, was in close touch with the rebels, and he was most anxious to have Hemingway's views on the prospects of their success. He made it clear that Hemingway was one of his heroes. Not only did he regard him as among the great writers of all time, but he had come to the conclusion through analysis of his writings, in particular his novel dealing with the Spanish Civil War, *For Whom the Bell Tolls*, that Hemingway had been in his day an extremely subtle and successful undercover agent, and probably still was one. He had written to Hemingway, but had received no reply. Uncharacteristically Fleming had forgiven him, and still hoped that contact could be made.

Hemingway's oldest friend in England was Jonathan Cape himself, who had been successfully publishing his books for thirty years, and Fleming, unable to make a direct approach to Jonathan, suggested that I should do so and persuade him to write to Cuba and ask Hemingway to see me. Jonathan agreed, and a favourable reply was received. There was a personal interest for Jonathan in this introduction, because Cape and the literary world in general had been waiting some years for any signs of a new book from the

Maestro, after the long pause in production following what had been hailed as his masterpiece, *The Old Man and the Sea*.

At the beginning of December, Fleming and I had a farewell lunch at the White Tower, after which we retired to his office for the briefing. Fleming said that it would be convenient for me to travel as a journalist, and the necessary accreditation was arranged with his paper. I was to take all the time I needed, and above all get out of Havana, and go into the country and see what was happening. He wanted to hear the viewpoints of Cubans of all kinds, from generals to waiters, and he still hoped that I might find some way of wheedling the fullest possible report out of the great Hemingway.

A few days later I flew to Havana, and, as suggested, took a room in the Seville Biltmore Hotel, in which Fleming's contact Edward Scott occupied a penthouse flat. We met within minutes of my arrival in the dark and icy solitude of the hotel's American bar. Scott was short, pink and rotund with a certain babyish innocence of expression that was wholly misleading. His manner, at first wary in the extreme, became congenial after he had read Fleming's letter.

Scott was the editor of the English language newspaper, the *Havana Post*, but appeared to have other, somewhat mysterious irons in the fire. He was a man Fleming much admired. Ian liked to have his friends ask him if his character James Bond was based upon any living person, and although he almost certainly believed Bond largely reflected his own personality, the standard reply was that he was a composite of several men of action he had known. When I asked the question that was expected of me, he agreed that Scott had contributed his share of the inspiration for his hero, while admitting that in Scott's case any physical similarities were excluded.

I mentioned to Scott that Fleming had asked me to see Hemingway and he seemed flabbergasted. The reason for his amazement was that he had just challenged Hemingway to a duel, following a fracas at a party given by the British Ambassador. Scott said that Hemingway had arrived in the company of the film actress Ava Gardner, who in a moment of high spirits had taken off her pants and waved them at the assembly. Scott, an ultra-patriotic New

Zealander, had objected to what he saw as an insult to the Crown, and, following a bellicose scene with Hemingway, the challenge had gone forth.

Leaving that situation aside, Scott's view was that Hemingway had withdrawn from the political scene, and no longer bothered himself with such uncomfortable things as wars or the rumours of wars and that, this being so, his views on the Castro revolt would have little value. Nevertheless, the briefing being what it was, I telephoned Hemingway's home to be told that both Hemingway and his wife were ill with influenza and were expected to be out of action for some days. I left my address and telephone number.

There seemed to be some uncertainty as to whether or not Hemingway would take up the challenge when he was on his feet again, and Scott, with whom I spent the first evenings in Havana, seeming to assume that any duel would be fought with pistols, always set aside a few minutes for target practice in a room fitted up like a range. He used a pistol employing CO_2 gas as the propellant for lead slugs. This fascinating and presumably lethal weapon was quite silent. We took turns to shoot at various small targets, but rarely hit anything.

Havana, the most beautiful city of the Americas, had quite suddenly become a dangerous place. Until the middle fifties, life there – at least as a tourist saw it – had seemed like a permanent carnival but, by the time of my visit in 1957, the spectacle of violence was commonplace. There was a good view from my hotel window of the Presidential palace and the garden-filled square in which it stood. The roads round the palace had been closed since March that year when twenty-one students had died in an attempt to shoot their way up to President Batista's office on the second floor. Now there were armed men everywhere.

I was standing at the window on the second evening of my visit, studying this scene, when machine-guns both in the square and on the palace roof opened fire, aiming it seemed in no particular direction, for a man standing on the balcony of a building across the street was hit, and fell, this being the first and last time in my life

I had actually seen anyone struck by a bullet. Such nightly alarms had become part of the existence of Havana. That same evening I had just returned from a visit to the city morgue, arranged by a reporter on the *Diario de la Marina*, where we saw the bodies of five murdered students, recovered from the streets during the previous night. It was an only slightly grimmer harvest than average, the victims being members of one or other of the left-wing groups opposed to the dictatorship. Several had been savagely handled either before or after death, in one case the victim's eyes having been gouged out. Batista's police were held responsible for these outrages. Outside Havana, the situation was worse, and in the province of Oriente, a private army, led by Rolando Masferrer, was busily torturing and extirpating 'Reds' – in other words any members of the peasantry objecting to the feudal conditions in which they lived.

The Batista regime was in its death throes. This ex-army sergeant who had assumed power twenty-three years earlier, had shown himself the most capable and, in his social measures, the most progressive president the country had ever known. His labour legislation had established Cuba as one of the most advanced nations in Latin America. He had fought big business over his social security laws, and still had the support of the trade unions and the organised urban workers, whose wages were at this time the highest in Cuban history.

But now, old and tired, he governed by force rather than flair, and he was losing control. He had forfeited the affection of most Cubans by his destruction of civil liberties, by press censorship, by the massive corruption, which he had ignored, and the ferocious repression of dissenters.

I was in Cuba to gather information, a task providing simple rules to be followed for obtaining the best results. In all nations there are sections of the population who know more than most about what is going on, and are usually happy, and often eager to discuss their experiences and opinions with anyone showing interest in them. These include most of those in positions of responsibility and power, and on a lower level, members of the legal and medical professions, journalists, and above all priests – who know about everything that happens in their parish.

In Havana I had excellent contacts including Ruby Hart Phillips of the *New York Times* who had arranged Herbert Matthew's visit to Castro in the Sierra, and who shared an office with Scott. Through Ruby, Scott, and others, I met bishops, disaffected senior officers, disgruntled politicians, student revolutionaries, a Batista torturer, the two legendary generals, Loynaz and Garcia Velez, both in their nineties who had led the last cavalry charges in the war against Spain, but above all the great capitalists, including Julio Lobo, chief of the sugar barons, without whose favour Batista's cause was lost.

From these encounters one certain fact emerged – that Castro's revolt, so far from being a proletarian revolution, knew nothing of Marxism and took little interest in the industrial workers. This was the middle class in action, and the hundred or so sons of good families who had taken to the mountains were not only not Communists, but they were at daggers drawn with them. How was it possible to believe, as our American friends had succeeded in believing, that Castro, who was receiving financial support from half the sugar magnates of Cuba, could have been the advocate of world-revolution and the dictatorship of the proletariat?

It was a moment when the United States was about to repeat its classic error in Latin America by renewing its assumption that any movement opposing a right-wing dictatorship must take its orders from Moscow. But in the case of Cuba this was not so. How was one to explain why the Cuban Communist Party should have sabotaged Castro's 26th July Movement in every possible way?

The antipathy shown at this time by Communists for the Castro movement sometimes took extreme forms, and was returned in full measure. The chief concern of a Castro agent from the Sierra Maestra whom I met in Havana was that any of his former comrades who had become Party members might spot him and denounce him to the police. It was an attitude that provoked talk of reprisals among Castro's men, including serious discussions as to whether the Communists should be granted legal existence after the Castro victory.

I discovered that three-quarters of the Cuban people were either openly or passively behind Castro, and it would have been logical for the United States to have thrown its weight behind him. In those days every declaration from the Sierra was underlined by assurances of the wholly democratic intentions of the rebels, their respect for private property and for foreign investments, and of their determination to hold elections within weeks of assuming power. As it was, the tottering dictator was supported by the Americans until the last. What little the majority of Castro's followers knew of Communism in December 1957, they distrusted or disliked. Three years later, largely through the success of the economic boycott organised by the United States, they had been herded into the Communist fold.

Fleming had said, 'Go into the country,' and I did so, travelling by bus from one end of the land to the other. The first discovery was that the mental attitudes of the countrymen were radically different from those of Cubans who worked for their living in the towns. The industrial worker had been converted to a kind of conservatism through his expectation of fairly steady employment the year round. The countryman enjoyed no such security. One quarter of Cuba grew nothing but sugar; and the single fact overshadowing the life of the Cuban peasant was that the sugar harvest occupied five months, to be followed by seven months of unemployment. He was therefore ready to support a revolution of any kind that would help to fill his stomach in the seven lean months, as well as relieving him from such feudal bullies as Masferrer and his thugs.

Santiago, capital of the sugar country, was necessarily a centre of action, and I went there to talk to cane-cutters and sugar magnates. I also went there on a strong recommendation to make contact with a famous clairvoyant, Tia Margarita, who was said to be consulted on occasion by Batista himself, and to know as much about what was going on as anyone in eastern Cuba. The astonishing statistic had been offered that one person in three in Cuba, regardless of colour, was a secret adherent of one of the cults introduced by the black slaves; and Tia Margarita happened to be high-priestess of Chango, Yoruba god of war, most powerful of the deities of the African interior.

She proved to be a comfortable-looking middle-aged black lady of compelling humour and charm, living in a small suburban house with a garden full of sweet-peas, with the usual attachment of a straw-thatched voodoo temple. Women of her kind were to be found in every town in Cuba, combining in their operations all the exciting mumbo-jumbo of horoscopy and divination with the real social service performed in solving personal problems of all kinds, as well as treating the sick from their wide repertoire of herbal remedies.

Tia Margarita ushered me into a chamber cluttered with the accessories of her profession, the skulls of small animals, the withered bats and the dusty salamanders, while gently kicking aside the live piglets and cockerels that would provide the material for future sacrifices. A faint culinary odour suggested the preparation of her celebrated remedy for nervous tension – a thick soup made from the bones of dogs. I added my contribution – a pair of dark spectacles – to the homely offerings, including roller skates, tubes of toothpaste, and a jar of Pond's Cold Cream, stacked under the war-god's altar. I noted the framed autographs, offered in gratitude by famous personalities: senators, baseball-players and motor racers who had come here with their troubles.

The mild maternal eyes scanned the print in the open book of my face, and her expression was one of slightly puzzled amusement. She expected to be called upon to demonstrate her speciality by forecasting the exact date of my death; instead of which I asked her what the people of Santiago thought about the war, and its likely outcome. If that was where my interest lay, she said, who better to discuss the matter with than Chango himself – surely the final authority on all such matters – who spoke through her mouth at seances held at the temple every Saturday night? Unfortunately this was a Monday, and when I asked Tia Margarita for an off-the-cuff opinion of the likely outcome of fighting in the Sierra, she was oracular and obscure. 'Chango says victory will be to whom victory is due,' she said. Still, something emerged from the interview, because Tia Margarita went into a kind of mini-trance, lasting perhaps ten seconds, then said that the war would be over in a year – which, give a day or two, it was.

In the few days I had been in Santiago, warlike activity had restarted. From the roof of the hotel in Cespedes Square, the night sparkled distantly where Castro partisans had gone into the cane-fields, to plant candles with their bases wrapped in paraffin-soaked rags. There was gunfire every night, usually when revolutionaries took on the police. By custom, the first shots were fired precisely at 10pm, giving the citizens a chance of a quiet stroll in the cool of the evening before the bullets began to fly. At 9:30pm, all the street lights ablaze, the promenaders would begin to stream out of the square and make for their homes, where they clustered at their doors like gophers ready to bolt for the shelter of their burrows if the shadow of an eagle fell upon them. Then, as the cathedral clock struck ten all the lights went out, and the streets would be cleared for battle.

Back in Havana a call came through from Ian Fleming in London. We had made a loose arrangement for a meeting in Jamaica, but there was a change in dates. He asked how things were going, and I told him that they were going fairly well, adding that there was not a lot more to be done.

'Have you talked to the Big Man?' he wanted to know.

By this I understood that he meant not the President, but Hemingway. I told him that Hemingway had been ill, adding that Scott did not seem to feel that a meeting would be specially rewarding.

'Never mind Scott,' Fleming said. 'Do your best to see him.'

I assured him that I would. Fleming said that he had just read *The Old Man and the Sea*, again, and had found it even better on second reading. He had the book open by the phone, and proceeded to read out a fairly long passage that he had found of special appeal.

A letter arrived from Hemingway next morning. It was neatly handwritten and formal in tone. He said he would be happy to see me at his farm, Finca Vigía, on the outskirts of the city, and would send his car to pick me up, suggesting the next afternoon for the visit.

Hemingway's concern for his privacy was in strong evidence at his farm, the roof of the building being screened by a high fence, with a gate secured by a chain and an enormous padlock. The driver got out to unlock the padlock, drove the car through the gate, then stopped to go back and chain and lock the gate again. I was ushered into a large room, furnished in the main with bookshelves, where I found Hemingway, in his pyjamas, seated on his bed. He pulled himself to his feet to mumble a lacklustre welcome.

I was stunned by his appearance. At sixty years old he looked like a man well into his seventies, and he was in wretched physical shape. He moved slowly under the great weight of his body to find the drinks, pouring himself, to my astonishment, a tumblerful of Dubonnet, half of which he immediately gulped down. Above all, it was his expression that shocked, for there was exhaustion and emptiness in his face. This was an encounter that might have been dangerous and undermining to any young man in the full enjoyment of ambition and hope, because it presented a parable on the subject of futility. Hemingway's mournful eyes urged you to accept your lot as it was, and be thankful for it.

Some people, and Fleming and I were among them, regarded this man as one of the great writers of the twentieth century, and at this time, three years after he had been awarded the Nobel Prize, he had only just overtopped the pinnacle of his fame. He was a man who had gained all that life had to offer. He had crammed himself with every satisfaction, driven his body to the utmost, loved so many women, dominated so many men, hunted so many splendid animals. It was hard to believe that anything Hemingway had set out to do had been left unachieved. Yet after all his conquests he seemed ready to weep with Alexander, and, looking into his face, it was hard to believe that he would ever smile again.

We talked in a desultory and spiritless fashion, and it was Hemingway who brought up the subject of his publishers, showing little affection for them, and ready with criticism. He found them parsimonious, nervous of spending money on publicity, and this, he said, had had an adverse effect on the sales of his books in Great Britain, which were disappointing compared with those in the United States. He disliked the dustjacket of the English edition of *The Old Man and the Sea*. A leading artist had been commissioned at great expense to produce the American version, which he showed me, and it had to be admitted that it was vivacious enough to increase sales.

The release of this unexpected grain of information about his literary affairs led to my undoing. It seemed, mistakenly, to open a suitable opportunity – although Jonathan Cape had warned me that this was a topic to be approached with extreme caution – to mention that his publishers were eager to know whether anything new from him could be expected in the near future. The reaction was instant and hostile. A wasted and watery eye swivelled to watch me with anger and suspicion. What had I come for? What was it I wanted of him? In the coldest manner he asked, 'Is this an interview?' and I hastened to reassure him that it was not.

There was something in the scene with the faint remembered flavour of an episode in *For Whom the Bell Tolls*, featuring Massart, 'one of France's great modern revolutionary figures', then Chief Commissar of the International Brigade – a 'symbol man' who cannot be touched. With time he had come to believe only in the reality of betrayals. With infallible discernment Hemingway had described this great old man's descent into pettiness; and now I was amazed that a writer who had understood how greatness could be pulled down by the wolves of weakness and old age, should – as it appeared to me – have been unable to prevent himself from falling into this trap.

Suddenly the talk was of Scott, and there was a note of harsh interrogation. Did I know him well, and had I heard about the challenge? I admitted that I had. I added quite sincerely that I regarded it as childish and absurd.

He seemed appeased, almost amicable. 'Take a look at this,' he said. He put in front of me a copy of a letter he had sent to the *Havana Post*. In this, couched in the most conciliatory language, he had taken note of the challenge to a duel made by its Editor, Edward Scott. This he had decided not to take up, in the belief that he owed it to his readers not to jeopardise his life in this way.

I nodded approval. It was the best thing in the circumstances that he could have done. For all that I was surprised, and in some way disappointed at the wording of the letter, as I felt that his readers might have been left out of it.

The problem now was how Fleming's demands – seeming more eccentric with every minute that passed – were to be satisfied. And yet Hemingway's opinions on Cuba *ought* to have been worth listening to. He had gone there in search of 'pay-dirt' for his post-war fiction many years before, and remembering his passionate involvement in the Spanish Civil War and in the politics of those days, it was hard to believe that suddenly he had torn himself free from all involvement with those times, and that Cuba for him was nothing but a tropical setting for the pursuit of visiting film actresses and gigantic fish. He downed another half-pint of Dubonnet, yawned, and I got up to go. He followed me to the waiting car. All his anger had passed, and I imagine that he felt little but boredom. 'A final word of advice,' he said. 'As soon as you get back to the hotel, I'd change that shirt.'

The shirt was a khaki affair, with convenient buttoning pockets of the kind it was hard to find in London at that time, and I had picked it up in an army-surplus shop in Oxford Street. 'By their standards that's a uniform,' he said. 'You could find yourself in a whole heap of trouble.'

I pointed out that I was wearing seersucker trousers with the shirt. 'It doesn't matter,' he said. 'It's still a uniform the way they see it, and they make the laws. A lot of cops on this island with itchy trigger-fingers. They have a rebellion on their hands.'

There was nothing to be lost. I took the plunge. 'How do you see all this ending?' I asked. Comrade Massart's cautious, watery, doubting eyes were on me again. 'My answer to that is I live here,' replied Hemingway.

In my letter to Fleming I wrote, 'Finally I saw the Great Man, as instructed. He told me nothing but taught me a lot.'

Acknowledgements

H UGE THANKS TO Lesley Lewis, who trusted me to make this collection and, remarkably, made no stipulations.

Special thanks must also go to Julian Evans, the author of Norman Lewis's biography. Authors can be difficult when one enters their territory, but instead Julian has been a rock of support, checking my facts, providing editing beyond the call of duty, and offering much needed encouragement.

Thanks must also go to Clare Morton for typing the entire book, and providing valuable comments; to David Vigar for spotting my errors; to Yvette Dickerson of the London Library, who laboured successfully to find lost articles; to Kevin Jackson who rescued me from computer traumas; and to my friends Nicholas Shakespeare and Mark Ellingham for valuable editorial help.

Finally, the greatest thanks go to the publishers, Rose Baring and Barnaby Rogerson, who have been exceptionally kind in allowing me to create this collection. They have never interfered, and instead have always been wonderfully encouraging. Nor have they complained when, too often, I have excused my tardiness by quoting Warren Buffet: 'You can't make a baby in a month by getting nine women pregnant.'

Permission to use the pieces in this collection was granted by Lesley Lewis and the Estate of Norman Lewis.

Norman Lewis (aged 72) photographed at home
by David Montgomery

About the Author

NORMAN LEWIS WAS BORN IN 1908. His early childhood was spent partly with his Welsh spiritualist parents in suburbs on the edge of north London, and partly with half-mad aunts in Wales. Forgoing a place at university for lack of funds, he used the income from wedding photography and petty trading – including selling lost umbrellas and cameras – to finance travels in Spain, Italy, Morocco, and the Balkans, as well as racing Bugattis. He then set up a successful camera business, which grew into a small chain.

One visitor to his High Holborn shop in London persuaded him to make an intrepid trip to southern Arabia and the Yemen, so that he could gather information for the intelligence services. The journey, which included a long voyage by dhow, greatly bolstered his long-held passion for the exotic.

During the Second World War he spent three years in the Intelligence Corps in North Africa and Southern Italy. His diaries from that time produced material for his masterpiece, *Naples '44*, written in 1977. Before the war, he had published two books, but his literary career only truly gained ground with the publication of *A Dragon Apparent* in 1951, which became an unexpected bestseller. Then followed several decades in which he wrote fifteen novels and fourteen works of non-fiction. In the year of its publication, an omnibus translation of his novels into Russian outsold Tolstoy. After travelling in Brazil in 1968 to research the ethnocide of the country's indigenous tribes, his later writings also included passionate campaigning on their behalf.

Until he reached his nineties Lewis was still enthusiastically travelling to offbeat parts of the world. He died in 2003.

For more information about Norman Lewis

The major source is the remarkable and thorough biography by Julian Evans, *Semi-Invisible Man: the life of Norman Lewis* (Jonathan Cape, 2008).

There are two autobiographies. Neither is complete or thorough, or completely satisfactory. But, of course, they both contain some great writing: *Jackdaw Cake* (Jonathan Cape, 1994) and *The World, the World* (Jonathan Cape, 1996).

Lewis wasn't knowledgeable or especially interested in music, but Michael Parkinson's interview on *Desert Island Discs* is well worth a listen. It can be found on BBC Sounds.

For a superb hour-long film about Lewis (*The Journeyman*, BBC Arena 1986) you can google: 'journeyman powell arena Norman Lewis'. The film's late director, Tristram Powell, brilliantly captures Lewis's character – not an easy thing to do.

ELAND

61 Exmouth Market, London EC1R 4QL
Email: info@travelbooks.co.uk

Eland was started in 1982 to revive great travel books which had fallen out of print. Although the list soon diversified into biography and fiction, all the titles are chosen for their interest in spirit of place. One of our readers explained that for him reading an Eland is like listening to an experienced anthropologist at the bar – she's let her hair down and is telling all the stories that were just too good to go into the textbook.

Eland books are for travellers, and for those who are content to travel in their own minds. We can never quite define what we are looking for, but they need to be observant of others, to catch the moment and place on the wing and to have a page-turning gift for storytelling. And they might do that while being, by turns, funny, wry, intelligent, humane, universal, self-deprecating and idiosyncratic. We take immense trouble to select only the most readable books and therefore many people collect the entire series.

Extracts from each and every one of our books can be read on our website, at www.travelbooks.co.uk. If you would like a free copy of our catalogue, please order it from the website, email us or send a postcard.